Lecture Notes in Artificial Intelligence 9728

Subseries of Lecture Notes in Computer Science

More information about this series at http://www.springer.com/series/1244

Petra Perner (Ed.)

Advances in Data Mining

Applications and Theoretical Aspects

16th Industrial Conference, ICDM 2016
New York, NY, USA, July 13–17, 2016
Proceedings

 Springer

Editor
Petra Perner
Institute of Computer Vision
  and applied Computer Sciences, IBaI
Leipzig, Saxony
Germany

ISSN 0302-9743   ISSN 1611-3349   (electronic)
Lecture Notes in Artificial Intelligence
ISBN 978-3-319-41560-4   ISBN 978-3-319-41561-1   (eBook)
DOI 10.1007/978-3-319-41561-1

Library of Congress Control Number: 2016942891

LNCS Sublibrary: SL7 – Artificial Intelligence

Printed on acid-free paper

This Springer imprint is published by Springer Nature
The registered company is Springer International Publishing AG Switzerland

Preface

The 16th event of the Industrial Conference on Data Mining ICDM was held in New York (www.data-mining-forum.de) running under the umbrella of the World Congress on "The Frontiers in Intelligent Data and Signal Analysis, DSA 2016" (www.worldcongressdsa.com).

After a peer-review process, we accepted 32 high-quality papers for oral presentation. The topics range from theoretical aspects of data mining to applications of data mining, such as in multimedia data, in marketing, in medicine, and in process control, industry, and society. Extended versions of selected papers will appear in the international journal *Transactions on Machine Learning and Data Mining* (www.ibai-publishing.org/journal/mldm).

In all, ten papers were selected for poster presentations and six for industry paper presentations, which are published in the ICDM Poster and Industry Proceedings by ibai-publishing (www.ibai-publishing.org).

A tutorial on "Data Mining," a tutorial on "Case-Based Reasoning," a tutorial on "Intelligent Image Interpretation and Computer Vision in Medicine, Biotechnology, Chemistry, and Food Industry," and a tutorial on "Standardization in Immunofluorescence" were held before the conference.

We were pleased to give out the best paper award for ICDM for the eighth time this year (three announcements are mentioned at www.data-mining-forum.de). The final decision was made by the Best Paper Award Committee based on the presentation by the authors and the discussion with the auditorium. The ceremony took place during the conference. This prize is sponsored by ibai solutions (www.ibai-solutions.de), one of the leading companies in data mining for marketing, Web mining, and e-commerce.

We would like to thank all reviewers for their highly professional work and their effort in reviewing the papers.

We also thank the members of the Institute of Applied Computer Sciences, Leipzig, Germany (www.ibai-institut.de), who handled the conference as secretariat. We appreciate the help and understanding of the editorial staff at Springer, and in particular Alfred Hofmann, who supported the publication of these proceedings in the LNAI series.

Last, but not least, we wish to thank all the speakers and participants who contributed to the success of the conference. We hope to see you in 2017 in New York at the next World Congress on "The Frontiers in Intelligent Data and Signal Analysis, DSA 2017" (www.worldcongressdsa.com), which combines under its roof the following three events: International Conferences Machine Learning and Data Mining (MLDM), the Industrial Conference on Data Mining (ICDM), and the International Conference on Mass Data Analysis of Signals and Images in Medicine, Biotechnology, Chemistry and Food Industry (MDA).

July 2016   Petra Perner

Organizations

Chair

Petra Perner   IBaI Leipzig, Germany

Program Committee

Ajith Abraham   Machine Intelligence Research Labs, USA
Brigitte Bartsch-Spörl   BSR Consulting GmbH, Germany
Orlando Belo   University of Minho, Portugal
Shirley Coleman   University of Newcastle, UK
Jeroen de Bruin   Medical University of Vienna, Austria
Antonio Dourado   University of Coimbra, Portugal
Geert Gins   KU Leuven, Belgien
Warwick Graco   ATO, Australia
Aleksandra Gruca   Silesian University of Technology, Poland
Hartmut Ilgner   Council for Scientific and Industrial Research,
                                  South Africa
Pedro Isaias   Universidade Aberta, Portugal
Piotr Jedrzejowicz   Gdynia Maritime University, Poland
Martti Juhola   University of Tampere, Finland
Janusz Kacprzyk   Polish Academy of Sciences, Poland
Mehmed Kantardzic   University of Louisville, USA
Mineichi Kudo   Hokkaido University, Japan
Eduardo F. Morales   INAOE, Ciencias Computacionales, Mexico
Armand Prieditris   Newstar Labs, USA
Rainer Schmidt   University of Rostock, Germany
Victor Sheng   University of Central Arkansas, USA
Kaoru Shimada   Fukuoka Dental College, Japan
Gero Szepannek   Santander Consumer Bank, Germany

Additional Reviewers

Alessandro Mammana   Max Planck Institute for Molecular Genetics
Juliane Perner   Cancer Research UK Cambridge Institute

Contents

Induction of Model Trees for Predicting BOD in River Water: A Data Mining Perspective

J. Alamelu Mangai[1] and Bharat B. Gulyani[2(✉)]

[1] Department of Computer Science, Presidency University,
Bengaluru 560089, Karnataka, India
alamelumaran@gmail.com
[2] Department of Chemical Engineering, BITS Pilani,
Dubai Campus, P.O. Box 345055, Dubai International Academic City, UAE
gulyanibb@gmail.com

Abstract. Water is a primary natural resource and its quality is negatively affected by various anthropogenic activities. Deterioration of water bodies has triggered serious management efforts by many countries. BOD is an important water quality parameter as it measures the amount of biodegradable organic matter in water. Testing for BOD is a time-consuming task as it takes 5 days from data collection to analyzing with lengthy incubation of samples. Also, interpolations of BOD results and their implications are mired in uncertainties. So, there is a need for suitable secondary (indirect) method for predicting BOD. A model tree for predicting BOD in river water from a data mining perspective is proposed in this paper. The proposed model is also compared with two other tree based predictive methods namely decision stump and regression trees. The predictive accuracy of the models is evaluated using two metrics namely correlation coefficient and RMSE. Results show that the model tree has a correlation coefficient of 0.9397 which is higher than the other two methods. It also has the least RMSE of 0.5339 among these models.

Keywords: Water quality parameters · Wastewater · BOD · Modeling and simulation · Data mining · Regression trees · Model trees

1 Introduction

Water is essential for the survival of all life-forms on earth, which makes it an important resource. Water resources are depleting fast because of rapid population growth. Also, water quality is deteriorating worldwide mainly due to human activities, rapid urbanization, discharge of new pathogens and chemicals into water from industries, etc. Impurities in water can be chemical, physical, and biological. Some impurities are benign while others are toxic. Ascertaining water quality is crucial before use for various intended purposes such as potable water, agricultural, industrial, etc. The difficulty of defining acceptable water quality is underscored [1] for specific cases. Water quality is defined in [2] as any characteristics of water, whether physical, chemical, and biological, that affects the survival, reproduction, growth, and management of fish. However, higher quality standards apply to water intended for human consumption than for other uses.

P. Perner (Ed.): ICDM 2016, LNAI 9728, pp. 1–13, 2016.
DOI: 10.1007/978-3-319-41561-1_1

The water quality of water resources and the assessment of long-term water quality changes is an important and challenging problem. During the past decades, there has been an increasing demand for monitoring water quality of water bodies such as rivers, ponds, lakes, underground water tables, and oceans, by regular measurements and/or prediction of various water quality variables. Some of the necessities of water quality monitoring are: (1) to monitor long-range trends in selected water quality parameters, (2) to detect actual or potential water quality problems; if such problems exist, (3) to determine specific causes, and (4) to devise solution strategies.

Various water analysis methods are employed to determine water quality parameters such as Dissolved Oxygen (DO), Chemical Oxygen Demand (COD), BOD, pH, Total Dissolved Solids (TDS), salinity, chlorophyll, coli form, and organic contaminants such as pesticides. The efficacy of treatment methods depends largely on assessment of incoming water contaminants levels. The list of potential water contaminants is exhaustive and impractical to test for in its entirety. Such water testing is sometimes costly and time consuming. In general, the organic pollution in an aquatic system is measured and expressed in terms of the biological oxygen demand (BOD) level. BOD is an important parameter [3] as it measures the amount of biodegradable organic matter in water. Testing for BOD is a time-consuming task as it takes 5 days from data collection to analyzing with lengthy incubation of samples [4]. There are various complicating factors such as oxygen demand resulting from the respiration of the algae in the sample and the possible oxidation of ammonia. Presence of toxic substances in samples may also affect microbial activity leading to a reduction in measured BOD value. The lab conditions for BOD determination differ from those in aquatic systems. It is also emphasized that there could be gross differences in test results, due to the approach adopted by laboratories in sample preservation, quality of chemicals used, and testing method applied [5]. So, there is a need for suitable secondary (indirect) method for predicting BOD.

Dissolved oxygen (DO), biochemical oxygen demand (BOD), and chemical oxygen demand (COD) are the important metrics used in pollution control. Oxygen dissolves in water by a purely physical process, proportional to its partial pressure in the gas in contact with the water. It is dependent on the temperature and the concentration of dissolved salts, notably chlorides. Dissolved oxygen is measured by a number of chemical techniques, the Winkler iodometric method and its several modifications, the choice depending on the type of water/wastewater and the kinds of interferences present. The BOD is the amount of DO required by microorganisms, mainly bacteria, for the oxidation of organic material in a waste under aerobic conditions. The BOD test is a bioassay technique involving the measurement of oxygen consumed by the bacteria while stabilizing the organic matter in the waste as they would normally do in nature but under normal laboratory conditions. By convention, the test is conducted for a period of 5 days at 20 °C. The level of dissolved oxygen present in a sample is limited to the saturation value of 9.2 mg/L at 20 °C, the temperature at which the test is normally run. However, the strength of typical wastes is such that several hundred mg/L is required for oxidation. In nature, this is accomplished by constant reaeration of the stream into which the waste is discharged. In the laboratory, a portion of the waste is diluted with oxygen saturated water to such an extent that the oxygen requirement is less than this saturation value, and reaeration is prevented. For wastes of unknown strength, several dilutions are necessary. This discussion shows the intricacies and uncertainties involved in BOD determination as well time and level of sophistication required from the laboratory personnel.

Data Mining is the process used to extract implicit, previously unknown, non trivial information from huge data repositories. Data Mining tasks are broadly classified into predictive and descriptive. The data base to be mined has a set of examples/observations where each instance/observation is defined using a fixed number of input variables and an output variable. The goal of predictive tasks is to learn a function of the output variable in terms of its input variables. The descriptive tasks generate a set of rules or clusters that identify the underlying relationship among the examples that exists in the data base. The two types of predictive tasks are classification and regression. In case of classification the target variable is a discrete label while it is continuous in case of regression. In this paper data mining model namely model trees is applied for predicting BOD in river water.

2   Related Work

Water quality variables (such as, temperature, pH, salinity, DO, BOD, COD, Chl-α, etc.) describe a complex process governed by a large number of hydrologic, hydro-dynamic, and ecological controls that operate over a wide range of spatiotemporal scales. Interactions among water quality variables make the modeling effort even more difficult [6]. There are two approaches to water quality modeling, broadly classified as - process based modeling (deterministic) and data driven modeling (stochastic). Classical process-based modeling approaches may provide good predictions, but they need cumbersome data calibration. They also rely on the approximation of various under-lying processes, thus limiting their applicability beyond the assumptions on which the developed model was based. Furthermore, model parameters may be far too many, making the model computation-intensive and slow. Limited water quality data and the high costs of water quality monitoring often pose serious problems for process-based modeling approaches. Data driven models offer a viable alternative as they require fewer input parameters and input conditions (than deterministic models) [6].

Most popular among data driven modeling approaches is the artificial neural net-work, ANN. A review of research dealing with the use of ANN in prediction and forecasting of water resources variables is provided by [7, 23]. Though ANN are increasingly being used for water quality prediction, the problems of assessing the optimality of the results still exists. Apart from the importance of preprocessing, specific mapping of ANN depends on network architecture, training techniques, and modeling parameters [4]. Another problem with ANN is that it is difficult to predict an unknown event that has not occurred in the training data.

Classification and Regression trees (CART) and neural networks have been used [8] to classify water quality of canals in Bangkok. However, the intended task is classification. Least squares support vector machines (SVM) with parameters tuned by Particle Swarm Optimization (PSO) has been used [9] to overcome the shortcomings of the MLP neural network model. A survey of data mining applications in water quality management is provided by [10]. The use of Model trees for predicting BOD in river water is less explored, hence the motivation for the work presented in this paper.

3 Tree Based Predictive Methods

Algorithms for building classification trees use a greedy strategy to grow the tree. They make a series of local optimum decisions on how to split an available subset of examples using a splitting condition. The TDIDT top down induction of decision trees is one such algorithm by Hunt's [11]. This forms the basis of other tree growing algorithms like ID3 [12] C4.5 [13] and CART [14]. At each level of the tree growing procedure the subset of examples that reach a node are further split into smaller and purer subsets based on an attribute that maximizes the gain after splitting the subset. A subset is said to be pure when all examples in it have the same class. Different measures of node impurity are used in various decision tree induction algorithms. Some of them are entropy, gini index and classification error (Tan 2006) as defined in Eqs. 1, 2, and 3. If S is a node that represents a subset of training examples, c is the number of class labels and P(i/s) is the probability of class i in node S, then

$$Entopy(S) = -\sum_{i=1}^{c} P(i|s)log_2^{P(i|S)}. \tag{1}$$

$$Gini(S) = 1 - \sum_{i=1}^{c} P(i|s)^2 \tag{2}$$

$$ClassificationError(S) = 1 - max_i P(i|s) \tag{3}$$

From a set of candidate splitting attributes, the best splitting attribute is identified using the information Gain as defined in Eq. 4. The attribute with the highest gain is the best splitting attribute.

$$Gain\ (S, A) = Entropy(S) - \sum_{i=1}^{|values(A)|} \frac{|S_i|}{|S|} Entropy\ (S_i) \tag{4}$$

where A is a candidate splitting attribute, $|values(A)|$ is the number of possible values of A, S_i is the subset of training examples where the value of the attribute A is 'i' and $|S_i|$ is the size of S_i. The disadvantage of Gain is that it favors attributes that result is large number of smaller but purer partitions. Classification and Regression Trees (CART) tree was invented by Breiman in 1984 [14]. This technique has been greatly studied in fields such as medicine, market research statistics, marketing and customer relations. However its application in chemical informatics is less explored. If the target variable is nominal, the resulting model is called a classification tree and for continuous valued numeric target variable, the tree is called a regression tree. The classification tree built by CART algorithm is same as that of ID3. However, unlike ID3, it uses Gini Index in selecting the best splitting attribute. If the target variable is continuous, CART builds a set of tree based regression equations to predict the target variable.

In C4.5, Gain ratio is used to determine the best splitting condition as defined in Eqs. 5 and 6.

$$\text{Gain ratio} = \frac{\text{Gain}_{\text{split}}}{\text{splitInfo}} \tag{5}$$

where,

$$\text{splitinfo} = -\sum_{i=1}^{|\text{values}(A)|} \frac{|S_i|}{|S|} \log \frac{|S_i|}{|S|} \tag{6}$$

3.1 Decision Stump

A decision stump [15] is a one level decision tree. It uses only one of the attributes for decision making. This is applied to examples that represent Boolean concepts. Each attribute A is assigned a score based on how well the attribute distinguishes the classes as given by Eq. 7.

$$\text{score (A)} = \frac{\max(|A \equiv C|, |A \neq C|)}{n} \tag{7}$$

where $|A \equiv C|$ is the examples where the value of the attribute and the class label are same, $|A \neq C|$ is the number of examples where the value of the attribute and the class label are different and n is the number of examples. The attribute with the best score is used for decision making. In case of a tie, it chooses at random.

3.2 Regression Trees

Regression trees are the alternative to statistical regression. They take the form of decision trees where the leaf nodes are numeric rather than categorical. Regression trees are constructed using the recursive partitioning algorithm [16]. It recursively partitions a subset of training samples into smaller subsets if a certain stopping condition is not met. The best split at each node is chosen using a local criteria. The most common approach for building a regression model is to identify the parameters that minimize the least squares error as defined in Eq. 8.

$$\text{Least Square error} = \frac{1}{n} \sum_{i=1}^{n} (\text{actual}_i - \text{predicted}_i)^2 \tag{8}$$

Where n is the number of training examples. The value of a leaf node l is the average of the target values of all examples that reach this node.

$$\text{value}_l = \frac{1}{n_l} \sum_{i=1}^{n_l} y_l \tag{9}$$

Where n_l is the number of examples in the leaf node l. The error at a leaf node l is calculated as

$$\text{Error (l)} = \frac{1}{n_l} \sum_{i=1}^{n_l} (y_i - \text{value}_l)^2 \tag{10}$$

If P(l) is the probability of a leaf node and nl is the number of leaf nodes, the error of a tree T is a weighted average of the error in all its leaves as illustrated in Eq. 11.

$$\text{Error (T)} = \sum_{l=1}^{nl} P(l)x \text{ error}(l) \tag{11}$$

A split that improves the error of the resulting tree after the split is chosen by the splitting criteria. The error of split 's' at a node t is the weighted average of the errors of the resulting subtrees as given in Eq. 12 where t_{left} and t_{right} are the left and right subtrees of node t after the split. The cardinality of t_{left} is $n_{t_{left}}$ and that of t_{right} is $n_{t_{right}}$.

$$\text{Error}(s, t) = \frac{n_{t_{left}}}{n_t} x \text{ Error}(t_{left}) + \frac{n_{t_{right}}}{n_t} x \text{ Error}(t_{right}) \tag{12}$$

With a set of candidate splits S, the best split s^* is that which maximizes

$$\Delta\text{Error}(s, t) = \text{Error}(t) - \text{Error}(s, t) \tag{13}$$

This greedy criteria is used to select the best split for all internal nodes. All possible splits of each input attribute is evaluated and the RP algorithm chooses the one with best ΔError.

3.3   Model Trees

Regression trees are sometimes difficult to interpret, although they are more accurate than linear regression for nonlinear data [17]. Hence regression trees are combined with linear regression to form model trees. Each leaf node of a model tree represents a linear regression equation instead of the average of the output values of all examples that reach it. The first step in building the model trees (M5 trees) [18] is to find the standard deviation of the target values of the examples T that reach a node. The set T may be linked with a leaf node or it may be further split into subsets based on the outcomes of a test. The process repeats with every subset until T has few examples or the values in T vary slightly. The expected reduction in error after splitting T into 'p' number of partitions based on a test condition is given as

$$\Delta\text{error} = \text{S.D (T)} - \sum_{i=1}^{p} P(i)x \text{ S.D}_i \tag{14}$$

From a set of candidate splits, M5 chooses the one that maximizes this reduction in error. Multivariate linear models are built at each node using standard regression techniques. The model is also simplified using pruning and smoothing techniques. M5 trees have been modified as M5' [19] to handle missing values and enumerated attributes a common characteristics of real life data sets.

3.4 Performance Metrics

The performance of the predictive model is evaluated using two metrics namely correlation coefficient and root mean square error RMSE. Correlation coefficient between actual (S_a) and predicted values (S_p) of an attribute is defined as [22].

$$\text{Correlation coefficient} = \frac{S_{pa}}{\sqrt{S_p S_a}} \tag{15}$$

Where $S_{pa} = \sum_i^N (p_i - \bar{p})(a_i - \bar{a}) \big/ (N - 1)$ where $\bar{p}, \bar{a}$ are the averages, respectively. And

$$S_p = \sum_i^N (p_i - \bar{p})^2 \big/ (N - 1) \text{ and } S_a = \sum_i^N (a_i - \bar{a})^2 \big/ (N - 1) \tag{16}$$

A correlation coefficient of 1 indicates that the values are perfectly correlated while that of 0 implies no correlation exists between them. If p_i is the predicted value for i^{th} instance, a_i is the actual value for i^{th} instance and N is the total number of instances in the given data set, the root mean square errorRMSE is given as,

$$\text{RMSE} = \sqrt{\sum_{i=1}^N \frac{(p_i - a_i)^2}{N}} \tag{17}$$

The smaller the RMSE the better is the performance of the model.

4 Methodology and Results

4.1 Data Set Description

The data set was obtained from the website of Department of Environment, Food and Rural Affairs, UK Government [20]. The descriptive statistics of the parameters used for water quality modeling in this paper is given in Table 1.

For this study a large data set from North-east region was considered with annual data available from 1980–2011. Parameters include - temperature (OC), pH, conductivity (μS/cm), suspended solids (mg/L), DO (mg/L), ammoniacal nitrogen (mg/L), nitrate (mg/L), nitrite (mg/L), chloride (mg/L), total alkalinity (mg/L), orthophosphate (mg/L), and BOD (mg/L). These attributes can be measured with the help of sampling in case of DO, temperature and conductivity; gravimetry for suspended solids and standard titration techniques using common chemicals for other parameters.

Table 1. Descriptive statistics of the data set

Attributes/statistics	Min	Max	Range	Mean	Std deviation	Variance
Temp	2.800	15.583	12.783	11.159	1.609	2.587
pH	7.331	7.983	0.652	7.632	0.166	0.027
Cond	149.644	1439.292	1289.647	609.365	352.087	123964.936
SS	2.100	71.968	69.868	16.666	11.149	124.295
DO	7.001	12.200	5.199	9.837	1.045	1.092
Amm	0.021	7.265	7.244	1.037	1.526	2.328
Nitrite	0.003	0.592	0.589	0.150	0.147	0.022
Nitrate	2.635	44.275	41.640	18.261	10.821	117.097
Chloride	10.825	265.632	254.807	76.186	60.992	3719.996
Alkaline	35.400	159.021	123.621	96.219	24.691	609.639
Orthop	0.011	2.003	1.992	0.599	0.498	0.248
BOD	1.163	6.797	5.633	3.062	1.567	2.455

4.2   Results and Analysis

The three decision tree models explored in this study are modeled using 10-fold cross validation using a data mining tool called WEKA [21]. The regression tree built for this data set is shown in the Fig. 1. As seen from the figure the size of the tree is 31 which is the depth of the tree.

The model trees that were induced for this data set allow minimum four number of instances to be kept at leaf nodes. The resulting unpruned model tree and pruned model tree are shown in Figs. 2 and 3. It can be observed from these figures that the unpruned tree is far more complex than the pruned tree. The pruned model uses just one attribute namely Ammonical nitrogen to predict BOD. Also the complexity of the pruned model tree is very simple when compared with the regression tree shown in Fig. 1.

In the model tree, each leaf node represents a regression model for the subset of examples that reach that node. Hence this model tree has identified three regression equations as shown in Fig. 4. The complexity of the tree in terms of the number of linear regression equations, depends on the linearity between the dependent and independent variables. The fewer the number of regression equations, the linear relationship between the dependent and independent variables is more. This hybrid model of combining regression trees with liner regression also helps in easy interpretation of the tree, since it is possible to achieve accurate results in fewer levels of the tree. The difference between the actual BOD and predicted BOD by the pruned model tree for each training example is shown in Fig. 5.

The proposed model is also compared with two other models namely decision stump and Regression Trees. Table 2 shows the correlation coefficient and the RMSE of all these models for the same data set.

It can be seen from Table 2 that the correlation coefficient of the Model Trees is higher than the other two models. For applying different algorithms on the same dataset,

Amm < 0.96
| Nitrite < 0.11
| | Nitrate < 13.8
| | | Chloride < 13.81 : 2.39 (2/0) [1/0.01]
| | | Chloride >= 13.81
| | | | Nitrite < 0.01 : 1.39 (5/0.01) [1/0.11]
| | | | Nitrite >= 0.01
| | | | | SS < 5.02 : 1.45 (4/0) [2/0.01]
| | | | | SS >= 5.02
| | | | | | Temp < 11.23
| | | | | | | Nitrite < 0.04
| | | | | | | | Temp < 9.96
| | | | | | | | Amm< 0.29 : 1.84 (3/0) [3/0.07]
| | | | | | | | Amm>= 0.29 : 2.19 (2/0) [1/0.53]
| | | | | | | Temp >= 9.96 : 1.67 (7/0.01) [3/0.07]
| | | | | | | Nitrite >= 0.04 : 2.21 (2/0) [0/0]
| | | | | | Temp >= 11.23 : 1.72 (6/0.01) [4/0.21]
| | Nitrate >= 13.8 : 2.17 (12/0.17) [8/0.25]
| Nitrite >= 0.11
| | Amm< 0.47 : 2.53 (10/0.07) [0/0]
| | Amm>= 0.47
| | | Temp < 12.38 : 3.51 (7/0.08) [8/0.14]
| | | Temp >= 12.38 : 2.8 (2/0.14) [0/0]
| | | | | | | | Temp >= 9.96 : 1.67 (7/0.01) [3/0.07]
| | | | | | | Nitrite >= 0.04 : 2.21 (2/0) [0/0]
| | | | | | Temp >= 11.23 : 1.72 (6/0.01) [4/0.21]
| | Nitrate >= 13.8 : 2.17 (12/0.17) [8/0.25]
| Nitrite >= 0.11
| | Amm< 0.47 : 2.53 (10/0.07) [0/0]
| | Amm>= 0.47
| | | Temp < 12.38 : 3.51 (7/0.08) [8/0.14]
| | | Temp >= 12.38 : 2.8 (2/0.14) [0/0]
Amm>= 0.96
| Nitrate < 26.07 : 5.59 (16/0.36) [9/0.43]
| Nitrate >= 26.07
| | DO < 8.96 : 5.24 (3/0.06) [1/1.15]
| | DO >= 8.96
| | | Cond < 1039.57 : 3.76 (2/0.01) [1/0.06]
| | | Cond >= 1039.57 : 4.19 (2/0.01) [1/0.05]

Fig. 1. Regression tree for BOD prediction

RMSE is a good measure for analyzing the performance of the model. Also, the RMSE is less for Model trees than the other two models. The difference between pruned and unpruned model trees in terms of the performance measures is statistically insignificant. However pruning helps in easy interpretation of the model, since the number of rules generated for pruned and unpruned model trees were 3 and 48, respectively.

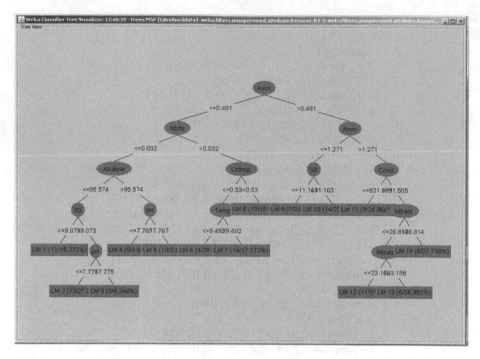

Fig. 2. The unpruned M5 model tree for BOD prediction

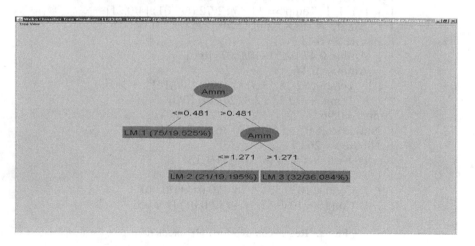

Fig. 3. The pruned M5 model tree for BOD prediction

For results of modeling effort to be practically important and usable, they must correlate favourably with real life situation and technical aspects of the problem at hand. The three regression equations given in the pruned M5 tree of Fig. 4, present BOD as a function of independent variables. For all three equations BOD is found to

```
LM num: 1
BOD =
        -0.092 * Temp
        - 0.2277 * pH
        + 0.0131 * SS
        - 0.0654 * DO
        + 0.0256 * Amm
        + 4.6082 * Nitrite
        - 0.0026 * Nitrate
        - 0.0014 * Alkaline
        + 0.466 * Orthop
        + 5.0125

LM num: 2
BOD =
        -0.0288 * Temp
        - 0.3014 * pH
        + 0.0284 * SS
        - 0.2202 * DO
        + 0.4619 * Amm
        + 2.1475 * Nitrite
        - 0.0272 * Nitrate
        - 0.0019 * Alkaline
      + 0.4352 * Orthop
        + 7.5833

LM num: 3
BOD =
        -0.0288 * Temp
        - 0.3014 * pH
        - 0.0005 * Cond
        + 0.0089 * SS
        - 0.189 * DO
        + 0.0339 * Amm
        + 1.9047 * Nitrite
        - 0.0691 * Nitrate
        - 0.0019 * Alkaline
        + 0.3736 * Orthop
        + 10.5721
```

Fig. 4. Linear regression models generated by the pruned M5 tree

strongly dependent on (1) nitrite, (2) orthophosphate and (3) pH. While nitrite and orthophosphate positively impact BOD, pH is shown as having negative impact. It indicates that the water contains fertilizer residues from leached water from agricultural activities. However, this cannot be treated as conclusive evidence.

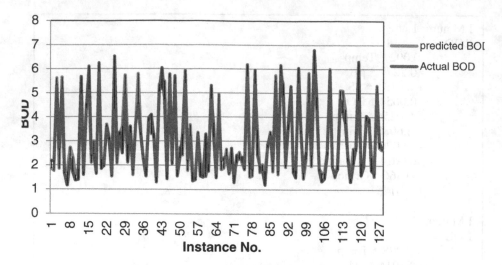

Fig. 5. Actual and predicted BOD by the pruned model tree (Color figure online)

Table 2. Performance Measures of the BOD Prediction Models

Model	Correlation coefficient	RMSE
Decision stump	0.8615	0.7937
Regression trees	0.9289	0.5803
Model trees (pruned)	0.9399	0.5339
Model trees (unpruned)	0.9438	0.5157

5 Conclusions

In this paper a prediction model for BOD in river water is proposed from a data mining perspective. It uses model trees which combines the best characteristics of regression trees and statistical linear regression. The proposed model is also compared with two other decision tree based methods namely decision stump and regression trees. The performance of the models is estimated using correlation coefficient and RMSE. Results show that the performance of the model trees is better than the other two methods.

References

1. Ajibade, W.A., Ayodele, I.A., Agbede, S.A.: Water quality parameters in the major rivers of Kainji Lake National Park. Niger. Afr. J. Environ. Sci. Technol. 2(7), 185–1996 (2008)
2. Boyd, C.E.: Water Quality in Warm Water Fish Ponds. Auburn University/Craftmaster Printers, Inc., Auburn/Opelika (1981)
3. Singh, K.P., Basant, A., Malik, A., Jain, G.: Artificial neural network modeling of the river water quality—a case study. Ecol. Model. **220,** 888–895 (2009)

4. Talib, A., Abu Hasan, Y., Abdul Rahman, N.N.: Predicting biochemical oxygen demand as indicator of river pollution using artificial neural networks. In: 18th World IMACS/MODSIM Congress, Cairns, Australia 13–17 July 2009
5. Alam, M.J.B., Islam, M.R., Muyen, Z., Mamun, M., Islam, S.: Water quality parameters along rivers. Int. J. Environ. Sci. Technol. **4**(1), 159–167 (2007)
6. Palani, S., Liong, S.-Y., Tkalich, P.: An ANN application for water quality forecasting. Mar. Pollut. Bull. **56**, 1586–1597 (2008)
7. Maier, H.R., Dandy, G.C.: Neural networks for the prediction and forecasting of water resources variables: a review of modelling issues and applications. Environ. Model Softw. **15**, 101–124 (2000)
8. Areerachakul, S., Sanguansintukul, S.: Classification and regression trees and MLP neural network to classify water quality of canals in Bangkok, Thailand. Int. J. Intell. Comput. Res. **1**(1), 43–50 (2010)
9. Xiang, Y., Jiang, L.: Water quality prediction using LS-SVM and particle swarm optimization. In: 2009 International Conference on Knowledge Discovery and Data Mining, pp. 900–904 (2009)
10. Dutta, P., Chaki, R.: A survey of data mining applications in water quality management. In: CUBE International Information Technology Conference, pp. 470–475 (2012)
11. Tan, P., Steinbach, M., Kumar, V.: Introduction to Data Mining. Pearson Education, Upper Saddle River (2006)
12. Quinlan, J.R.: Induction of decision trees. Mach. Learn. **1**, 81–106 (1986)
13. Quinlan, J.R.: C4.5: Programs for Machine Learning. Morgan Kaufmann, Burlington (1993)
14. Breiman, L., Friedman, J.H., Olshen, R.A., Stone, C.J.: Classification and Regression Trees. Chapman and Hall/CRC, London (1984)
15. Wayne, I., Pat, L.: Induction of one-level decision trees. In: Proceedings of the Ninth International Conference on Machine Learning, Aberdeen, Scotland, 1–3 July 1992, pp. 233–240. Morgan Kaufmann, San Francisco (1992)
16. Soman, K.P., Diwakar, S.: Insight into Data Mining: Theory and Practise. PHI, Delhi (2006)
17. Roiger, R.J., Geatz, M.W.: Data Mining: A Tutorial Based Primer. Addison Wesley, Boston (2003)
18. Quinlan, J.R.: Learning with continuous classes. In: Proceedings of 5th Australian Joint Conference on Artificial Intelligence, pp. 343–348. World Scientific, Singapore (1992)
19. Wang, Y., Witten, I.H.: Induction of Model Trees for Predicting Continuous Classes. Working Paer Series. University of Waikato, New Zealand (1996)
20. Department of Environment, Food and Rural Affairs (DEFRA). UK Government website-http://data.gov.uk/dataset/river-water-quality-regions
21. Hall, M., Frank, E., Holmes, G., Pfahringer, B., Reutemann, P., Witten, I.H.: The WEKA data mining software: an update. SIGKDD Explor. **11**(1), 10–18 (2009)
22. Witten, I.H., Frank, E.: Data Mining-Practical Machine Learning Tools and Technology with Java Implementations. Morgan Kauffman Publications, San Francisco (2000)
23. Jain, J., Alamelu Mangai, J., Gulyani, B.B.: Water quality modeling using LM and BR based ANN with sensitivity analysis. In: Proceedings of the International Conference on Computational Methods and Software Engineering, 28–30 December 2015, pp. 73–88. Anna University, Chennai (2015)

Efficient Mining of High Average-Utility Itemsets with Multiple Minimum Thresholds

Jerry Chun-Wei Lin[1]([✉]), Ting Li[1], Philippe Fournier-Viger[2],
Tzung-Pei Hong[3,4], and Ja-Hwung Su[5]

[1] School of Computer Science and Technology,
Harbin Institute of Technology Shenzhen Graduate School, Shenzhen, China
jerrylin@ieee.org, tingli@ikelab.net
[2] School of Natural Sciences and Humanities,
Harbin Institute of Technology Shenzhen Graduate School, Shenzhen, China
philfv@hitsz.edu.cn
[3] Department of Computer Science and Information Engineering,
National University of Kaohsiung, Kaohsiung, Taiwan
tphong@nuk.edu.tw
[4] Department of Computer Science and Engineering,
National Sun Yat-sen University, Kaohsiung, Taiwan
[5] Department of Information Management, Cheng Shiu University,
Kaohsiung, Taiwan
bb0820@ms22.hinet.net

Abstract. High average-utility itemsets mining (HAUIM) is a key data mining task, which aims at discovering high average-utility itemsets (HAUIs) by taking itemset length into account in transactional databases. Most of these algorithms only consider a single minimum utility threshold for identifying the HAUIs. In this paper, we address this issue by introducing the task of mining HAUIs with multiple minimum average-utility thresholds (HAUIM-MMAU), where the user may assign a distinct minimum average-utility threshold to each item or itemset. Two efficient IEUCP and PBCS strategies are designed to further reduce the search space of the enumeration tree, and thus speed up the discovery of HAUIs when considering multiple minimum average utility thresholds. Extensive experiments carried on both real-life and synthetic databases show that the proposed approaches can efficiently discover the complete set of HAUIs when considering multiple minimum average-utility thresholds.

Keywords: High average-utility itemsets · Multiple thresholds · Data mining · Downward closure · Utility

1 Introduction

The main purpose of knowledge discovery in database (KDD) is to discover implicit and useful information in a collection of data. Association-rule mining (ARM) or frequent itemset mining (FIM) plays an important topic in KDD,

© Springer International Publishing Switzerland 2016
P. Perner (Ed.): ICDM 2016, LNAI 9728, pp. 14–28, 2016.
DOI: 10.1007/978-3-319-41561-1_2

which has been extensively studied [1, 3]. A major limitation of traditional ARM and FIM is that they focus on mining association rules or frequent itemsets in binary databases, and treat all items as having the same importance without considering factors. To address this limitation, the problem of high utility itemset mining (HUIM) [4, 10, 18, 19] was introduced. An important limitation of traditional HUIM is that the utility of an itemset is generally smaller than the utility of its supersets. Hence, traditional HUIM tends to be biased toward finding itemsets of greater length (containing many items), as these latter are more likely to be high utility itemsets. The utility measure used in traditional HUIM thus does not provide a fair measurement of the utility of itemsets.

To alleviate the influence of an itemset's length on its utility, and find more useful high utility itemset for recommendation, Hong et al. [7] proposed the average utility measure, and the problem of high average utility itemset mining (HAUIM). The average utility of an itemset is defined as the total utility of its items in transactions where the itemset appears, divided by the number of items in the itemset. Numerous algorithms have been designed to more efficiently mine high average-utility itemsets (HAUIs) [11, 13, 14, 16] but most of them rely on a single minimum average-utility threshold to mine HAUIs. In real-life situations, each item or itemset may be more or less important to the user. It is thus unfair to measure the utility of all items in a database using the same minimum utility threshold.

To address this issue, this paper proposes a novel framework for high average-utility itemset mining with multiple minimum average-utility thresholds (HAUIM-MMAU). Based on the proposed framework, a two-phase algorithm named HAUI-MMAU is proposed to discover HAUIs. To improve the performance of the proposed algorithm, two efficient pruning strategies called IEUCP and PBCS are designed to prune unpromising itemsets early, thus reducing the search space and speeding up the discovery of HAUIs. Extensive experiments were conducted on both real-life and synthetic datasets to show that the proposed algorithm can efficiently mine the complete and correct set of HAUIs in databases, while considering multiple minimum average-utility thresholds to assess the utility of itemsets.

2   Related Work

In recent years, HUIM [4, 10, 18, 19] has become a key research topic in the field of data mining. Chan et al. [4] presented a framework to mine the top-k closed utility patterns based on business objectives. Yao et al. [18, 19] defined the problem of utility mining while considering both purchase quantities of items in transactions (internal utility) and their unit profits (external utility). Liu et al. [10] introduced the transaction-weighted utility (TWU) model and the transaction-weighted downward closure (TWDC) property. Lin et al. [11] adopted the TWU model to design the high-utility pattern (HUP)-tree for mining HUIs using a condensed tree structure called HUP-tree. Liu and Qu [15] proposed the HUI-Miner algorithm to discover HUIs without generating candidates using a designed utility-list structure. Fournier-Viger et al. [5] then presented the FHM algorithm and the Estimated Utility Co-occurrence Structure (EUCS) to mine HUIs.

In HUIM, the utility of an itemset is defined as the sum of the utilities of its items, in transactions where the itemset appears. An important drawback of this definition is that it does not consider the length of the itemset. Hong et al. [7] first proposed the average utility measure, and the stated the problem of HAUIM. The average-utility of an itemset is the sum of the utilities of its items, in transaction where it appears, divided by its length (number of items). The average-utility model provides an alternative measure to assess the utility of itemsets. Because the average utility measure considers the length of itemsets, it is more suitable and applicable in real-life situations, than the traditional measure. Lin et al. [11] then developed a high average-utility pattern (HAUP)-tree structure to mine HAUIs more efficiently. Lan et al. [13] developed a projection-based average-utility (PBAU) mining algorithm to mine HAUIs. Lan et al. [14] then also extended the PBAU algorithm and designed a PAI approach using an improved strategy for mining the HAUIs. Lu et al. [16] then developed a HAUI-tree structure to mine HAUIs without candidate generation.

The above algorithms were designed to mine HAUIs using a single minimum average-utility threshold (count). But in real-life situations, items are often regarded as having different importance to the user. Hence, it is unfair to assess the utility of all items using a same minimum average utility threshold (count), to determine if itemsets are HAUIs. In the past, the MSApriori [9] was first proposed to mine FIs under multiple minimum support thresholds. The CFP-growth algorithm [8] was then designed to build the MIS-tree and perform a recursive depth-first search to output the FIs. The MHU-Growth algorithm [17] extends CFP-Growth, to mine high utility frequent itemsets with multiple minimum support thresholds. Lin et al. [12] then developed the HUIM-MMU model for discovering HUIs with multiple minimum utility thresholds. Besides, two improved TID-index and EUCP strategies were proposed to prune unpromising itemsets early, thus speeding up the discovery of HUIs.

3   Preliminaries and Problem Statement

Let $I = \{i_1, i_2, \ldots, i_r\}$ be a finite set of r distinct items occurring in a database D, and $D = \{T_1, T_2, \ldots, T_n\}$ be a set of transactions, where for each transaction $T_q \in D$, T_q is a subset of I and has a unique identifier q, called its *TID* (transaction identifier). For each item i_j and transaction T_q, a positive number $q(i_j, T_q)$ represents the purchase quantity of i_j in transaction T_q. Moreover, a profit table $ptable = \{p(i_1), p(i_2), \ldots, p(i_r)\}$ is defined, where $p(i_m)$ is a positive integer representing the unit profit of item i_m $(1 \leq m \leq r)$. A set of k distinct items $X = \{i_1, i_2, \ldots, i_k\}$ such that $X \subseteq I$ is said to be a k-itemset, where k is the length or level of the itemset. An itemset X is said to be contained in a transaction T_q if $X \subseteq T_q$. An example quantitative database is shown in Table 1. It consists of five transactions and six items, represented using letters from (a) to (f). The profit of each item in Table 2.

Table 1. A quantitative database.   **Table 2.** A profit table.

TID	Items
1	b:7, c:2, d:3, e:1
2	b:4, c:3, d:3
3	a:2, d:1
4	a:1, c:6, f:4
5	b:2, c:3, d:1, f:2

Item	Profit
a	5
b	2
c	1
d	2
e	4
f	1

Definition 1. The minimum average-utility threshold to be used for an item i_j in a database D is a positive integer denoted as $mau(i_j)$. A *MMAU-table* is used to store the minimum average-utility thresholds of all items in D, and is defined as:

$$MMAU - table = \{mau(i_1), mau(i_2), \ldots, mau(i_r)\}, \tag{1}$$

In the following, it will be assumed that the minimum average-utility thresholds of all items for the running example are defined as: $\{mau(a):8, mau(b):8, mau(c):13, mau(d):14, mau(e):20, mau(f):9\}$

Definition 2. For a k-itemset X, the minimum average-utility threshold of X is denoted as $mau(X)$, and is defined as:

$$mau(X) = \frac{\sum\limits_{i_j \in X} mau(i_j)}{|X|} = \frac{\sum\limits_{i_j \in X} mau(i_j)}{k}. \tag{2}$$

Definition 3. The average-utility of an item i_j in a transaction T_q is denoted as $au(i_j, T_q)$, and is defined as:

$$au(i_j, T_q) = \frac{q(i_j, T_q) \times p(i_j)}{1}, \tag{3}$$

where $q(i_j, T_q)$ is the purchase quantity of item i_j in T_q, and $p(i_j)$ is the unit profit of item i_j.

Definition 4. The average-utility of a k-itemset X in a transaction T_q is denoted as $au(X, T_q)$, and defined as:

$$au(X, T_q) = \frac{\sum\limits_{i_j \in X \wedge X \subseteq T_q} q(i_j, T_q) \times p(i_j)}{|X|} = \frac{\sum\limits_{i_j \in X \wedge X \subseteq T_q} q(i_j, T_q) \times p(i_j)}{k}, \tag{4}$$

where k is the number of items in X.

Definition 5. The average-utility of an itemset X in a database D is denoted as $au(X)$, and is defined as:

$$au(X) = \sum_{X \subseteq T_q \wedge T_q \in D} au(X, T_q). \tag{5}$$

Problem Statement: The purpose of HAUIM-MMAU is to efficiently discover the set of all high average-utility itemsets, where an itemset X is said to be a HAUI if its average utility is no less than its minimum average-utility threshold $mau(X)$ as:

$$HAUI \leftarrow \{X | au(X) \geq mau(X)\}. \tag{6}$$

4   Proposed HAUIM-MMAU Framework

Based on the proposed HAUIM-MMAU framework, a baseline algorithm for mining high-utility itemsets with multiple minimum average-utility thresholds (HAUI-MMAU) is proposed in this section. Thereafter, two improved algorithms based on two novel pruning strategies named IEUCP and PBCS are introduced to further increase mining efficiency.

4.1   Proposed Downward Closure Property

In HAUIM, the downward closure (DC) property does not hold for the average-utility measure. But a DC property can be restored using the average-utility upper bound ($auub$) model [7]. This allows to prune the search space early for discovering the HAUIs.

Definition 6. The average-utility upper bound of an itemset X is defined as the sum of the maximum utilities of transactions where X appears:

$$auub(X) = \sum_{X \subseteq T_q \wedge T_q \in D} mu(T_q), \tag{7}$$

where $mu(T_q)$ is the maximum utility of transaction T_q, defined as $mu(T_q) = max(q(i_j, T_q) \times p(i_j)) \forall i_j \in I$.

Definition 7. An itemset X is a high average-utility upper-bound itemset (HAUUBI) if its $auub$ is no less than its minimum average-utility threshold. The set of all HAUUBIs is thus defined as:

$$HAUUBI \leftarrow \{X | auub(X) \geq mau(X)\}. \tag{8}$$

Property 1. Let X^k and X^{k-1} be two itemsets respectively containing k and $k - 1$ items, such that $X^{k-1} \subset X^k$. The $auub$ property of HAUIM states that $auub(X^k) \leq auub(X^{k-1})$. Thus, if an itemset X^{k-1} is not a HAUUBI, then any superset X^k of X^{k-1} is also not a HAUUBI, nor a HAUI.

In HAUIM, the *auub* property was shown to be very effective at reducing the number itemsets to be considered in the search space. However, for the problem of high average-utility itemset mining with multiple minimum average-utility thresholds (HAUIM-MMAU), the *auub* property does not hold. The reason is that itemsets in different levels (itemsets of different lengths) may be evaluated using different minimum average-utility thresholds. Thus, the *auub* property of the HAUIM framework cannot be directly adopted into the designed HAUIM-MMAU framework. To address this limitation, this paper proposes the concept of least minimum average utility (*LMAU*), which is defined as follows.

Strategy 1. The items in the *MMAU-table* are sorted in ascending order of their *mau* values.

For example, items in the *MMAU-table* of the running example are sorted as $\{mau(a):8, mau(b):8, mau(f):9, mau(c):13, mau(d):14, mau(e):20\}$, that is by descending order of their *mau* values.

Based on the designed Strategy 1, the following transaction-maximum-utility downward closure (TMUDC) property holds for the HAUIM-MMAU framework.

Theorem 1 (Transaction- Maximum-Utility Downward Closure Property, TMUDC Property). *Without loss of generality, assume that items in itemsets are sorted in ascending order of mau values. Let X^k be a k-itemset $(k \geq 2)$, and X^{k-1} be a subset of X^k of length $k-1$. If X^k is a HAUUBI, then X^{k-1} is also a HAUUBI.*

Although the TMUDC property can guarantee the anti-monotonicity for HAUUBIs, some HAUIs may still be missed. This can happen when a HAUUBI X of length 1 is evaluated using its $mau(X)$ value. For example, assume that the *auub* value of an itemset (i_j) is less than $mau(i_j)$. This itemset will thus be considered as not being an HAUUBI. Moreover, according to the TMUDC property, any extension Y of the itemset (i_j) is also not a HAUUBI. Consider that $Y = \{i_j, i_z\}$ such that the item i_z has a lower *mau* value than i_j. Then Y would still be a HAUI if $au(Y) \geq \frac{mau(i_j)+mau(i_z)}{2}$. Thus, if the TMUDC is directly applied to obtain the HAUUBIs of length 1, Y would not be considered as a HAUUBI since $auub(i_j) < mau(i_j)$. As a result, Y would not be included into the final set of HAUIs. Hence, it is incorrect to apply the TMUDC property to determine the HAUUBIs of length 1. To solve this problem, the concept of *LMAU* value is introduced, which guarantee the discovery of the complete set of HAUIs in the designed HAUIM-MMAU framework.

Definition 8 (Least Minimum Average-Utility, LMAU). The least minimum average-utility (LMAU) in the *MMAU-table* is defined as:

$$LMAU = min\{mau(i_1), mau(i_2), \ldots, mau(i_r)\}. \qquad (9)$$

where r is the total number of items.

The *LMAU* in the running example is calculated as $LMAU = min\{mau(a), mau(b), mau(f), mau(c), mau(d), mau(e)\} = min\{8, 8, 9, 13, 14, 20\}(= 8)$.

Theorem 2 (HAUIs ⊆ HAUUBIs). *Let be an itemset X^{k-1} of length $k-1$ and X^k be one of its supersets. If X^{k-1} has an auub value lower than the LMAU, X^{k-1} is not a HAUUBI nor a HAUI, as well as all its supersets. Hence, X^{k-1} and its supersets can be discarded.*

The above theorem indicates that any superset of a non-HAUUBI cannot be a HAUUBI. That is, only the combination of two HAUUBIs may generate a potential HAUI. By using this approach for candidate generation, many unpromising candidates can be eliminated, and it becomes unnecessary to calculate their actual average-utility value in the later mining process.

4.2 The Proposed HAUI-MMAU Algorithm

The designed baseline HAUI-MMAU algorithm consists of two phases. In the first phase, the designed HAUI-MMAU algorithm performs a breadth-first search to mine the HAUUBIs. In the second phase, an additional database scan is performed to identify the actual high average-utility itemsets from the set of HAU-UBIs discovered in the first phase. The pseudocode of the proposed algorithm is given in Algorithm 1.

Algorithm 1. HAUI-MMAU

Input: D, a quantitative transactional databases; *ptable*, a profit table;
 MMAU-table, the user predefined multiple minimum average utility
 threshold table.

Output: The set of complete high average-utility itemsets (HAUIs).

1 find the *LMAU* in the *MMAU-table*;
2 scan D to find $auub(i_j)$;
3 **for** *each item i_j* **do**
4 | **if** $auub(i_j) \geq LMAU$ **then**
5 | |_ $HAUUBI^1 \leftarrow HAUUBI^1 \cup i_j$;

6 sort items in $HAUUBI^1$ in ascending order of their *mau* values;
7 set $k \leftarrow 2$;
8 **while** $HAUUBI^{k-1} \neq null$ **do**
9 | $C_k = generate\_candidate(HAUUBI^{k-1})$;
10 | **for** *each k-itemset $X \in C_k$* **do**
11 | | scan D to calculate $auub(X)$;
12 | | **if** $auub(X) \geq mau(X)$ **then**
13 | | |_ $HAUUBI^k \leftarrow HAUUBI^k \cup X$;

14 |_ set $k \leftarrow k + 1$;

15 $HAUUBIs \leftarrow \bigcup HAUUBI^k$;
16 **for** *each itemset X in HAUUBIs* **do**
17 | scan D to calculate $au(X)$;
18 | **if** $au(X) \geq mau(X)$ **then**
19 | |_ $HAUIs \leftarrow HAUIs \bigcup X$;

20 return $HAUIs$;

The proposed HAUI-MMAU algorithm takes as input: (1) a quantitative transactional database D (2) a profit table *ptable* indicating the unit profit of each item, and (3) a multiple minimum average-utility threshold table, *MMAU-table*. First, the least minimum average-utility value (*LMAU*) is found in the *MMAU-table* (Line 1). Then, the database is scanned to find the *auub* values of all 1-items (Line 2). For each 1-item, the algorithm check if its *auub* value is no less than the *LMAU*, to determine if it is an *HAUUBI* of length 1 (Lines 3 to 5). After obtaining the set of all *HAUUBIs* of length 1 ($HAUUBI^1$), the items in $HAUUBI^1$ are sorted in ascending order of their *mau* values (Line 6). This process is necessary to ensure that all *HAUIs* will be retrieved, as explained in the previous section. The parameter k is then set to 2 (Line 7), and a loop is performed to mine all *HAUUBIs* in a level-wise way, starting from itemsets of length $k = 2$ (Lines 8 to 14). During the $(k-1)$-th iteration of the loop, the set $HAUUBI^k$, containing the *HAUUBIs* of length k, is obtained by the following process. First, pairs of $(k-1)$-itemsets in $HAUUBI^{k-1}$ are combined to generate their supersets of length k by applying the *generate_candidate* procedure (Line 9). The result is a set named C_k, containing potential *HAUUBIs* of length k. The database is then scanned again to calculate the *auub* value of each itemset X in C_k. If the itemset X has an *auub* value that is no less than its maximum average-utility (*mau*), it is added to the set $HAUUBI^k$ (Lines 11 to 13). This loop terminates when no more candidates can be generated.

In the second phase, the database is scanned again to find the high average-utility itemsets (*HAUIs*) in the set of *HAUUBIs*, found in first phase (Lines 16 to 19). Based on the proposed theorems, the designed algorithm is correct and complete, and thus returns the full set of *HAUIs* (Line 20).

4.3 Improved IEUCP Strategy

The TMUDC property proposed in the designed HAUI-MMAU algorithm can considerably reduce the search space and speed up the discovery of HAUIs. However, the designed HAUI-MMAU algorithm still suffers from the combinational explosion of the number of itemsets in the search space, since it generates candidates in a level-wise way, and many candidates may be generated at each level. To address this limitation, the Improved estimated utility co-occurrence pruning strategy (IEUCP) is designed to reduce the number of join operations (the operations of combining pairs of itemsets to generate larger itemsets) by pruning unpromising itemsets early. The IEUCP strategy is applied for the generation of k-itemsets ($k \geq 2$). It relies on the following theorem to ensure the correctness and the completeness of the HAUI-MMAU algorithm with the IEUCP strategy, for discovering HAUIs.

Theorem 3 (Improved Estimated Utility Co-occurrence Pruning Strategy, IEUCP). *Let there be an itemset X^{k-1} and an itemset X^k that is an extension of X^{k-1} (that was obtained by appending an item to X^{k-1}). Without loss of generality, assume that items in itemsets are sorted by ascending order of mau values. A sorted prefix of X^{k-m} is a subset of X^{k-1} containing* the m first items of X^{k-1} according to ascending order of mau values, where $1 \leq m < k - 1$. For example, consider the itemset (bcde), which is a 4-itemset ($k = 4$). Its sorted prefixes are (b), (bc), and (bcd). If there exists a sorted prefix X^{k-m} of X^k that is not a HAUUBI, then X^k is also not a HAUUBI.

It is important to note that an itemset X can only be safely pruned by Theorem 3 if X has a sorted prefix that is not a HAUUBI, and that this pruning condition should not be applied by considering subsets of X that are not sorted prefixes of X. Based on Theorem 3 and the proposed TMUDC property, the completeness and correctness of the HAUI-MMAU algorithm is preserved, when applying the IEUCP strategy.

4.4   Pruning Before Calculation Strategy, PBCS

Although the developed IEUCP strategy is effective at reducing the number of unpromising candidates in the first phase, calculating the set of actual HAUIs in the second phase remains a very time-consuming process. To address this issue, this article further proposes an efficient pruning before calculation strategy (PBCS) to prune itemsets without performing a database scan.

Theorem 4. Let X, X^a, and X^b be itemsets such that $X^a \subset X$, $X^b \subset X$, and $X^a \cup X^b = X$, $X^a \cap X^b = \varnothing$. If both X^a and X^b are not HAUIs, the itemset X is also not a HAUI.

Thus, if it is found that two subsets X^a and X^b of an itemset X are not HAUIs and respect the condition of Theorem 4, it can be concluded that X is not a HAUI. It is thus unnecessary to scan the database to calculate the actual utility of X. Based on this PBCS, the cost of calculating the actual utility of HAUUBIs in the second phase can be greatly reduced.

5   Experimental Evaluation

Substantial experiments were conducted to evaluate the effectiveness and efficiency of the proposed algorithms. The performance of the baseline HAUI-MMAU algorithm was compared with two versions named HAUI-MMAU$_{\text{IEUCP}}$ and HAUI-MMAU$_{\text{PBCS}}$, which respectively integrates the proposed IEUCP strategy, and both the designed IEUCP and PBCS strategies. Note that no previous studies have been done for mining HAUIs with multiple average-utility thresholds, and it is thus unnecessary to compare the designed algorithms with the previous algorithms for HAUIM with a single minimum average-utility threshold. Experiments were carried on both real-life and synthetic dataset, having various characteristics. The foodmart, retail, and chess datasets were obtained from the SPMF website [6]. The IBM Quest Synthetic Dataset Generator [2] was used to generate a synthetic dataset named T40I10D100K. Parameters and characteristics of the datasets used in the experiments are respectively shown in Tables 3 and 4.

Table 3. Parameters of the datasets.

#$|D|$	Transaction count
#$|I|$	Number of distinct items
AvgLen	Average transaction length
MaxLen	Maximal transaction length
Type	Dataset type

Table 4. Characteristics of the datasets.

Dataset	#$|D|$	#$|I|$	AvgLen	MaxLen	Type
Foodmart	21,556	1,559	4.4	10	Sparse
Retail	88,162	16,470	10.3	76	Sparse
Chess	3,196	75	36	36	Dense
T40I10D100K	100,000	942	39.6	77	Dense

Furthermore, the method proposed in [9] for assigning the multiple thresholds to items was adapted to automatically set the *mau* value of each item in the proposed HAUIM-MMAU framework. The following equation is thus used to set the *mau* value of each item i_j:

$$mau(i_j) = max\{\beta \times p(i_j), GLMAU\}, \qquad (10)$$

where β is a constant used to set the *mau* values of an item as a function of its unit profit. To ensure randomness and diversity in the experiments, β was respectively set in different interval for varied datasets. The constant $GLMAU$ is user-specified and represents the global least average-utility value. Lastly, $p(i_j)$ represents the external utility (unit profit) of the item i_j. If β is set to zero, a single minimum average-utility threshold $GLMAU$ is used for all items. In that case, the task of HAUI-MMAU would become the same as traditional HAUIM.

5.1 Runtime

In this section, the runtimes of three proposed approaches are respectively compared. Figure 1 shows the runtime of the proposed algorithms when β is varied with a fixed $GLMAU$ value within the predefined interval for each dataset.

It can be observed in Fig. 1 that the improved HAUI-MMAU$_{\text{IEUCP}}$ and HUAI-MMAU$_{\text{PBCS}}$ algorithms outperform the baseline HAUI-MMAU algorithm. For example, when $GLMAU$ is set to 168,000 and β is set in the [200, 400] interval for the chess dataset, the runtime of HAUI-MMAU, HAUI-MMAU$_{\text{IEUCP}}$ and HAUI-MMAU$_{\text{PBCS}}$ are respectively 116, 133 and 58 seconds. It can also be observed that HAUI-MMAU$_{\text{PBCS}}$ always has the best performance among the three algorithms. The reason is that HAUI-MMAU$_{\text{PBCS}}$ applies both the IEUCP and PBCS strategies to improve the performance of the mining process,

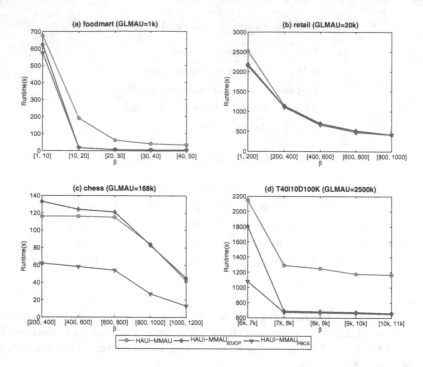

Fig. 1. Runtime for a fixed *GLMAU* and varied β.

while HAUI-MMAU$_{\text{IEUCP}}$ only applies the IEUCP strategy. But both strategies are complementary to each other. The IEUCP strategy can prune a large amount of unpromising itemsets to avoid several database scans when calculating the *auub* values of itemsets in the first phase, while the PBCS strategy can prune unpromising HAUIs from the remaining *HAUUBIs* to avoid calculating their actual average-utility values, in the second phase. Hence, HAUI-MMAU$_{\text{PBCS}}$ has the best performance among the three algorithms. In Fig. 1, it can also be observed that the performance gap between HUAI-MMAU$_{\text{PBCS}}$ and the other algorithms is larger for the chess and T40I10D100K datasets, especially when β is set lower. The reason is that both chess and T40I10D100K are dense datasets with relatively few distinct items. As a result, the discovered HAUUBIs are highly related to each other and the designed PBCS strategy can thus be used to prune the unpromising HAUIs early from the remaining HAUUBIs.

5.2 Candidate Analysis

We further compared the number of candidates generated by each of the three algorithms. Here, an itemset is said to be a candidate if its actual average-utility value is calculated in the first phase, and its exact average-utility is calculated in the second phase. Figure 2 shows the number of candidates for the three algorithms when β is varied with a fixed *GLMAU* value.

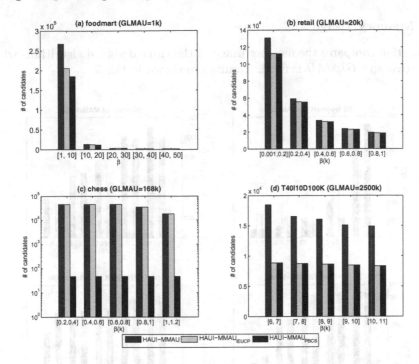

Fig. 2. Number of candidates for a fixed *GLMAU* and varied β

In Fig. 2, it can be observed that the HAUI-MMAU algorithm always considers more candidates than the other two algorithms. The reason is that HAUI-MMAU$_{\text{IEUCP}}$ can efficiently prune some unpromising HAUUBIs using its designed pruning strategy, while HAUI-MMAU$_{\text{PBCS}}$ can prune unpromising HAUIs from the remaining HAUUBIs for revealing the actual HAUIs. It can also be found in Fig. 2(a) that when β is set in the [1, 10] interval, more candidates are generated than for the other intervals of β such as [10, 20], [20, 30], [30, 40] and [40, 50]. The reason is that foodmart is a very sparse dataset having a large number of distinct items. In the [1, 10] interval, the generated candidates have similar average-utility values. Since the minimum average-utility threshold is set lower, the amounts of candidates are thus revealed in this interval. When the minimum average-utility threshold is increased, itemsets with similar average-utility values may be lower than the certain threshold; a large amount of candidates can thus be pruned. It can be observed in Fig. 2(b) and (d), that the number of candidates considered by HAUI-MMAU$_{\text{IEUCP}}$ and HAUI-MMAU$_{\text{PBCS}}$ is almost the same. This explains why they have similar runtimes, as shown in Fig. 1(b) and (d). But in Fig. 2(c), it is obvious that the number candidates considered by HAUI-MMAU$_{\text{PBCS}}$ is much less than the two other algorithms, which shows that the proposed PBCS strategy can greatly reduce the search space in the second phase for the chess dataset.

5.3   Memory Usage

This section compares the memory usage of the three designed algorithms when β is varied and $GLMAU$ is fixed. Results are shown in Fig. 3.

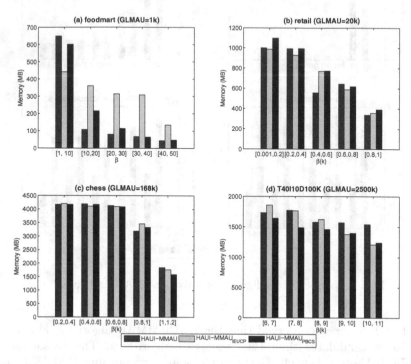

Fig. 3. Memory usage for a fixed $GLMAU$ and various values of β.

It can be seen in Fig. 3 that HAUI-MMAU sometimes consumes less memory than HAUI-MMAU$_{\text{IEUCP}}$ when β is varied and $GLMAU$ is fixed. In most cases, the HAUI-MMAU$_{\text{PBCS}}$ algorithm consumes more memory than the other two algorithms since both the IEUCP and PBCS strategies are adopted to prune unpromising candidates, as it can be observed in Fig. 3(b) and (c). Although the IEUCP strategy requires additional memory to store its structure, pruning itemsets reduces the number of itemsets to then be considered as potential HAUIs, and thus the runtime and memory usage. Thus, the HAUI-MMAU$_{\text{PBCS}}$ algorithm consumes less memory than the HAUI-MMAU and HAUI-MMAU$_{\text{IEUCP}}$ algorithms, in most cases.

6   Conclusion

In this paper, the high average-utility itemset mining with multiple minimum average-utility thresholds (HAUIM-MMAU) framework was designed to mine high average-utility itemsets (HAUIs) with multiple minimum average-utility thresholds. The baseline HAUI-MMAU algorithm is a two phases algorithm, which relies on several designed theorems to find the HAUIs. The first IEUCP strategy is designed to prune the search space, and thus increasing the efficiency of HAUI mining. The second PBCS pruning strategy is used to reduce the number of HAUUBIs at the beginning of the second phase, for revealing the actual HAUIs. An extensive experimental study was conducted on both synthetic and real datasets to evaluate the performance of the algorithms in terms of runtime, number of candidates, and memory usage. Results show that the designed algorithms can efficiently discover the HAUIs and that the two pruning strategies can effectively reduce the number of candidates in the first and second phase.

References

1. Agarwal, R., Imielinski, T., Swami, A.: Database mining: a performance perspective. IEEE Trans. Knowl. Data Eng. **5**(6), 914–925 (1993)
2. Agrawal, R., Srikant, R.: Quest synthetic data generator (1994). http://www. Almaden.ibm.com/cs/quest/syndata.html
3. Agrawal, R., Srikant, R.: Fast algorithms for mining association rules in large databases. In: International Conference on Very Large Data Bases, pp. 487–499 (1994)
4. Chan, R., Yang, Q., Shen, Y.D.: Mining high utility itemsets. In: IEEE International Conference on Data Mining, pp. 19–26 (2003)
5. Fournier-Viger, P., Wu, C.-W., Zida, S., Tseng, V.S.: FHM: faster high-utility itemset mining using estimated utility co-occurrence pruning. In: Andreasen, T., Christiansen, H., Cubero, J.-C., Raś, Z.W. (eds.) ISMIS 2014. LNCS, vol. 8502, pp. 83–92. Springer, Heidelberg (2014)
6. SPMF: an open-source data mining library. http://www.philippe-fournier-viger. com/spmf/
7. Hong, T.P., Lee, C.H., Wang, S.L.: Effective utility mining with the measure of average utility. Expert Syst. Appl. **38**(7), 8259–8265 (2011)
8. Kiran, R.U., Reddy, P.K.: Novel techniques to reduce search space in multiple minimum supports-based frequent pattern mining algorithms. In: ACM International Conference on Extending Database Technology, pp. 11–20 (2011)
9. Liu, B., Hsu, W., Ma, Y.: Mining association rules with multiple minimum supports. In: ACM SIGKDD International Conference on Knowledge Discovery and Data Mining, pp. 337–341 (1999)
10. Liu, Y., Liao, W., Choudhary, A.K.: A two-phase algorithm for fast discovery of high utility itemsets. In: Ho, T.-B., Cheung, D., Liu, H. (eds.) PAKDD 2005. LNCS (LNAI), vol. 3518, pp. 689–695. Springer, Heidelberg (2005)
11. Lin, C.-W., Hong, T.-P., Lu, W.-H.: Efficiently mining high average utility itemsets with a tree structure. In: Nguyen, N.T., Le, M.T., Świątek, J. (eds.) ACIIDS 2010. LNCS, vol. 5990, pp. 131–139. Springer, Heidelberg (2010)
12. Lin, J.C.W., Gan, W., Fournier-Viger, P., Hong, T.P.: Mining high-utility itemsets with multiple minimum utility thresholds. In: International C* Conference on Computer Science and Software Engineering, pp. 9–17 (2015)
13. Lan, G.C., Hong, T.P., Tseng, V.S.: A projection-based approach for discovering high average-utility itemsets. J. Inf. Sci. Eng. **28**(1), 193–209 (2012)

14. Lan, G.C., Hong, T.P., Tseng, V.S.: Efficiently mining high average-utility itemsets with an improved upper-bound strategy. Int. J. Inf. Technol. Decis. Making **11**(5), 1009–1030 (2012)
15. Liu, M., Qu, J.: Mining high utility itemsets without candidate generation. In: ACM International Conference on Information and Knowledge Management, pp. 55–64 (2012)
16. Lu, T., Vo, B., Nguyen, H.T., Hong, T.-P.: A new method for mining high average utility itemsets. In: Saeed, K., Snášel, V. (eds.) CISIM 2014. LNCS, vol. 8838, pp. 33–42. Springer, Heidelberg (2014)
17. Ryang, H., Yun, U., Ryu, K.: Discovering high utility itemsets with multiple minimum supports. Intell. Data Anal. **18**(6), 1027–1047 (2014)
18. Yao, H., Hamilton, H.J., Butz, C.J.: A foundational approach to mining itemset utilities from databases. In: SIAM International Conference on Data Mining, pp. 211–225 (2004)
19. Yao, H., Hamilton, H.J.: Mining itemset utilities from transaction databases. Data Knowl. Eng. **59**(3), 603–626 (2006)

Data Mining in Medicine: Relationship of Scoliotic Spine Curvature to the Movement Sequence of Lateral Bending Positions

Athena Jalalian[1], Francis E.H. Tay[1(✉)], and Gabriel Liu[2]

[1] Department of Mechanical Engineering, Faculty of Engineering,
National University of Singapore,
9 Engineering Drive 1, Singapore 117575, Singapore
athena@u.nus.edu, mpetayeh@nus.edu.sg
[2] Department of Orthopedic Surgery,
National University of Singapore, Singapore, Singapore
gabriel_liu@nuhs.edu.sg

Abstract. We aim to determine relationships between scoliotic spine curvatures in movement sequence from left bending to erect to right bending positions in the frontal plane. A multi-body kinematic modelling approach is utilized to reconstruct the curvatures and study the relationships. The spine is considered as a chain of micro-scale motion-segments (MMSs). Linear regression method is adopted to identify relationships between angles of MMSs in erect and lateral bending positions. Excellent linear relationships ($R^2 = 0.93 \pm 0.09$) were identified between angles of MMSs placed between each two successive vertebrae. We showed that these relationships give good estimates of the curvatures (Root-mean-square-error = 0.0172 ± 0.0114 mm) and the key parameters for scoliosis surgery planning; estimation errors for Cobb angle, spinal mobility, and flexibility were $0.0016 \pm 0.0122°$, $0.0010 \pm 0.086°$, and 0.0002 ± 0.0002 respectively. This paper provides an important insight: scoliotic spine curvatures in lateral bending positions and the key parameters for surgery planning can be predicted using spine curvature in erect position.

Keywords: Adolescent idiopathic scoliosis · Lateral bending positions · Multi-body kinematic model · Spine curvature · Spine movement sequence

1 Introduction

Scoliosis is a complex 3-dimensional structural deformity of the human spine. Surgery is required for correction of severe scoliosis (i.e. Cobb angle[1] [1] >45°) [2]. To plan the surgery (e.g. selection of fusion levels), several X-ray images are taken from the scoliotic spine in different spine positions in the frontal plane of the human body. Considering the X-ray images of the scoliotic spine in the erect and lateral bending positions as the representative of a sequence of spine movement from left bending to erect to right bending positions, the motion data acquired from this sequence is used to

---

[1] The definition is provided in Sect. 2.

© Springer International Publishing Switzerland 2016
P. Perner (Ed.): ICDM 2016, LNAI 9728, pp. 29–40, 2016.
DOI: 10.1007/978-3-319-41561-1_3

measure the key parameters (e.g. Cobb angle, spinal mobility[1], spinal flexibility[1]) for planning the surgery [3]. The scoliotic spine movement sequence is represented by the sequence of the first, intermediate, and end spine positions.

Radiological imaging exposes the scoliotic patients to harmful radiation. To mitigate radiation risks, reducing the number of the required X-ray images is potentially helpful. In this regard, motion capture [5], and inclinometer [6], methods have been adopted to obtain the required data from the sequence of the spine movement in order to measure the key parameters: spinal mobility and flexibility. However, Hresko et al. [7] found no significant statistical correlation between the measurements by using these methods and the X-ray images. It should be noted that the measurements on X-ray images are the gold standard for the values of the parameters [8, 9]. Another alternative to measurement of the parameters can be prediction of the scoliotic spine curvatures in the erect and lateral bending positions by using the information acquired from the movement sequence of the spine. However, to the best of our knowledge, study on such a prediction is considerably lacking in the existing literature.

In our attempt to study such a prediction, this paper aims to determine whether mathematical relationships exist between the scoliotic spine curvatures in the movement sequence from left bending to erect to right bending positions in the frontal plane. For this end, we utilize a multi-body kinematic modelling approach to reconstruct the spine curvatures in the positions. This study focuses on adolescent idiopathic scoliosis (AIS), which requires surgical correction. AIS is the most common form of the scoliosis ($\approx$80 % [10]), and patients are predominantly female [11].

2 Scoliotic Spine

In this section, we briefly explain the scoliotic spine and its evaluation. Scoliosis is a side-to-side curvature of the spine with a twisting of the vertebral column about its axis (Fig. 1a) [12]. It has been diagnosed in 1.5 % to 3 % of the population [2]. The scoliotic deformity affects the thoracic and lumbar regions of the spine from the fifth vertebra (L5) of the lumbar region to the first vertebra (T1) of the thoracic region.

Scoliosis is normally evaluated by measuring Cobb angles on a 2-dimensional radiograph taken in the erect position in the frontal plane [1]. Cobb angles are measured between the inflection vertebra[2] and the end vertebrae[3] (Fig. 1b) [13]. The severity of the scoliosis is determined by Cobb angle; Cobb angle greater than 45° corresponds to the severe scoliosis.

We define the terms below to describe the scoliotic spine curvatures in the frontal plane in order to study the relationships between the curvatures. The definitions are based on the concept of 'vertebral body line' proposed by Scoliosis Research Society (SRS) (term #2 in [14]).

---

[2] Inflection vertebra is where the spine curvature changes direction from convex to concave and vice versa [15].

[3] The vertebrae that define the ends of the spine curvature in a certain plane [15].

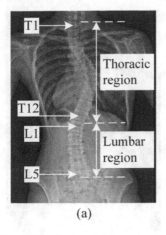 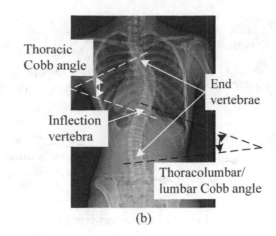

(a)   (b)

Fig. 1. Scoliotic spine in erect position in frontal plane (a) spine regions, and (b) Cobb angles

1. **Global coordinate system (G).** G is represented by XYZ defined on the lowest vertebra in the spine (Fig. 2a). G has its origin at the mid-point of the vertebral body of the lowest vertebra. The mid-point of a vertebra is the intersection of the line drawn from the upper left corner to the lower right of the vertebral body and the line drawn from the upper right to the lower left of the vertebral body (the white circles in Fig. 2a) [15]. Y-axis of G is parallel to the line joining the left and right pedicles of the lowest vertebra. Z-axis is parallel to the line that passes through the center of the upper and lower endplates of the lowest vertebra. X-axis defines the anterior direction. The plane YZ is the frontal plane of the human body.
2. **Scoliotic spine curvature in the frontal plane.** It is the curved line that passes through the mid-points of the vertebral bodies projected on the frontal plane (Fig. 2b).
3. **Location of a vertebra in the frontal plane (LOC).** It is the location of the mid-point (Fig. 2a) of the vertebral body in the frontal plane. LOC is given by the ordered pair of (Z,Y) according to the definition of G.
4. **Orientation of a vertebra in the frontal plane (Θ).** It is the angle between Z-axis and a line (Fig. 2a) passing through the center of the upper and lower endplates of the vertebra in the frontal plane.
5. **Spinal mobility in the frontal plane.** It is the angle between the lines connecting the mid-points of the lowest and uppermost vertebrae in the erect and left/right bending positions. It can be defined for a part of the spine, e.g. thoracic region.
6. **Spinal flexibility in the frontal plane.** It is the difference of spinal movement (excursion) from the erect to lateral bending positions [16]. A scoliosis flexibility index is calculated by (1) according to [7]. It can be defined for each spine region.

$$\text{Flexibility index} = \left| \frac{\text{Cobb}_{\text{erect}} - \text{Cobb}_{\text{left/right}}}{\text{Cobb}_{\text{erect}}} \right| \tag{1}$$

Where, $|\cdot|$ is the absolute operation.

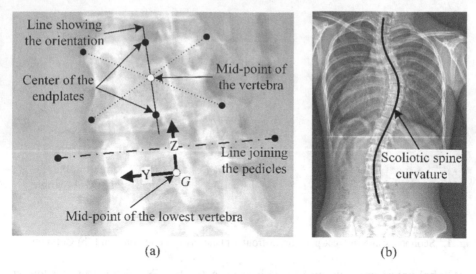

(a)   (b)

Fig. 2. The global coordinate system of the spine, the spine curvature, and the geometrical parameters, i.e. location and orientation of vertebrae

3   Methods

3.1   Subjects and Data Collection

After obtaining domain specific review board approval, pre-operative X-ray images of 18 patients with severe AIS were used for the study (Table 1). The images were taken in three posterior-anterior positions in the frontal plane: one erect position and two prone positions (left and right bending positions). These X-ray images represent the sequence of the scoliotic spine movement from the left bending to erect to right bending positions.

The patients had no neurologic deterioration, and they were admitted to hospital for surgical treatment. There were 12 female and six male patients between the ages of 12 and 19 years (mean age of 15 years). The thoracolumbar/lumbar and thoracic Cobb angles of the patients were $46.28 \pm 13.94°$ (mean $\pm$ SD) and $49.78 \pm 12.25°$ respectively.

To obtain the spine curvature in a position, the respective X-ray image was analyzed to measure LOC and Θ of the vertebrae from L4 to T2. LOC and Θ of a vertebra give the translation and rotation of the vertebra, respectively, with respect to its inferior vertebra when the spine moves from the erect to a bending position. It should be noted that L5 and T1 were excluded since they were not visible in the majority of the images. Section 3.2 explains how the curvatures are obtained by using LOC and Θ.

The accuracy limit of the measurement of LOC and Θ was 0.1 mm and 0.1° respectively. LOC and Θ were measured three times by two experts familiar with X-ray images of the scoliotic spine. The mean values of the measurements were considered for the parameters. The measurements were supervised by GL (one of the authors) who is an experienced scoliosis surgeon at National University Hospital, Singapore.

Table 1. Descriptive data of the patients

Patient	Gender	Age (year)	Lenke classification [3]	Thoracolumbar/lumbar Cobb angle (°)	Thoracic Cobb angle (°)
1	Female	13	1A	24	49
2	Female	15	1B	39	53
3	Female	16	1C	46	32
4	Female	12	1C	48	33
5	Female	13	2A	53	38
6	Female	16	2B	35	53
7	Female	19	2C	43	48
8	Female	14	2C	54	55
9	Female	15	3A	59	54
10	Female	13	4C	63	86
11	Female	12	5C	62	49
12	Female	14	6C	59	51
13	Male	14	2A	47	59
14	Male	19	2B	26	48
15	Male	18	2B	28	61
16	Male	18	3B	24	46
17	Male	14	3C	53	31
18	Male	19	3C	70	50

The repeatability and reliability of the measurements were evaluated by using Pearson correlation coefficient. The intra-observer coefficients were 0.95 ± 0.04 and 0.93 ± 0.05 for expert-1 and expert-2 respectively. The inter-observer coefficient was 0.91. These agreements are excellent and can show the repeatability and reliability of the measurements according to [17].

3.2   Reconstruction of the Spine Curvatures

The spine curvatures are obtained by fitting polynomials to LOC and Θ. The polynomials (2) have no first-degree and constant terms because the location and orientation of L4 are zero according to definition of G.

$$P(Z) = \lambda_t Z^t + \lambda_{t-1} Z^{t-1} + \ldots + \lambda_2 Z^2 \tag{2}$$

The linear least squares method is adopted to fit the polynomial. The coefficients (λ) of P are given in terms of the measured parameters by

$$\lambda = \left(\mathbf{Z}^\mathrm{T} \mathbf{Z}\right)^{-1} \mathbf{Z}^\mathrm{T} \mathbf{Y} \tag{3}$$

Where, the $^\mathrm{T}$ sign stands for the transpose operation. λ and $\mathbf{Y}$ are the vectors of the coefficients and measured parameters respectively. $\mathbf{Z}$ is the design matrix. λ, $\mathbf{Y}$, and $\mathbf{Z}$ are given by (4).

The upper and lower halves of **Y** and **Z** correspond to the measured locations and orientations respectively. The order of each polynomial is adjusted in order to find the best fitting, i.e. the least RMSE for LOC and Θ.

$$
\lambda = \begin{bmatrix} \lambda_t \\ \lambda_{t-1} \\ \vdots \\ \lambda_2 \end{bmatrix}, \mathbf{Y} = \begin{bmatrix} Y_{L4} \\ Y_{L3} \\ \vdots \\ Y_{T2} \\ \hline \tan\Theta_{L4} \\ \tan\Theta_{L3} \\ \vdots \\ \tan\Theta_{T2} \end{bmatrix}, \mathbf{Z} = \begin{bmatrix} Z_{L4}^{t} & Z_{L4}^{t-1} & \cdots & Z_{L4}^{2} \\ Z_{L3}^{t} & Z_{L3}^{t-1} & \cdots & Z_{L3}^{2} \\ \vdots & \vdots & \ddots & \vdots \\ Z_{T2}^{t} & Z_{T2}^{t-1} & \cdots & Z_{T2}^{2} \\ tZ_{L4}^{t-1} & (t-1)Z_{L4}^{t-2} & \cdots & 2Z_{L4}^{1} \\ tZ_{L3}^{t-1} & (t-1)Z_{L3}^{t-2} & \cdots & 2Z_{L3}^{1} \\ \vdots & \vdots & \ddots & \vdots \\ tZ_{T2}^{t-1} & (t-1)Z_{T2}^{t-2} & \cdots & 2Z_{T2}^{1} \end{bmatrix} \tag{4}
$$

Multi-body modelling approach considers the spine as a chain of motion segments comprising joints and links [18, 19]. By employing this approach, we consider the spine curvature as a chain of MMSs laying on the curvature (Fig. 3). An MMS consists of a rigid link and a 1-DOF rotary joint. The links of MMSs are equal in length. The chain is constrained at the first MMS attached to the mid-point of L4, and cannot translate with respect to this mid-point. The last MMS corresponds to the mid-point of the uppermost vertebrae considered in the spine curvature (T2 in this study).

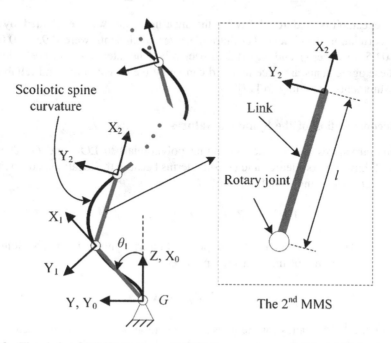

Fig. 3. The chain of MMSs that lays on the spine curvature and the coordinate systems

Denavit-Hartenberg convention [20] is adopted to represent the chain of MMSs. According to Denavit-Hartenberg convention, the angle (θ) of the rotary joints of MMSs must be identified to reconstruct the spine curvature in a certain position. θ of an MMS is defined with respect to X-axis attached to its inferior MMS in the counterclockwise direction (Fig. 3). For example, θ_1 is the angle between X_1 and X_0, and it is measured with respect to X_0 in the counterclockwise direction.

The chain of MMSs is fully characterized to a given patient by specification of the number of MMSs and the length of their links. In this study, the total number of MMSs is considered 1000 according to our previous work [21], and the length of the links is 0.37 ± 0.03 mm. After the characterization, MMSs nearest (the shortest Euclidian distance) to the mid-point of the vertebrae in the erect position are considered as the representative of LOC and Θ of the vertebrae in all the positions.

4 Results

Fifty-four curvatures (3 positions $\times$ 1 movement sequence $\times$ 18 patients) were reconstructed by using the chain of MMSs. The RMSEs of curvatures estimated by the chain were 0.0000 mm. Besides, the RMSEs of Cobb angles, spinal mobility, and spinal flexibility were 0.0002°, 0.0004°, and 0.0000 respectively. These errors were calculated between the curvatures and parameters estimated by using the polynomial (the actual spine curvature) and chain. It should be noted that the estimates by using the polynomials are the reference in this study because we adopted Analytic Cobb method [22, 23] that measures the parameters on the spine curvature. The small RMSEs show that the curvatures estimated by the chain are good estimates of the actual spine curvatures, and the estimated curvatures give good estimates of the key parameters for planning scoliosis surgery.

For each patient, the chain gave θs for MMSs of the spine in the movement sequence from the left bending to erect to right bending positions. By utilizing linear regression method, it was found that a linear relationship (5) exists between θ_{erect} and $\theta_{\text{left/right}}$ of MMSs placed between the mid-points of each two successive vertebrae.

$$\theta_{\text{left/right}} \approx \hat{\theta}_{\text{left/right}} = \alpha_{\text{left/right}}(i) \cdot \theta_{\text{erect}} + \beta_{\text{left/right}}(i) \qquad (5)$$

Where, $\hat{\theta}$ represents the estimated θ. α and β are the slopes and intercepts of the linear relationships respectively. i corresponds to i^{th} vertebra.

In total, 504 linear relationships (14 vertebrae $\times$ 2 positions $\times$ 18 patients) were identified. The coefficient of determination (R^2) for the linear relationships was 0.93 ± 0.09 (Fig. 4a). The confidence interval of the mean was between 0.92 and 0.95, with the confidence level of 95 %. According to [17] and [24], these high R^2 show that there exists an excellent linear relationship between θ_{erect} and $\theta_{\text{left/right}}$ of MMSs between each two successive vertebrae. As an example, Fig. 4b shows the linear relationship identified between θ_{erect} and θ_{left} of MMSs placed between the mid-points of L3 and L4 for one of the patient.

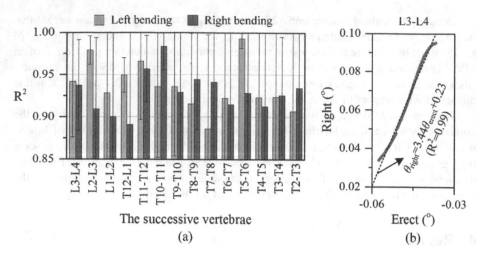

Fig. 4. (a) Mean ± SD of R^2 of the relationships between θ_{erect} and $\theta_{left/right}$ of MMSs placed between each two successive vertebrae, (b) an example of the linear relationship identified for L3-L4 of one of the patients in the movement from the erect to right bending position.

The curvatures of 36 spines (2 lateral bending positions × 18 patients) were reconstructed by $\hat{\theta}_{left/right}$, and compared with the actual curvatures (Fig. 5). The small RMSEs (0.0172 ± 0.0114 mm) demonstrate that the scoliotic spine curvatures estimated by using the linear relationships are good estimates of the actual curvatures.

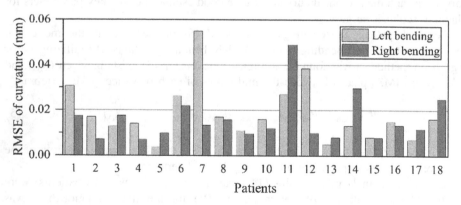

Fig. 5. RMSEs between the estimated and actual curvatures for each patient in the left and right bending positions

As mentioned in Sect. 1, the estimated curvatures are supposed to replicate the X-ray images taken in the lateral bending positions. Besides, Cobb angle and spinal mobility and flexibility are the key parameters measured on the X-ray images to plan the scoliosis surgery [3]. Therefore, the parameters measured on the estimated

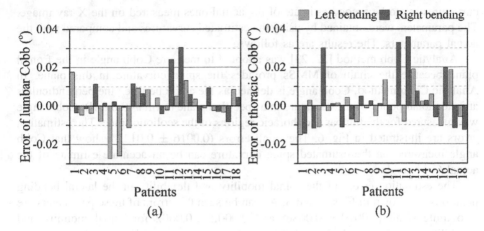

Fig. 6. The error between the actual and estimated Cobb angles for the individual patients

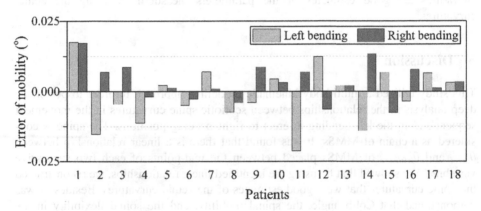

Fig. 7. The error between the actual and estimated spinal mobility

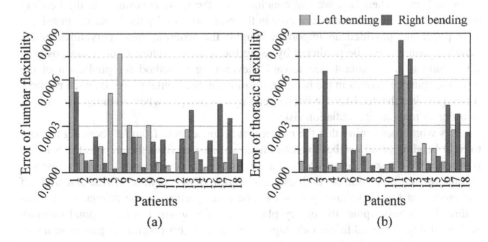

Fig. 8. The error between the actual and estimated spinal flexibility

curvatures are to be a good estimate of the actual ones measured on the X-ray image. The parameters were obtained by using the estimated curvatures and compared with the actual parameters. The results are as follows.

Analytic Cobb method [22, 23] was adopted to measure Cobb angle in the frontal plane because the chain of MMSs provides the spine curvature in this plane. In Analytic Cobb method, Cobb angle is defined as the angle between the perpendiculars at inflectional points of the curvature in a specified plane (term #48 in [4]). Cobb angle was measured for the lumbar and thoracic regions of the scoliotic spine. The estimation errors are illustrated in Fig. 6. The small errors ($0.0016 \pm 0.0122°$) show that Cobb angle measured on the estimated spine curvature can be an accurate estimate of the actual Cobb angle.

The estimation errors of the spinal mobility and flexibility in the lateral bending positions are shown in Figs. 7 and 8. As can be seen the errors of these parameters are also quite small ($0.0010 \pm 0.0086°$ and 0.0002 ± 0.0002 for spinal mobility and flexibility respectively) showing that the parameters measured by using the estimated curvatures are good estimates of the parameters measured by using the actual curvatures.

5 Discussion

This study utilized a multi-body kinematic modelling approach in order to perform a deep analysis in the relationships between scoliotic spine curvatures in the movement sequence from the left bending to erect to right bending positions. The spine is considered as a chain of MMSs. It was found that there is a linear relationship between θ_{erect} and $\theta_{left/right}$ of MMSs placed between the mid-points of each two successive vertebrae. We showed that by using the identified linear relationships, we reconstructed the spine curvatures that were good estimates of the actual curvatures. Besides, it was demonstrated that Cobb angle, the spinal mobility, and the spinal flexibility in the bending positions obtained from the estimated curvatures were also good estimates of the actual ones. Therefore, we can conclude that the spine curvatures in the bending positions can be related to the curvature in the erect position by the linear relationships. This paper has provided an important insight: the scoliotic spine curvatures in the bending positions can be predicted by using the spine curvature in the erect position. This study is an important step towards developing a method for prediction of the scoliotic spine curvatures in the bending positions in the frontal plane. By such method, we can obviate the need for the lateral bending radiographs when planning the surgery in order to mitigate the radiation risks.

It is worth mentioning that the movement sequence was from left bending to erect to right bending positions, but the movement was studied from the erect to the bending positions. We chose the erect position as the start position of the movement sequence, and the reference position for both characterization of the chain of MMSs and study of the prediction for the following reasons. The erect position is the reference for evaluating the scoliotic spine, its surgery planning, and assessment of the spinal mobility and flexibility [1]. In addition, radiological imaging of this position is part of scoliosis routine standard care [7]. Moreover, the evaluation of the scoliosis surgery outcome is mainly made by comparing the pre- and post-operative X-ray images taken in the erect position. Therefore, the X-ray image of the spine in the erect position is required and cannot be removed from the scoliosis routine standard care. As such, the erect position is an excellent choice to provide the information for predicting the spine curvature in the lateral bending positions.

Indeed, to predict the spine curvatures, α (slope) and β (intercept) of the identified linear relationships should be estimated. Estimation of α and β is out of the scope of this paper. Further studies should be done on the estimation of α and β. We are in the process of estimating α and β to fully characterize the identified relationships to develop a method for prediction of scoliotic spine curvatures in the bending positions.

The data was collected from Asian subjects. Thus, the results of this study may not be applicable to other races. In addition, the majority of the Lenke types in this study are types 1, 2, and 3. More information relating to patients with the Lenke types of more than three can help to better generalize the results to all Lenke types.

6 Conclusions

This paper studied the relationships between the scoliotic spine curvatures in the movement sequence from the left bending to erect to right bending positions in the frontal plane. We identified the relationships that give good estimates of the spine curvatures in the bending positions by using the curvature in the erect position. We also demonstrated that by using the estimated curvatures, good estimates of Cobb angle, spinal mobility, and spinal flexibility in the bending positions were obtained. The paper provides an important insight: scoliotic spine curvatures in lateral bending positions and the key parameters for surgery planning can be predicted using spine curvature in erect position. This study is an ideal starting point for developing a method for prediction of the scoliotic spine curvatures in the lateral bending positions in the frontal plane. By such method, we can reduce the number of the X-ray images required for planning the scoliosis surgery (i.e. X-ray of the bending positions) in the scoliosis standard care, and thus, reduce the radiation risks.

References

1. Cobb, J.: Outline for the study of scoliosis. Instr. Course Lect. 5, 261–275 (1948)
2. Duke, K., et al.: Biomechanical simulations of scoliotic spine correction due to prone position and anaesthesia prior to surgical instrumentation. Clin. Biomech. 20(9), 923–931 (2005)
3. Lenke, L.G., et al.: Adolescent idiopathic scoliosis a new classification to determine extent of spinal arthrodesis. J. Bone Joint Surg. 83(8), 1169–1181 (2001)
4. Stokes, I.: Three-dimensional terminology of spinal deformity (2001). http://www.srs.org/professionals/glossary/SRS_3D_terminology.htm. Accessed 29 Dec 2014
5. Wilk, B., et al.: The effect of scoliosis fusion on spinal motion: a comparison of fused and nonfused patients with idiopathic scoliosis. Spine 31(3), 309–314 (2006)

6. Poussa, M., Mellin, G.: Spinal mobility and posture in adolescent idiopathic scoliosis at three stages of curve magnitude. Spine **17**(7), 757–760 (1992)

7. Hresko, M.T., et al.: A comparison of methods for measuring spinal motion in female patients with adolescent idiopathic scoliosis. J. Pediatr. Orthop. **26**(6), 758–763 (2006)

8. Perret, C., et al.: Pelvic mobility when bending forward in standing position: validity and reliability of 2 motion analysis devices. Arch. Phy. Med. Rehabil. **82**(2), 221–226 (2001)

9. Wong, K.W., et al.: The flexion–extension profile of lumbar spine in 100 healthy volunteers. Spine **29**(15), 1636–1641 (2004)

10. Reamy, B.V., Slakey, J.B.: Adolescent idiopathic scoliosis: review and current concepts. Am. Fam. Physician **64**(1), 111–116 (2001)

11. Keenan, B.E., et al.: Segmental torso masses in adolescent idiopathic scoliosis. Clin. Biomech. **29**(7), 773–779 (2014)

12. Tan, K.-J., et al.: Curve progression in idiopathic scoliosis: follow-up study to skeletal maturity. Spine **34**(7), 697–700 (2009)

13. O'Brien, M.F., et al.: Spinal Deformity Study Group Radiographic Measurement Manual. Medtronic Sofamor Danek, Memphis (2004)

14. Stokes, I.A.: Three-dimensional terminology of spinal deformity: a report presented to the scoliosis research society by the scoliosis research society working group on 3-D terminology of spinal deformity. Spine **19**(2), 236–248 (1994)

15. Lenke, L.: SRS Terminology committee and working group on spinal classification revised glossary of terms (2000). http://www.srs.org/professionals/glossary/SRS_revised_glossary_of_terms.htm. Accessed 21 July 2015

16. Bridwell, K.H., DeWald, R.L.: The Textbook of Spinal Surgery. Wolters Kluwer Health, New York (2012)

17. Richards, B.S., et al.: Assessment of trunk balance in thoracic scoliosis. Spine **30**(14), 1621–1626 (2005)

18. Jalalian, A., Gibson, I., Tay, E.H.: Computational biomechanical modeling of scoliotic spine: challenges and opportunities. Spine Deformity **1**(6), 401–411 (2013)

19. Jalalian, A., et al.: A review of computer simulation of spine biomechanics for the treatment of scoliosis. In: The 5th TSME International Conference on Mechanical Engineering. The Empress, Chiang Mai, Thailand (2014)

20. Denavit, J.: A kinematic notation for lower-pair mechanisms based on matrices, trans, of the ASME. J. Appl. Mech. **22**, 215–221 (1955)

21. Jalalian, A., Tay, E.H., Arastehfar, S., Liu, G.: A patient-specific multibody kinematic model for representation of the scoliotic spine movement in frontal plane of the human body. Multibody Syst. Dyn. Accepted

22. Jeffries, B., et al.: Computerized measurement and analysis of scoliosis: a more accurate representation of the shape of the curve. Radiology **134**(2), 381–385 (1980)

23. Koreska, J., Smith, J.: Portable desktop computer-aided digitiser system for the analysis of spinal deformities. Med. Biol. Eng. Comput. **20**(6), 715–726 (1982)

24. Colton, T.: Statistics in Medicine, p. 164. Little Brown, Boston (1974)

A Data Mining-Based Approach for Exploiting the Characteristics of University Lecturers

Thuc-Doan Do, Thuy-Van T. Duong$^{(\boxtimes)}$, and Ngoc-Phien Nguyen

Ton Duc Thang University, Ho Chi Minh City, Vietnam
{dothucdoan, duongthithuyvan,
nguyenngocphien}@tdt.edu.vn

Abstract. The faculty evaluation forms can be considered as valuable data source to exploit knowledge which helps to improve the quality of teaching and learning in universities. In this paper, we analyze previous studies on exploiting faculty evaluation forms according to major problems and their solutions. On that basis, we propose and solve the problem of mining useful knowledge about human resource of Ton Duc Thang University using a data mining-based approach. The experimental data are collected from the online faculty evaluation system of our university, with more than 140,000 evaluation forms. We apply the solution to analyze the data set and draw meaningful comments for the characteristics of the lecturers so that human resource can be exploited and constructed appropriately and efficiently. The results obtained are compared to a previous study on clustering lecturers based on performance and correlation coefficient analysis method.

Keywords: Student feedback · Faculty performance · Teaching performance evaluation · Faculty evaluation form · Clustering

1 Introduction

The quality of education has always been considered as the foundation of the long-term development of all countries. In order to provide people with sufficient knowledge and skills to labor market and enhance their reputation, universities must constantly improve the quality of teaching and learning. Many strategies have been applied to measure the faculty performance, including: student ratings, peer ratings, self-evaluation, videos, student interviews, exit and alumni ratings, employer ratings, administrator ratings, teaching scholarships, teaching awards, learning outcome measures and teaching portfolio [2]. Among these strategies, student ratings are considered as the most popular evaluation tool [4].

In this paper, based on the analysis of previous studies on the exploitation of knowledge from faculty evaluation forms to improve the quality of teaching and support stakeholders such as administrators, lecturers and students in making decisions, we propose a new problem that exploits evaluation forms to obtain useful knowledge about human resource of our university and the method to solve that problem. On that basis, administrators can make decisions in salary increase and task assignment; students can choose appropriate lecturers; lecturers realize their strengths and weaknesses.

© Springer International Publishing Switzerland 2016
P. Perner (Ed.): ICDM 2016, LNAI 9728, pp. 41–53, 2016.
DOI: 10.1007/978-3-319-41561-1_4

We apply the proposed solution on a real data set including 143,117 forms from the online faculty evaluation system of Ton Duc Thang University. The results obtained are compared to the only study on clustering lecturers based on performance and correlation coefficient analysis method. It provides an overview of human resource in our university, laying the foundation for the exploitation and development of human resource efficiently.

The main contributions of our work are the following:

- Construct a new faculty evaluation form suitable for our university
- Analyze previous studies in terms of main problems solved
- Propose a new problem and the solution to tackle that problem
- Apply the solution to analyze the data set collected from the online faculty evaluation system of Ton Duc Thang University and discuss about results obtained
- Compare the results obtained to the only study on clustering lecturers based on performance and correlation coefficient analysis method

The rest of the paper is organized as follows. Section 2 presents the studies of exploiting faculty evaluation forms in terms of solved problem and proposed method. Section 3 proposes new problem and its solution. Section 4 presents the experiments and results obtained. Section 5 is used for a discussion. Section 6 draws the conclusion.

2   Related Works

To the best of our knowledge, there are a few studies on exploiting faculty evaluation forms to improve teaching quality and support stakeholders in making decisions. This section is divided into three parts according to the main problems solved [15].

2.1   Identifying Determining Factors of Faculty Performance

Regression analysis was applied to find the relationship between one dependent variable, which was the faculty performance in this case, and one or more independent variables such as subject knowledge, communication skills, etc. In [12], the authors analyzed the 4,589 evaluation forms about an online MBA program of a university in 2007 to identify determining factors of faculty performance and course satisfaction. Each form consists of many questions divided into three groups of criteria: personal attributes, learner facilitation and quality of feedback. Two overall evaluation factors are overall performance of the lecturer and overall satisfaction of the course. The result obtained shows that personal attributes are determining factors. In [1], evaluation forms were collected from Management Information System department's courses at Bogazici University between 2004 and 2009 and some other lecturer and course characteristics drawn from the Student Evaluation of Teaching research (SET). Stepwise regression method was used to identify the determining factors of faculty performance. The experimental results show that five factors consisting of the attitudes of the lecturer, the attendance of the student, the ratio of students filled the questionnaire to the class size, the lecturer is a part-time laborer and the workload of the course largely determine the faculty performance.

Statistical tests such as Chi-square test, Info Gain test, Gain Ratio test were used to analyze the impact of each factor on faculty performance. In [7], the empirical data are the faculty evaluation forms from the graduates of a faculty at an engineering university in 3 years. The evaluation factors include: teacher name, speed of delivery, content arrangement, presentation, communication, knowledge, content delivery, explanation power, doubts clearing, discussion of problems, overall completion of course and regularity, students attendance, and result. The result is that content arrangement is the determining factor of faculty performance.

Apriori algorithm was used to find the association rule with the form $A \rightarrow B$ in which A was evaluation factor and B was faculty performance. In [8], the empirical data were collected from a faculty evaluation system in spring semester of 2007–2008. The experimental result shows that the teaching content and teaching attitude have the strongest relationship with the faculty performance. In [3], the authors collected data from a personnel management system and educational evaluation system. Apriori algorithm was used to find the relationship between the personal information of lecturers namely gender, age, certification and overall rating; the relationship between the evaluation factors namely teaching attitude, teaching ability, teaching content, teaching organization, teaching methods and faculty performance. The factors having strong relationship with the faculty performance should be focused to improve the quality of teaching.

Some algorithms were applied to build the model to classify faculty performance based on evaluation factors. In [5], the empirical data were collected from the evaluation forms of an online system based on four groups of factors: subject knowledge, teaching skills and assessment methods, behavior towards students, communication skills. Models for classifying faculty performance using those factors obtained from M5P [18] and REP [19] algorithms were used to identify the determining factors of faculty performance. In particular, the factor at the root of the tree is the determining factor because it helps to split the data into groups with the lowest entropy. The lower level in the tree the factor appears at, the less impact on the faculty performance it has. REP algorithm builds the tree faster and achieves higher accuracy than M5P algorithm in the data set. Subject knowledge is the determining factor of faculty performance in both algorithms. In [1], two CHAID and CART algorithms were used to identify the determining factors of faculty performance. Experimental results generated two different trees. Factors appearing at all levels in the tree are considered as the set of the important factors to faculty performance, in which the attitudes of the lecturer at the root of both trees is the most important factor. In [7], the empirical data were collected from the graduates of a faculty at an engineering university in 3 years. Classification methods consisting of four algorithms: Naive Bayes, ID3, CART, LAD tree were used to build faculty performance classification model based on evaluation factors. These factors include: teacher name, speed of delivery, content arrangement, presentation, communication, knowledge, content delivery, explanation power, doubts clearing, discussion of problems, overall completion of course and regularity, students attendance, and result. The result obtained shows that Naïve Bayes algorithm has the highest accuracy.

2.2   Finding the Relationship Among Evaluation Factors

Apriori algorithm was used to find the relationship among the evaluation factors in [11], including: subject knowledge, teaching with new aids, motivating self and students, communication skills, class control, punctuality and regularity, knowledge beyond syllabus, and aggregate.

2.3   Adjusting Faculty Performance Based on Clustering Evaluation Forms

Some algorithms were applied to cluster evaluation forms then recalculate the faculty performance based on clusters obtained. In [6], the evaluation factors consist of clear and understandable presentation, methodical and systematic approach, tempo of lecturers, preparedness for a lecture, the accuracy of arrival to the lecture, encouraging students to participate in classes, informing students about their work, considering student comments and answering questions, availability (through individual teacher/student meetings or via e-mail). The authors partitioned students into several clusters based on the similarity on evaluation forms using k-means algorithm then analyzed the faculty performance in each cluster. In [10], the empirical data obtained from the 3,000 student feedbacks about 77 factors to assess 50 Information Technology lecturers of a university. Expectation Maximization algorithm [20] was used to cluster data according to four levels of performance evaluation: very good, good, satisfactory and poor. The number of clusters is 14. The average value of faculty performance was calculated for each aforementioned level based on results obtained from the clusters.

3   Problem and Solution

3.1   Problem Definition

In terms of the main problems solved as described in the previous section, the studies are divided into three groups: identifying determining factors of faculty performance, finding the relationship among evaluation factors, and adjusting faculty performance based on clustering evaluation forms. In terms of problem-solving methods, the studies on exploiting knowledge from faculty evaluation forms can be divided into three groups: using statistical methods, using machine learning methods, and combining both statistical methods and machine learning methods. While statistical methods are suitable for identifying important factors that influence faculty performance, using machine learning methods in finding relationship among evaluation factors are relevant. However, in general, the exploitation of useful knowledge from evaluation forms is still limited. Therefore we propose the problem of exploiting faculty evaluation forms to obtain characteristics of the human resource in our university.

Let $F_{ijkl} = \ <f_{ijkl}^1, f_{ijkl}^2, \ldots, f_{ijkl}^n>$ be an evaluation form of student i about lecturer j, after studying course k in semester l, in which f_{ijkl}^m is the mth factor of the form and $domain(\ f_{ijkl}^m) = \{1, 2, 3, 4, 5\}$, equivalent to a Likert-scale with intervals of 1 to 5

(5 = Strongly satisfied, 4 = Satisfied, 3 = Neither, 2 = Dissatisfied, 1 = Strongly Dissatisfied). The form consists of n questions or n evaluation factors, in which first $n-1$ factors are specific factors while the last factor is the overall rating. A database D contains a set of all evaluation forms.

Let $T_{jl} = <t_{jl}^1, t_{jl}^2, ..., t_{jl}^n>$ be average rating of lecturer j in semester l, in which t_{jl}^m is the average rating of the mth factor. This feature vector describes specialized features of each lecturer based on all of the evaluation forms about him/her.

Let $I(j,l)$ be a set of students taught by lecturer j in semester l, $K(j,l)$ be the set of courses taught by lecturer j in semester l.

3.2 Method

Our solution is a 3-stage process as follows:

- Stage 1 - Pre-process data:
  - Step 1.1: Firstly, we eliminated inconsistent evaluation forms with the deviation between average rating of specific factors and overall rating being greater than δ because the reason for the lack of consistence may be that the students did not pay attention to the content of the questions completely and seriously.
  - Step 1.2: We then calculated the feature vectors of all lecturers.

The pseudo code of the stage 1 is as follows:

	Step 1.1: Eliminate all inconsistent evaluation forms
1:	for l = 1 to *number of semesters* do
2:	for j = 1 to *number of lecturers* do
3:	for i ∈ I(j,l) do
4:	for k ∈ K(j,l) do
5:	//calculate the sum of rating of student i
6:	for lecturer j after studying course k in semester l
7:	sum(i,j,k,l) = 0
8:	for m = 1 to n-1 do
9:	sum(i,j,k,l) = sum(i,j,k,l) + f_{ijkl}^m
10:	end for
11:	avg(i,j,k,l) = sum(i,j,k,l) / (n-1)
12:	if (|avg(i,j,k,l)- f_{ijkl}^n| >= δ) then
13:	exclude F_{ijkl} from D
14:	end if
15:	end for
16:	end for
17:	end for
18:	end for

	Step 1.2: Calculate feature vectors
1:	**for** l = 1 **to** *number of semesters* **do**
2:	**for** j = 1 **to** *number of lecturers* **do**
3:	**for** m = 1 **to** n **do**
4:	//calculate the sum of rating for lecturer j
5:	in semester l in terms of mth factor
6:	acc_sum(j,l,m) = 0
7:	acc_count(j,l,m) = 0
8:	**for** i ∈ I(j,l) **do**
9:	**for** k ∈ K(j,l) **do**
10:	acc_sum(j,l,m) = acc_sum(j,l,m) + f_{ijkl}^{m}
11:	acc_count(j,l,m) = acc_count(j,l,m) + 1
12:	**end for**
13:	**end for**
14:	t_{jl}^{m} = acc_sum(j,l,m) / acc_count(j,l,m)
15:	**end for**
16:	**end for**
17:	**end for**

- Stage 2 - Process data: We divided lecturers into different clusters according to the similarity of the feature vectors using k-means [17] and X-means [13] methods. We chose k-means as it is the most common clustering algorithm. With each k, we calculated the sum of the squared error measure (SSE) [16] to find the most suitable value of k:

$$SSE = \sum_{i=1}^{K} \sum_{x \in C_i} dist^2(m_i, x)$$

- x is a vector which belongs to cluster C_i
- m_i is a representative vector for cluster C_i (the mean of all vectors in cluster C_i)
- $dist$ is Euclidean distance between each vector and representative vector

We also chose X-means algorithm which is extended from k-means and able to estimate the optimal number of clusters and more efficient in terms of computational cost than traditional k-means algorithm.

- Stage 3 - Post-process data: We analyzed results obtained from stage 2 and drew conclusions.

3.3 Comparison

To the best of our knowledge, there is one study on clustering lecturers based on performance. In [9], the authors identified 77 factors which influence faculty performance. The empirical data include information about 50 Information Technology lecturers of a university. These lecturers were clustered according to performance, using k-means algorithm. The result shows that there are two clusters: cluster 1 consists of the lecturers assessed distinctively while cluster 2 consists of the lecturers who are similar to the others. We implemented this method with our data and then compared results obtained with those of our method.

4   Experiments and Results

We have collected data from the online faculty evaluation system of Ton Duc Thang University for the second semester 2014–2015. The total number of evaluation forms obtained is 143,117. The form consists of 13 closed questions (12 specific questions and a question about overall satisfaction) and two open questions. The form was constructed on the following basis:

- SEEQ evaluation form consists of 33 closed questions and one open question [14] which is widely used in the world
- Evaluation form of the first semester 2014–2015 in our university
- Evaluation form of the second semester 2013–2014 in our university
- Suggestion from departments in our university
- Characteristics of Vietnamese students and our university's students
- Requirements and current situation of our university

For closed questions, we use the Likert scale as mentioned before. The specific evaluation factors were divided into 12 specific questions as presented in Table 1. Specific questions or specific factors in the faculty evaluation form, corresponding to detailed evaluation factors about the lecturers. Thus, each evaluation form can be considered as a student's perspective on specialized features or the strengths and the weaknesses of a lecturer.

Table 1. Specific questions or specific factors in the faculty evaluation form

ID	Question
Q1	Are you satisfied with the specialised knowledge/skills of the lecturers
Q2	Lecturers can inspire students
Q3	Are you satisfied with the enthusiasm of the lecturers
Q4	Lecturers often discuss and answer the questions of students
Q5	Lecturers prepare complete and updated course materials
Q6	Lively, clear, easy to understand and take notes lectures
Q7	Lecturers encourage students to give questions, situations, new issues and discuss in class
Q8	Lecturer present and discuss about the development trends and applications of the subject
Q9	Individual assignments and group assignments are given to help students grasp the subject
Q10	Lecturers instruct students the methods of self-study and deeply exploiting the subject
Q11	Lecturers clearly present the forms of examination and assessment to students
Q12	Contents of the lectures are suitable for the tests

In the preprocessing stage, we eliminated the evaluation forms with the deviation between the average rating of 12 specific factors and overall satisfaction being greater than one ($\delta = 1$). The number of remaining forms after this stage is 139,994 (97.82 %). The value of each faculty evaluation factor is the average of corresponding factor from all relevant forms, rounded to the nearest unit. The results obtained are 647 12-dimensional vectors describing specialized features of 647 lecturers of the whole university.

We applied X-means algorithm for clustering the vectors. The number of clusters obtained is 4. The number of members in each cluster and the values of cluster centroid are presented in Tables 2 and 3, respectively.

Figure 1 illustrates the values of cluster centroid on the graph.

Table 2. Number of members in each clusters

Cluster	Number of members
1	54
2	507
3	28
4	58

Table 3. Values of cluster centroid

Attribute	Cluster 1	Cluster 2	Cluster 3	Cluster 4
Q1	4.777778	4.005906	3.392857	4
Q2	4.351852	3.984252	3	3.052632
Q3	4.981481	4.021654	3.285714	3.982456
Q4	4.796296	4.007874	3.392857	4
Q5	4.333333	4.001969	3.535714	4
Q6	4.277778	3.988189	3	3.017544
Q7	4.055556	3.980315	3.071429	3.789474
Q8	4.037037	3.990157	3.107143	3.824561
Q9	4.037037	3.994094	3.285714	3.789474
Q10	4.037037	3.992126	3	3.614035
Q11	4.407407	4.005906	3.75	4
Q12	4.314815	4.003937	3.571429	3.982456

We used k-means algorithm implemented by Rapid-Miner Studio 6.4 and analyzed results from this tool using SSE. Figure 2 illustrates the relationship between the number of clusters k and SSE. We chose k in the range [2, 17] with 17 being the number of departments in our university. It can be seen from the line graph that the value $k = 4$ creates an elbow, where SSE value starts declining much more slowly. In other words, from the point $k = 4$ the clusters begin to be split into smaller clusters without improving SSE significantly. Therefore the relevant number of clusters is 4.

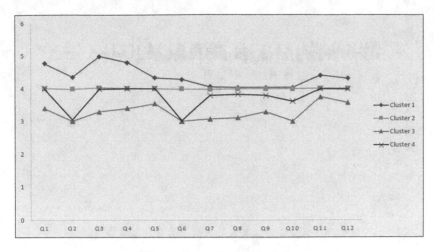

Fig. 1. Centroid of each cluster

In general, the result we obtained with this value matches the result from X-means algorithm presented.

We implemented the method in [9] which proposed to cluster lecturers according to their performance by using X-means algorithm. The result shows that lecturers are distributed into two clusters as shown in Fig. 3. We then analyzed each cluster obtained. Cluster 1 (centroid value: 4.248) consists of 385 lecturers with overall rating being greater than or equal to 4, who are in cluster 1 and a part of cluster 2 (with high overall rating) in our method. Cluster 2 (centroid value: 3.728) consists of 262 lecturers with overall rating being less than 4, who are in the remaining part of cluster 2 (with low overall rating), cluster 3 and cluster 4 in our method.

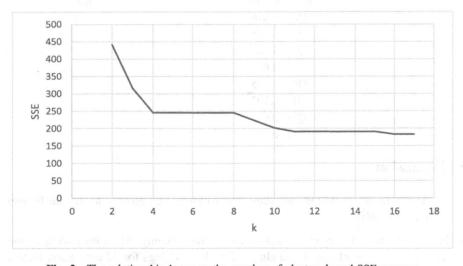

Fig. 2. The relationship between the number of clusters k and SSE measure

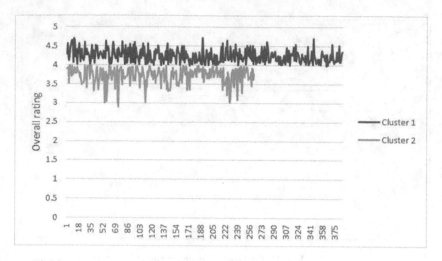

Fig. 3. Lecturers distributed into two clusters according to the method in [9]

In order to examine influence of each evaluation factor on the overall rating in more details, we conducted the analysis about correlations among them. We calculated Pearson correlation by using SPSS software. The result obtained is presented in Table 4.

Table 4. Correlation coefficient between specific factors and overall rating

ID	Correlation coefficient
Q1	0.738
Q2	0.745
Q3	0.728
Q4	0.709
Q5	0.676
Q6	0.745
Q7	0.665
Q8	0.675
Q9	0.685
Q10	0.702
Q11	0.677
Q12	0.709

5   Discussion

From the results of clustering, we drew the following comments about the human resource in our university:

- In general, the lecturers in our university are rated highly. More than 86 % of the lecturers belong to cluster 1 and cluster 2 with the ratings for 12 evaluation factors being greater than or equal to four. The factors getting the highest satisfaction are enthusiasm (Q3, Q4) and knowledge conveyed by the lecturer (Q1). It is quite reasonable for a university that was founded only 18 years ago and the majority of lecturers are young. On the other hand, the ability to inspire students (Q2) and give lively lectures (Q6) is considered as weaknesses of all lecturers. The administrators should pay attention to this problem and try to remedy the situation. In addition, some combinatorial aspects of lecturer characteristics can also be derived from the clustering result. For example, lecturers that are evaluated as not being clear and easy to understand (Q6) are also evaluated as being less able to inspire students (Q2). Another example is that lecturers often discuss and answer the questions of students (Q4) are also evaluated as having good knowledge or skills (Q1).

- Most lecturers belong to cluster 2 (78.4 %) and are assessed uniformly for all criteria (4/5 in Likert scale), showing that there is no significant difference in the quality of teaching among the lecturers in the university.
- Cluster 1 consists of the lecturers with the highest rating (8.3 %). There is no remarkable difference between the lecturers of cluster 1 and that of cluster 2 except evaluation factors Q1, Q3 and Q4 in which Q3, Q4 assess the enthusiasm of the lecturers. More than 97 % (38 out of 39) of the lecturers with the average ratings of the overall satisfaction being equal to 5 belong to cluster 1. Therefore it can be seen that the enthusiasm plays an important role in improving the overall satisfaction. The remaining criteria such as the ability to inspire students (Q2) and give lively lectures (Q6), the expansion of lectures (Q7, Q8), and applications and deeply exploiting the subject (Q9, Q10) are not appreciated compared to the aforementioned criteria. It can be explained by the fact that as the lecturers are young, they do not have much practical experience, wisdom and ability to apply academy knowledge.
- Cluster 3 consisting of 4.3 % of the lecturers is assessed almost similar to the lecturers of cluster 2 except two factors: the ability to inspire (Q2) and give lively lectures (Q6).
- Cluster 4 includes the lecturers with the lowest ratings, accounting for 9 %. These lecturers were rated higher in objective factors such as preparing complete and updated course materials (Q5), presenting clearly the forms of examination and assessment (Q11), contents of the lectures are suitable for the tests (Q12). Therefore, they need to pay attention to improve a variety of factors including specialized knowledge and ability to convey knowledge.

With regards to the method proposed in [9], it is clear that it only partitions lecturers according to overall rating, not based on their specific features. Therefore it can not provide valuable knowledge about characteristics of lecturers belonging to each cluster.

When it comes to the correlation coefficient analysis method, it can be seen that all correlation coefficients are greater than 0.6, which are considered as strong correlations. Among 12 factors, Q2 and Q6 are the factors which have the strongest correlation to the overall rating. These are also the weaknesses of lecturers in our university as analyzed before. The next important factors are Q1 and Q3. The interesting thing is that they are also the strengths of our lecturers. Overall, the results obtained by analyzing the correlation coefficient are consistent to comments drawn from clustering characteristics of the lecturers.

6 Conclusion and Future Works

In this paper, we analyzed the previous studies on exploiting faculty evaluation forms in terms of the main problems solved and their solutions. In general, these studies only focus on solving a few problems such as identifying factors that have the largest influence on faculty performance or seeking dependencies among evaluation factors. On that basis, we proposed a new problem which clusters evaluation forms according to the similarity in specialized features of the lecturers in order to build an overall picture of human resource in our university. We have applied the solution in analyzing real data collected from the online evaluation system of Ton Duc Thang University. We drew useful comments about the strengths and weaknesses of the lecturers in the university as well as those of the lecturers belonging to each cluster, gave some explanations, and identified evaluation factors which influence the overall satisfaction of the students. These results obtained after comparing to applying correlation coefficient analysis. We also compared results obtained to those of the previous study on clustering lecturers and proved that our method provides more valuable information about the characteristics of lecturers.

In future, we continue to exploit the data source to predict the faculty performance based on personal characteristics of the lecturers such as qualifications, age, gender, etc. In addition, we will also investigate the change of assessment trend over time as well as mining knowledge from open questions in the evaluation forms.

References

1. Badur, B., Mardikyan, S.: Analyzing teaching performance of instructors using data mining techniques. Inform. Educ. **10**(2), 245–257 (2011)
2. Berk, R.A.: Survey of 12 strategies to measure teaching effectiveness. Int. J. Teach. Learn. High. Educ. **17**(1), 48–62 (2005)
3. Geng, S., Guo, Z.: Application of association rule mining in college teaching evaluation. In: Electrical, Information Engineering and Mechatronics 2011 Lecture Notes in Electrical Engineering, vol. 138, pp. 1609–1615 (2012)
4. Kelly, M.: Student Evaluations of Teaching Effectiveness: Considerations for Ontario Universities. COU Academic Colleagues Discussion Paper (2012)
5. Kumar, S.A., Vijayalakshmi, M.N.: A naïve based approach of model pruned trees on learner's response. Int. J. Adv. Res. Comput. Sci. Softw. Eng. **4**(9), 52–57 (2012)
6. Kuzmanovic, M., et al.: A new approach to evaluation of university teaching considering heterogeneity of students' preferences. High. Educ. **66**(2), 153–171 (2013)
7. Pal, A.K., Pal, S.: Evaluation of teacher's performance: a data mining approach. IJCSMC **2** (12), 359–369 (2013)
8. Qingxian, P., Linjie, Q., Lanfang, L.: Data mining and application of teaching evaluation based on association rules. In: Proceedings of 4th International Conference on Computer Science and Education, pp. 1404–1407 (2009)
9. Singh, C., Gopal, A.: Performance Analysis of Faculty using Data Mining Techniques. International Journal of Computer Science and Application, pp. 140–144 (2010)

10. Singh, C., Gopal, A., Mishra, S.: Performance assessment of faculties of management discipline from student perspective using statistical and mining methodologies. Int. J. Data Eng. (IJDE) 1(5), 63–69 (2011)
11. Singh, C., Gopal, A., Mishra, S.: Extraction and analysis of faculty performance of management discipline from student feedback using clustering and association rule mining techniques. In: Proceedings of 3rd International Conference on Electronics Computer Technology, ICECT, pp. 94–96 (2011)
12. Wong, A., Fitzsimmons, J.: Student Evaluation of Faculty: An Analysis of Survey Results. U21GlobalWorking Paper Series, no. 003/2008 (2008)
13. Pelleg, D., Moore, A.: X-means: extending K-means with efficient estimation of the number of clusters. In: Proceedings of 17th International Conference on Machine Learning, ICML 2000, pp. 727–734 (2000)
14. Marsh, H.W.: SEEQ: a reliable, valid and useful instrument for collecting students' evaluations of university teaching. Br. J. Educ. Psychol. 52(1), 77–95 (1982)
15. Duong, T.-V.T., Do, T.-D., Nguyen, N.-P.: Exploiting faculty evaluation forms to improve teaching quality: an analytical review. In: Proceedings of 3rd Science and Information Conference, SAI 2015, pp. 457–462 (2015)
16. Tan, P.N., Steinbach, M., Kumar, V.: Introduction to Data Mining, 3rd edn. Pearson Education, New Delhi (2009)
17. Aggarwal, C.C.: Data Mining: The Textbook. Springer, Switzerland (2015)
18. Wang, Y., Witten, I.H.: Induction of model trees for predicting continuous classes. In: Proceedings of the Poster Papers of the European Conference on Machine Learning, pp. 128–137 (1997)
19. Quinlan, J.R.: Simplifying decision trees. Int. J. Man Mach. Stud. 27(3), 221–234 (1987)
20. Dempster, A.P., Laird, N.M., Rubin, D.B.: Maximum likelihood from incomplete data via the EM algorithm. J. Roy. Stat. Soc. B 39(1), 1–38 (1977)

Incremental Response Modeling Based on Segmentation Approach Using Uplift Decision Trees

Sankara Prasad Kondareddy[1]([⊠]), Shruti Agrawal[2], and Shishir Shekhar[3]

[1] Specialist, Fidelity Investments, Bangalore, India
k.shankarprasad@gmail.com
[2] M.SC, Indian Institute of Technology Kanpur, Kanpur, India
Shruti.agrawal271@gmail.com
[3] Lead, Fidelity Investments, Bangalore, India
shishir.shekhar@gmail.com

Abstract. Data mining methods have been successfully used in direct marketing to model the behavior of responders. But these response models do not take in account, the behavior of customers who would take an action irrespective of marketing action. Redundant marketing communications sometimes annoy the customer and reduce the brand value of the company. Accurate targeting of customers also reduces direct marketing cost. Incremental response modeling aims to predict the behavior of customers who respond positively only in the case of marketing. In this paper, we propose a two-step approach for incremental response modeling. In the first step, we segment the data using uplift decision trees using traditional and modified divergence metrics. Then, in the second step we use the standard incremental response modeling methods. Experiments on real world direct marketing campaign data showed that the proposed method outperforms the uplift decision trees.

Keywords: True-lift modeling · Net lift · Uplift decision trees

1 Introduction

Direct marketing cost is one of the biggest contributors of overall marketing cost. Most companies use data mining tools to develop predictive models to identify customers who are likely to respond to direct marketing campaigns. But improper use of these models may lead to unexpectedly low response rates. This is mainly because many of the customers get annoyed due to repetitive phone calls, emails, etc. Companies thus want to target their customers precisely to retain the brand value and improve customer satisfaction.

Traditionally, response models try to predict the likelihood of response for a particular customer, given a marketing action. These models do not account for likelihood of response irrespective of marketing action. Incremental response modeling as shown in Fig. 1 (also known as uplift modeling or differential response modeling or true uplift modeling) removes this drawback and works towards choosing the customers

© Springer International Publishing Switzerland 2016
P. Perner (Ed.): ICDM 2016, LNAI 9728, pp. 54–63, 2016.
DOI: 10.1007/978-3-319-41561-1_5

effectively. Incremental response models are used to identify the customers where there is a true uplift [5]. We want to identify customers who respond to the direct marketing campaign positively and remove the customers from the targeting list who will take an action irrespective of marketing or not.

		Respond if targeted	
		Yes	No
Respond if not targeted	Yes	Spontaneous response (Targeting not required here)	Don't target at all (customers getting annoyed, leading to negative response)
	No	Customers to be targeted	No effect of targeting (only increasing the cost to the company)

Fig. 1. True response uplift.

In true uplift modeling, historical treatment and control campaign datasets are used to build a predictive model. The treatment dataset consists of customers to whom the marketing communication was sent and the control set are the ones who are not reached out but kept aside for measurement purpose. Still, we get some response from control set which is called as natural response. The response from our treatment set consists of natural response and uplift in response due to campaign as shown in Fig. 2.

TREATMENT DATA	CONTROL DATA
Response= Yes (Natural Response)	Response =Yes (Natural Response)
Response= Yes (Uplift due to campaign)	Response=No
Response = No	

Fig. 2. Potential incremental response [10].

This uplift due to campaign is of importance to marketing division. For a particular customer, true uplift can be predicted by modeling the difference between probability of response, given treatment and probability of response, given no treatment (control). The customers with positive uplift can then be reached out. Standard classification models cannot account for the control set and thus cannot be used directly. In this work, we tried to model uplift using decision trees [1] based on various modified splitting criteria. We propose a model which constructs an uplift decision tree and then applies logistic regression based uplift models on the observations at each terminal node to get the prediction. Experimental results showed that the proposed method outperforms various other uplift modeling techniques available.

2   Literature Review

Despite the importance of the topic, surprisingly there is not much work done on uplift modeling. Basically, there are two ways of building uplift models. The first method is called two-model approach. It is based on creating different classification models on treatment set and control set separately, to predict probability of response for each set. Each of the treatment and control dataset is divided into training and test data. Models are constructed separately using standard machine learning classifiers on each training data. The difference of the two predicted probabilities from treatment and control model gives us the predicted uplift. Models are then used to predict probability of response given treatment and probability of response given no treatment (control) on the test data (consisting of treatment as well as control data) for evaluating the model. If uplift follows different pattern than the probabilities of response in treatment and control then, this approach may not perform well.

The second modeling technique predicts the uplift directly from treatment and control dataset using a single model. The complete dataset is divided into training and test data, each containing both treatment and control. A single model is developed using the training data and further evaluated on test data. Some of the papers that addressed decision tree based models for uplift prediction are [2, 3, 7–9, 11]. [7] presented the first paper which explicitly discussed the topic of uplift modeling. [7] provided various illustrations relating to uplift based on real world data and came up with modified decision tree based algorithm which was further detailed in [8]. [3] used splitting criterion based on incremental response rate ($\Delta\Delta P$) for construction of decision trees. This algorithm did not account for population size and thus leading to unwanted weightage on small populations with more uplift. In [9], trees are constructed on information theory based splitting criterion which is more in line with modern machine learning algorithms. The divergence measure (Euclidean or Kullback-Leiber) between the distribution of treatment and control response rate is used as a splitting criteria. Also, they have presented generalized versions for calculation of gain, normalization factor and used variance based pruning technique to construct tree. [11] further accounted for multiple treatment case. They also confirmed through experiments that decision trees based uplift model show significant improvement over previous uplift modeling techniques.

The regression based technique for uplift modeling were also been addressed in literature [4–6]. [5] developed a model by first adding interaction terms between independent variable and treatment-control flag variable. He then created model on the dataset and used it with great success for years in marketing at Fidelity Investments. In [12], two model based approach is followed using logistic regression. [4] proposed a model based on net weight of evidence. He also proposed the modified version of Information value for variable selection for uplift modeling. Most papers in literature used artificially created data to carry out their experiments. In this paper, we have proposed a two-step approach for incremental response modeling and used real time direct marketing campaign datasets to carry out our experiments.

The paper is further arranged into following sections. Section 3 describes the notation and definition of various divergence measures used. Section 4 presents the approach of work, along with the new proposed method. Section 5 shows the results of various experiments done. Section 6 concludes the paper with the result of proposed method. Section 7 provides the scope for future research.

3 Notations and Definitions

Let us introduce and formalize the notation used in the following sections. $Y \in \{1, 0\}$ denotes the response variable for a marketing campaign, which is binary (1 for response = yes and 0 for no), flag $F \in \{1, 0\}$ denotes the binary variable which is 1 for treatment input and 0 for control input and $X_1, \ldots, X_m \in \mathbb{R}$ denotes the set of independent variables. $P^T(.)$ and $P^C(.)$ denote the probability with respect to treatment and control dataset respectively. We want to predict the uplift as difference in predicted probabilities between treatment and control group i.e. $P(Y = 1 | F = 1, X) - P(Y = 1 | F = 0, X)$, which is denoted as $P^T(Y = 1 | X) - P^C(Y = 1 | X)$. We used Laplace correction while estimating the probabilities. For any test A of a numeric variable X at split value v at a node, let a_l and a_r denote the two outcomes $(X \leq v)$ and $(X > v)$ of the test and N(left) and N(right) denote the corresponding number of outcomes for each. As we have only considered binary splits, for any categorical variable also, we will have correspondingly two outcomes a_L and a_R for any test of split. Also, N = N(left) + N(right) denote the number of observations at the node where split is to occur.

4 Proposed Method

In our work, we tried various new splitting criteria for uplift decision tree along with those presented in [11], on real world direct marketing campaign data. The problem with decision tree based approach is that the predicted posterior uplift values will not be a true representation of the actual uplift, because the terminal nodes will have one value as predicted posterior uplift.

We propose a two-step model for incremental response modeling as shown in Fig. 3. In the first step, we build an uplift decision tree using traditional and modified divergence measure as splitting criteria and then in the second step we build uplift models at each of the terminal node using traditional uplift regression methods. This is equivalent to segmenting the complete data (treatment & control) using uplift decision trees and building uplift models in each of the segments. This is same as using clustered weighted modeling approach for uplift modeling. In this paper, we used two-model approach in the second step. We use a logistic model for treatment and control dataset at each terminal node separately and predicted uplift probability is calculated as the difference in individual predicted probabilities.

In the first step, we generated the uplift tree up to some significant depth, so that there are enough observations at the terminal nodes to build two-model logistic regression. For applying logistic regression, variable selection was first done using net information value [10] of variables and then, removing some of those variables while checking for multi-collinearity. As the logistic regression model is created on treatment and control samples separately at each terminal node, we constrained the number of observations that can be present in treatment and control set at a node after each split in the uplift decision tree. The difference between the predicted probabilities of two models on treatment and control dataset at each terminal node is the predicted posterior uplift probability. Applying logistic regression on terminal nodes is expected to give better results as it assigns posterior probabilities to each input observation that comes at terminal node, rather than giving the same value to entire node. As expected, this proposed model outperformed all the decision tree based models.

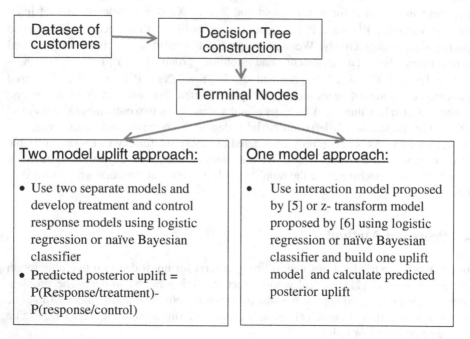

Fig. 3. Proposed hybrid model or two step model.

We implement the decision tree as described in [9] using various other splitting criteria along with Euclidean and Kullback-Leibler divergence measures.

For any split test A at a node, gain using divergence measure D is defined as follows:

$$D_{gain}(A) = D_{gain}(P^T(Y) : P^C(Y)/A) - D_{gain}(P^T(Y) : P^C(Y)) \qquad (1)$$

Where

$$D_{gain}(P^T(Y) : P^C(Y)/A) = \frac{N(Left)}{N}D[P^T(Y/a_L) : P^C(Y/a_L)] + \frac{N(Right)}{N}D[P^T(Y/a_R)$$
$$: P^C(Y/a_R)]$$

(2)

Let $P^T(Y) = [a_1 = P^T(Y = 1), a_0 = P^T(Y = 0)]$ & $P^C(Y) = [a_1 = P^C(Y = 1), a_0 = P^C(Y = 0)]$. The divergence measures defined for probability distributions $P^T(Y)$ and $P^T(Y)$ are defined in (3), (4), (5), (6), (7) and (8).

$$\text{Euclidean divergence measure } ED = \sum\nolimits_i (a_i - b_i)^2$$

(3)

$$\text{Hellinger divergence measure } HD = SQRT[\sum\nolimits_i (a_i - b_i)^2]$$

(4)

$$\text{Kullback - Liebler divergence measure } KL = \sum\nolimits_i \left(a_i Log\left(\frac{a_i}{b_i}\right)\right)$$

(5)

$$\text{J - divergence measure } JD = \sum\nolimits_i \left((a_i - b_i)Log\left(\frac{a_i}{b_i}\right)\right)$$

(6)

$$\text{Modified Euclidean divergence measure } MED = (a_1 - b_1)^2$$

(7)

$$\text{Modified J - divergence measure } MJD = (a_1 - b_1)Log\left(\frac{a_1}{b_1}\right)$$

(8)

For modified Euclidean Divergence and modified J divergence, we have incorporated only the $P(Y = 1)$ while calculating the divergence. The normalization factors considered to penalize the uneven splits are shown in (9) and (10).

$$I(A) = H\left(\frac{N^T}{N},\frac{N^C}{N}\right) * KL(P^T(A) : P^C(A)) + \frac{N^T}{N}H(P^T(A)) + \frac{N^C}{N}H(P^C(A)) + 0.5$$

(9)

$$J(A) = G\left(\frac{N^T}{N},\frac{N^C}{N}\right) * D(P^T(A) : P^C(A)) + \frac{N^T}{N}G(P^T(A)) + \frac{N^C}{N}G(P^C(A)) + 0.5$$

(10)

Where $H(A)$ denotes the entropy function that is $H(A) = -\sum_i a_i * log(a_i)$ and $G(A) = \sum_i a_i(1 - a_i)$ for probability distribution $A = (a_1, \ldots, a_n)$. Our final splitting criterion is given in (11). Label to the leaf node (predicted posterior uplift probability) is given as in (12).

$$\frac{ED_{gain}}{J(A)}, \frac{HD_{gain}}{J(A)}, \frac{KL_{gain}}{I(A)}, \frac{JD_{gain}}{I(A)}, \frac{MED_{gain}}{J(A)}$$

(11)

$$\text{Label value to leaf node} = P^T(Y = 1/L) - P^C(Y = 1/L) \tag{12}$$

Where $P^T(Y = 1/L)\&P^C(Y = 1/L)$ denote treatment and control target class distributions at leaf node L. Some modified versions based on changes in normalization factors and computation of gain for each split were also tried out, but they did not perform very well. We found that the euclidean divergence and modified euclidean divergence measure performed best on the considered dataset.

$$a = P^T(X \leq v/Y = 1) * P^C(X \leq v/Y = 0) \tag{13}$$

$$b = P^T(X \leq v/Y = 0) * P^C(X \leq v/Y = 1) \tag{14}$$

$$c = P^T(X > v/Y = 1) * P^C(X > v/Y = 0) \tag{15}$$

$$d = P^T(X > v/Y = 0) * P^C(X > v/Y = 1) \tag{16}$$

Using (13), (14), (15) and (16), Modified Net weight of evidence (NWOE) and Modified Net Information Value (MNIV) as splitting criterion is defined as follows:

$$\text{NWOE for left node NWOE}_{\text{Left}} = \log\left(\frac{a}{b}\right) \tag{17}$$

$$\text{NWOE for right node NWOE}_{\text{Right}} = \log\left(\frac{c}{d}\right) \tag{18}$$

$$\text{MNIV} = (a - b) * \text{NWOE}_{\text{Left}} + (c - d) * \text{NWOE}_{\text{Right}} \tag{19}$$

We also implemented decision tree based on modified net information value [10] but it could not outperform the measures based on euclidean divergence. Along with two separate logistic regression model approach, we also tried a logistic regression model based on z transformation (single linear model) as described in [6], but it performed very poorly on the dataset considered.

5   Experimental Results

We have used two real time marketing campaign datasets to evaluate the performance of the uplift decision trees as proposed in [11] and the proposed method. Dataset 1 has overall uplift response rate of 4.2 % (population treatment response rate-control response rate) and Dataset 2 has 4.1 % overall uplift response rate. Since we have treatment and control groups, evaluation of the model performance is different from standard classification models procedure. For evaluating the uplift model, we followed the same approach as of [5] and this method is adopted by many industrial practitioners.

We first create the model predicted posterior uplift probability for each observation in validation sample. We use the decile report based on the predicted posterior uplift probability to evaluate the uplift model. The deciles are created by rank ordering all the observations of the validation sample by predicted posterior uplift probability. In each decile, actual uplift is calculated as the difference in observed responses rates of treatment and control group. We checked the correlation of actual uplift (treatment response-control response) with predicted posterior uplift and also rank ordering of actual uplift. Since the application is from direct marketing, we have used the lift in uplift response (top decile uplift response rate/overall data uplift response rate) in the top decile while comparing different methods.

Table 1. Comparison of different uplift models.

Method	Technique	Lift in uplift response rate	
		Dataset 1	Dataset 2
One step	Decision tree using euclidean divergence	2.72	1.49
One step	Decision tree using modified euclidean divergence	2.76	1.90
One step	Decision tree using net information value	1.88	0.76
Two step	Decision tree using euclidean divergence + logistic regression	3.30	1.98
Two steps	Decision tree using modified euclidean divergence + logistic regression	3.09	1.94

Table 1 shows the comparison of different uplift models. Uplift decision trees based on modified Euclidean Divergence is performing better than Uplift decision trees based on Euclidean Divergence. But the proposed two-step method is working better than the one step uplift decision trees.

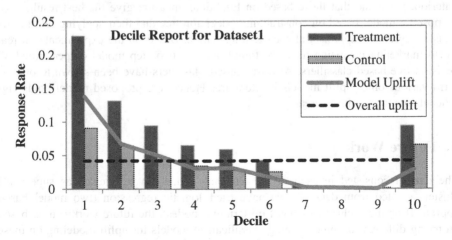

Fig. 4. Response rate by decile using two step method (decision tree using euclidean divergence + logistic regression) on validation dataset1.

The validation dataset decile reports using the proposed two step method: Decision Tree using Euclidean Divergence + Logistic Regression is shown in Figs. 4 and 5. Clearly we can see that the proposed method outperforms the other uplift decision trees.

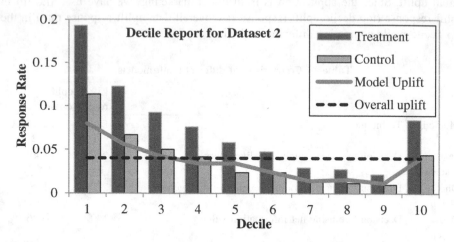

Fig. 5. Response rate by decile using two step method (decision tree using euclidean divergence + logistic regression) on validation dataset2.

6 Conclusion

Incremental response modeling or true lift modeling has got little attention in literature and is getting importance among marketing practitioners in recent years. We have implemented decision trees based classifiers for incremental response modeling on real world marketing campaign data. Out of various divergence measures used for splitting criterion, we found that those based on Euclidean measure give the best results. We proposed a model based on constructing a decision tree and then applying two-model logistic regression approach at each terminal node. Based on the experiments on real world marketing campaign data, we found that the two step model outperformed all decision tree based classifiers. Also tree based classifiers have been shown to outperform various other uplift models in literature. Hence, the proposed model holds significant importance in its area.

7 Future Work

The dataset consisted in each of the terminal nodes can be viewed as supervised clusters created from data. As we have tried logistic regression (two model based approach) on these clusters (dataset on terminal nodes), the future work can be based on trying different machine learning classification models for uplift modeling on these nodes.

References

1. Breiman, L., Friedman, J., Stone, C.J., Olshen, R.A.: Classification and Regression Trees. CRC Press, New York (1998)
2. Chickering, D.M., Heckerman, D.: A decision theoretic approach to targeted advertising. In: Proceedings of the 16th Conference on Uncertainty in Artificial Intelligence, Stanford, CA, pp. 82–88 (2000)
3. Hansotia, B., Rukstales, B.: Incremental value modeling. J. Interact. Mark. **16**(3), 35–46 (2002)
4. Larsen, K.: Net lift models: optimizing the impact of your marketing. In: Predictive Analytics World (2011)
5. Lo, V.S.: The true lift model - a novel data mining approach to response modeling in database marketing. SIGKDD Explor. **4**(2), 78–86 (2002)
6. Jaskowski, M., Jaroszewicz, S.: Uplift modeling for clinical trial data. In: ICML 2012 Workshop on Clinical Data Analysis, Edinburgh, Scotland (2012)
7. Radcliffe, N.J., Surry, P.D.: Differential response analysis: modeling true response by isolating the effect of a single action. In: Credit Scoring and Credit Control VI, Edinburgh, Scotland (1999)
8. Radcliffe, N.J., Surry, P.D.: Real-world uplift modeling with significance-based uplift trees. White Paper TR-2011-1, Stochastic Solutions (2011)
9. Rzepakowski, P., Jaroszewicz, S.: Decision trees for uplift modeling. In: 2010 IEEE International Conference on Data Mining, Sydney, Australia, pp. 441–450 (2010)
10. Lee, T., Zhang, R., Meng, X., Ryan, L.: Incremental response modeling using SAS enterprise miner. In: Proceedings SAS Global Forum Conference, San Francisco, p. 96 (2013)
11. Rzepakowski, P., Jaroszewicz, S.: Decision trees for uplift modeling with single and multiple treatments. Knowl. Inf. Syst. **32**, 303–327 (2012)
12. Vansteelandt, S., Goetghebeur, E.: Causal inference with generalized structural mean models. J. Roy. Stat. Soc. B **65**(4), 817–835 (2003)

PHM: Mining Periodic High-Utility Itemsets

Philippe Fournier-Viger[1(✉)], Jerry Chun-Wei Lin[2], Quang-Huy Duong[3],
and Thu-Lan Dam[3,4]

[1] School of Natural Sciences and Humanities,
Harbin Institute of Technology Shenzhen Graduate School, Shenzhen, China
philfv@hitsz.edu.cn
[2] School of Computer Science and Technology,
Harbin Institute of Technology Shenzhen Graduate School, Shenzhen, China
jerrylin@ieee.org
[3] College of Computer Science and Electronic Engineering,
Hunan University, Changsha, China
huydqyb@gmail.com
[4] Faculty of Information Technology, Hanoi University of Industry,
Hanoi, Vietnam
lanfict@gmail.com

Abstract. High-utility itemset mining is the task of discovering high-utility itemsets, i.e. sets of items that yield a high profit in a customer transaction database. High-utility itemsets are useful, as they provide information about profitable sets of items bought by customers to retail store managers, which can then use this information to take strategic marketing decisions. An inherent limitation of traditional high-utility itemset mining algorithms is that they are inappropriate to discover recurring customer purchase behavior, although such behavior is common in real-life situations (for example, a customer may buy some products every day, week or month). In this paper, we address this limitation by proposing the task of periodic high-utility itemset mining. The goal is to discover groups of items that are periodically bought by customers and generate a high profit. An efficient algorithm named PHM (Periodic High-utility itemset Miner) is proposed to efficiently enumerate all periodic high-utility itemsets. Experimental results show that the PHM algorithm is efficient, and can filter a huge number of non periodic patterns to reveal only the desired periodic high-utility itemsets.

Keywords: High-utility itemset · Periodic itemset · Average periodicity

1 Introduction

High-utility itemset mining (HUIM) [4,7,10–12,15] is a popular data mining task. It has attracted a lot of attention in recent years. It extends the traditional problem of *Frequent Itemset Mining* (FIM) [1]. This latter consists of discovering frequent itemsets, i.e. groups of items (itemsets) appearing frequently in a transaction database [1]. FIM has many applications. However, an important

© Springer International Publishing Switzerland 2016
P. Perner (Ed.): ICDM 2016, LNAI 9728, pp. 64–79, 2016.
DOI: 10.1007/978-3-319-41561-1_6

limitation of FIM is that it assumes that each item cannot appear more than once in each transaction and that all items have the same importance (e.g. weight, unit profit or value). *High-Utility Itemset Mining* (HUIM) addresses this issue by considering that each item may have non binary purchase quantities in transactions and that each item has a weight (e.g. unit profit). The goal of HUIM is to discover itemsets having a high utility (e.g. yielding a high profit) in a transaction database. Besides, market basket analysis, HUIM has several other applications such as website click stream analysis, and biomedical applications [11,15]. Mining high-utility itemsets is widely recognized as more challenging than FIM because the utility measure used in HUIM is not anti-monotonic, i.e. a high utility itemset may have supersets or subsets having lower, equal or higher utilities [4]. Thus, techniques for reducing the search space in FIM cannot be directly reused in HUIM. Though several algorithms have been proposed for HUIM [4,7,10–12,15], an inherent limitation of these algorithms is that they are inappropriate to discover recurring customer purchase behavior, although such behavior is common in real-life situations. For example, in a retail store, some customers may buy some set of products on approximately a daily or weekly basis. Detecting these purchase patterns is useful to better understand the behavior of customers and thus adapt marketing strategies, for example by offering specific promotions to cross-promote products such as reward or points to customers who are buying a set of products periodically. In the field of FIM, algorithms have been proposed to discover periodic frequent patterns (PFP) [2,3,8,9,13,14] in a transaction database. However, these algorithms are inadequate to find periodic patterns that yield a high profit, as they only select patterns based on their frequency. Hence, these algorithms may find a huge amount of periodic patterns that generate a low profit and miss many rare periodic patterns that yield a high profit.

To address this limitation of previous work, this paper proposes the task of periodic high-utility itemset mining. The goal is to efficiently discover all groups of items that are bought together periodically and generate a high profit, in a customer transaction database. The contributions of this paper are fourfold. First, the concept of periodic patterns used in FIM is combined with the concept of HUIs to define a new type of patterns named *periodic high-utility itemsets* (PHIs), and its properties are studied. Second, novel measures of a pattern's periodicity named *average periodicity* and *minimum periodicity* are introduced to provide a flexible way of assessing the periodicity of patterns. Third, an efficient algorithm named PHM (Periodic High-utility itemset Miner) is proposed to efficiently discover the periodic high-utility itemsets. Fourth, an extensive experimental evaluation is carried to compare the efficiency of PHM with the state-of-the-art FHM algorithm for HUIM. Experimental results show that the PHM algorithm is efficient, and can filter a huge number of non periodic patterns to reveal only the desired itemsets. The rest of this paper is organized as follows. Sections 2, 3, 4 and 5 respectively present preliminaries related to HUIM, related work, the PHM algorithm, the experimental evaluation and the conclusion.

2   Related Work

This section reviews related work in high-utility itemset mining and periodic frequent pattern mining.

2.1   High-Utility Itemset Mining

Definition 1 (Transaction database). Let I be a set of items (symbols). A *transaction database* is a set of transactions $D = \{T_1, T_2, ..., T_n\}$ such that for each transaction T_c, $T_c \in I$ and T_c has a unique identifier c called its Tid. Each item $i \in I$ is associated with a positive number $p(i)$, called its external utility (e.g. unit profit). For each transaction T_c such that $i \in T_c$, a positive number $q(i, T_c)$ is called the internal utility of i (e.g. purchase quantity).

Example 1. Consider the database of Table 1, which will be used as running example. This database contains seven transactions $(T_1, T_2...T_7)$. Transaction T_3 indicates that items a, b, c, d, and e appear in this transaction with an internal utility of respectively 1, 5, 1, 3 and 1. Table 2 indicates that the external utility of these items are respectively 5, 2, 1, 2 and 3.

Table 1. A transaction database

TID	Transaction
T_1	$(a, 1), (c, 1),$
T_2	$(e, 1)$
T_3	$(a, 1), (b, 5), (c, 1), (d, 3), (e, 1)$
T_4	$(b, 4), (c, 3), (d, 3), (e, 1)$
T_5	$(a, 1), (c, 1), (d, 1)$
T_6	$(a, 2), (c, 6), (e, 2)$
T_7	$(b, 2), (c, 2), (e, 1)$

Table 2. External utility values

Item	a	b	c	d	e
Unit profit	5	2	1	2	3

Definition 2 (Utility of an item/itemset). The utility of an item i in a transaction T_c is denoted as $u(i, T_c)$ and defined as $p(i) \times q(i, T_c)$. The utility of an itemset X (a group of items $X \subseteq I$) in a transaction T_c is denoted as $u(X, T_c)$ and defined as $u(X, T_c) = \sum_{i \in X} u(i, T_c)$. The utility of an itemset X (in a database) is denoted as $u(X)$ and defined as $u(X) = \sum_{T_c \in g(X)} u(X, T_c)$, where $g(X)$ is the set of transactions containing X.

Example 2. The utility of item a in T_6 is $u(a, T_6) = 5 \times 2 = 10$. The utility of the itemset $\{a, c\}$ in T_6 is $u(\{a, c\}, T_6) = u(a, T_6) + u(c, T_6) = 5 \times 2 + 1 \times 6 = 16$. The utility of the itemset $\{a, c\}$ (in the database) is $u(\{a, c\}) = u(a) + u(c) = u(a, T_1) + u(a, T_3) + u(a, T_5) + u(a, T_6) + u(c, T_1) + u(c, T_3) + u(c, T_5) + u(c, T_6) = 5 + 5 + 5 + 10 + 1 + 1 + 1 + 6 = 34$.

Definition 3 (High-utility itemset mining). The *problem of high-utility itemset mining* is to discover all high-utility itemsets [4,7,10–12,15]. An itemset *X is a* high-utility itemset if its utility $u(X)$ is no less than a user-specified minimum utility threshold *minutil* given by the user.

Example 3. If *minutil* = 30, the complete set of HUIs is $\{a,c\}$: 34, $\{a,c,e\}$: 31, $\{b,c,d\}$: 34, $\{b,c,d,e\}$: 40, $\{b,c,e\}$: 37, $\{b,d\}$: 30, $\{b,d,e\}$: 36, and $\{b,e\}$: 31, where each HUI is annotated with its utility.

In HUIM, the utility measure is not monotonic or anti-monotonic [12,15], i.e., an itemset may have a utility lower, equal or higher than the utility of its subsets. Several HUIM algorithms circumvent this problem by overestimating the utility of itemsets using the *Transaction-Weighted Utilization (TWU)* measure [12,15], which is anti-monotonic, and defined as follows.

Definition 4 (Transaction weighted utilization). The *transaction utility (TU)* of a transaction T_c is the sum of the utility of all the items in T_c.i.e. $TU(T_c) = \sum_{x \in T_c} u(x, T_c)$. The *transaction-weighted utilization (TWU)* of an itemset X is defined as the sum of the transaction utility of transactions containing X, i.e. $TWU(X) = \sum_{T_c \in g(X)} TU(T_c)$.

Example 4. The TUs of $T_1, T_2, T_3, T_4, T_5, T_6$ and T_7 are respectively 6,3, 25, 20, 8, 22 and 9. The TWU of single items a, b, c, d, e are respectively 61, 54, 90, 53 and 79. $TWU(\{c,d\}) = TU(T_3) + TU(T_4) + TU(T_5) = 25 + 20 + 8 = 53$.

Theorem 1 (Pruning search space using the TWU). *Let X be an itemset, if $TWU(X) <$ minutil, then X and its supersets are low utility.* [12]

Algorithms such as Two-Phase [12], BAHUI [10], PB [7], and UPGrowth+ [15] utilizes the above property to prune the search space. They operate in two phases. In the first phase, they identify candidate high utility itemsets by calculating their TWUs. In the second phase, they scan the database to calculate the exact utility of all candidates found in the first phase to eliminate low utility itemsets. Recently, an alternative algorithm called *HUI-Miner*[11] was proposed to mine HUIs directly using a single phase. Then, a faster depth-first search algorithm FHM [4] was proposed, which extends HUI-Miner. In FHM, each itemset is associated with a structure named *utility-list* [4,11]. Utility-lists allow calculating the utility of an itemset quickly by making join operations with utility-lists of shorter patterns. Utility-lists are defined as follows.

Definition 5 (Utility-list). Let $\succ$ be any total order on items from I. The *utility-list* of an itemset X in a database D is a set of tuples such that there is a tuple $(tid, iutil, rutil)$ for each transaction T_{tid} containing X. The *iutil* element of a tuple is the utility of X in T_{tid}. i.e., $u(X, T_{tid})$. The *rutil* element of a tuple is defined as $\sum_{i \in T_{tid} \wedge i \succ x \forall x \in X} u(i, T_{tid})$.

Example 5. Assume that $\succ$ is the alphabetical order. The utility-list of $\{a\}$ is $\{(T_1, 5, 1), T_3, 5, 20), (T_5, 5, 3), (T_6, 10, 12)\}$. The utility-list of $\{d\}$ is $\{(T_3, 6, 3), (T_4, 6, 3), (T_5, 2, 0)\}$. The utility-list of $\{a, d\}$ is $\{(T_3, 11, 3), (T_5, 7, 0)\}$.

To discover HUIs, FHM performs a single database scan to create utility-lists of patterns containing single items. Then, longer patterns are obtained by performing the join operation of utility-lists of shorter patterns. The join operation for single items is performed as follows. Consider two items x, y such that $x \succ y$, and their utility-lists $ul(\{x\})$ and $ul(\{y\})$. The utility-list of $\{x, y\}$ is obtained by creating a tuple $(ex.tid, ex.iutil + ey.iutil, ey.rutil)$ for each pair of tuples $ex \in ul(\{x\})$ and $ey \in ul(\{y\})$ such that $ex.tid = ey.tid$. The join operation for two itemsets $P \cup \{x\}$ and $P \cup \{y\}$ such that $x \succ y$ is performed as follows. Let $ul(P)$, $ul(\{x\})$ and $ul(\{y\})$ be the utility-lists of P, $\{x\}$ and $\{y\}$. The utility-list of $P \cup \{x, y\}$ is obtained by creating a tuple $(ex.tid, ex.iutil + ey.iutil - ep.iutil, ey.rutil)$ for each set of tuples $ex \in ul(\{x\})$, $ey \in ul(\{y\})$, $ep \in ul(P)$ such that $ex.tid = ey.tid = ep.tid$. Calculating the utility of an itemset using its utility-list and pruning the search space is done as follows.

Property 1 (Calculating utility of an itemset using its utility-list). The utility of an itemset is the sum of $iutil$ values in its utility-list [11].

Theorem 2 (Pruning search space using utility-lists). *Let X be an itemset. Let the extensions of X be the itemsets that can be obtained by appending an item y to X such that $y \succ i$, $\forall i \in X$. If the sum of $iutil$ and $rutil$ values in $ul(X)$ is less than minutil, X and its extensions are low utility* [11].

FHM is very efficient. However, an important limitation of current HUIM algorithms is that they are not designed for discovering periodic patterns.

2.2   Periodic Frequent Pattern Mining

In the field of FIM, algorithms have been proposed to discover periodical frequent patterns (PFP) [2,3,8,9,13,14] in a transaction database. Discovering PFP has applications in many domains such as web mining, bioinformatics, and market basket analysis [14]. The concept of PFP is defined as follows [14].

Definition 6 (Periods of an itemset). Let there be a database $D = \{T_1, T_2, ..., T_n\}$ containing n transactions, and an itemset X. The set of transactions containing X is denoted as $g(X) = \{T_{g_1}, T_{g_2}..., T_{g_k}\}$, where $1 \leq g_1 < g_2 < ... < g_k \leq n$. Two transactions $T_x \supset X$ and $T_y \supset X$ are said to be *consecutive with respect to* X if there does not exist a transaction $T_w \in g(X)$ such that $x < w < y$. The *period* of two consecutive transactions T_x and T_y in $g(X)$ is defined as $pe(T_x, T_y) = (y - x)$, that is the number of transactions between T_x and T_y. The *periods of an itemset* X is a list of periods defined as $ps(X) = \{g_1 - g_0, g_2 - g_1, g_3 - g_2, ...g_k - g_{k-1}, g_{k+1} - g_k\}$, where g_0 and $g_k + 1$ are constants defined as $g_0 = 0$ and $g_k + 1 = n$. Thus, $ps(X) = \bigcup_{1 \leq z \leq k+1} (g_z - g_{z-1})$.

Example 6. For the itemset $\{a, c\}$, The list of transactions containing $\{a, c\}$ is $g(\{a, c\}) = \{T_1, T_3, T_5, T_6\}$. Thus, the periods of this itemset are $ps(\{a, c\}) = \{1, 2, 2, 1, 1\}$.

Definition 7 (Periodic frequent pattern). The maximum periodicity of an itemset X is defined as $maxper(X) = max(ps(X))$ [14]. An itemset X is a periodic frequent pattern (PFP) if $|g(X)| \geq minsup$ and $maxper(X) < maxPer$, where $minsup$ and $maxPer$ are user-defined thresholds [14].

The first algorithm for mining PFPs is PFP-Tree [14]. It utilizes a tree-based and pattern-growth approach for discovering PFPs. Then, the MTKPP algorithm [2] was proposed for discovering the k most frequent PFPs in a database, where k is a user-specified parameter. MTKPP utilizes a vertical structure to maintain information about itemsets in the database A variation of the PF-Tree algorithm named the ITL-Tree was also introduced [3] to reduce the time for mining PFPs by approximating the periodicity of itemsets. Another approximate algorithm for PFP mining was recently proposed [9]. Other extensions of the PF-Tree algorithm named MIS-PF-tree [8] and MaxCPF [13] were respectively proposed to mine PFPs using multiple $minsup$ thresholds, and multiple $minsup$ and $minper$ thresholds. An important limitation of traditional algorithms for PFP mining is that they are inadequate to find periodic patterns that yield a high profit, since they only consider the support (frequency) of patterns. Hence, they may find a huge amount of periodic patterns that yield a low profit and miss many rare periodical patterns that yield a high profit.

3 The PHM Algorithm

To address the aforementioned limitation of HUI and PFP mining algorithms, this section introduces the concept of periodic high-utility itemsets (PHUIs). The first subsection present novel measures to assess the periodicity of HUIs, while the second subsection presents and efficient algorithm named PHM (Periodic High-Utility Itemset Miner) to discover PHUIs efficiently.

3.1 Measuring the Periodicity of High-Utility Patterns

A drawback of the maximum periodicity measure used by most PFP algorithms is that an itemset is automatically discarded if it has a single period of length greater than the $maxPer$ threshold. Thus, this measure may be viewed as too strict. To provide a more flexible way of evaluating the periodicity of patterns, the concept of *average periodicity* is introduced in the proposed algorithm.

Definition 8 (Average periodicity of an itemset). The average periodicity of an itemset X is defined as $avgper(X) = \sum_{g \in ps(X)} /|ps(X)|$.

Example 7. The periods of itemsets $\{a, c\}$ and $\{e\}$ are respectively $ps(\{a, c\}) = \{1, 2, 2, 1, 1\}$ and $ps(\{e\}) = \{2, 1, 1, 2, 1, 0\}$. The average periodicities of these itemsets are respectively $avgper(\{a, c\}) = 1.4$ and $avgper(\{e\}) = 1.16$.

Lemma 1 (Relationship between average periodicity and support). *Let X be an itemset appearing in a database D. An alternative and equivalent way of calculating the average periodicity of X is $avgper(X) = |D|/(|g(X)| + 1)$.*

Proof. Let $g(X) = \{T_{g_1}, T_{g_2}, \ldots, T_{g_k}\}$ be the set of transactions containing X, such that $g_1 < g_2 < \cdots < g_k$. By definition, $avgper(X) = \sum_{g \in ps(X)} /|ps(X)|$. To prove that the lemma holds, we need to show that $\sum_{g \in ps(X)} /|ps(X)| = |D|/(|g(X)| + 1)$.

(1) We first show that $\sum_{g \in ps(X)} = |D|$, as follows:

$\sum_{g \in ps(X)} = (g_1 - g_0) + (g_2 - g_1) + \cdots (g_k - g_{k-1}) + (g_{k+1} - g_k)\}$
$= \sum_{g \in ps(X)} = g_0 + (g_1 - g_1) + (g_2 - g_2) + \cdots (g_k - g_k) + (g_{k+1})$
$= g_{k+1} - g_0 = |D|$.

(2) We then show that $|ps(X)| = |g(X)| + 1$, as follows:

By definition, $ps(X) = \bigcup_{1 \le z \le k+1} (g_z - g_{z-1})$. Thus, the set $ps(X)$ contains $k+1$ elements. Since X appears in k transactions, $sup(X) = k$, and thus $|ps(X)| = |g(X)| + 1$.

Since (1) and (2) holds, the lemma holds. □

The above lemma is important as it provides an efficient way of calculating the average periodicity of itemsets in a database D. The term $|D|$ can be calculated once, and thereafter the average periodicity of any itemset X can be obtained by only calculating $|g(X)| + 1$, and then dividing $|D|$ by the result. This is more efficient than calculating the average periodicity using Definition 8. Besides, this lemma is important as it shows that there is a relationship between the support used in FIM and the average periodicity of a pattern. Although the average periodicity is useful as it measures what is the typical period length of an itemset, it should not be used as the sole measure for evaluating the periodicity of a pattern because it does not consider whether an itemset has periods that vary widely or not. For example, the itemset $\{b, d\}$ has an average periodicity of 2.33. However, this is misleading since this itemset only appears in transaction T_3 and T_4, and its periods $ps(\{T_3, T_4\}) = \{3, 1, 4\}$ vary widely. Intuitively, this pattern should not be a periodic pattern. To avoid finding patterns having periods that vary widely, our solution is to combine the average periodicity measure with other periodicity measure(s). The following measures are combined with the average periodicity to achieve this goal. First, we define the *minimum periodicity* of an itemset as $minper(X) = min(ps(X))$ to avoid discovering itemsets having some very short periods. But this measure is not reliable since the first and last period of an itemset are respectively equal to 1 or 0 if the itemset respectively appears in the first or the last transaction of the database. For example, the last period of itemset $\{e\}$ is 0, because it appears in the last transaction (T_7), and thus its minimum periodicity is 0. Our solution to this issue is to exclude the first and last periods of an itemset from the calculation of the minimum periodicity. Moreover, if the set of periods is empty as a result of this exclusion, the minimum periodicity is defined as ∞. In the rest of this paper, we consider this definition. Second, we consider the *maximum periodicity* of an itemset $maxper(X)$ as defined in the previous section. The rationale for using this measure in combination with the average periodicity is that it can avoid discovering periodical patterns that do not occur for long periods of time. In terms of calculation costs, a reason for choosing the minimum periodicity, maximum periodicity and average periodicity as measure is that they can be calculated very efficiently for an itemset X by scanning the list of transactions $g(X)$ only once. That is, calculating these measures do not require to store the set of periods $ps(X)$ in memory. Conversely, other measures such as the standard deviation would require to calculate all periods of an itemset beforehand. Thus, we define the concept of periodic high-utility itemsets by considering the minimum periodicity, maximum periodicity and average periodicity measures.

Definition 9 (Periodic high-utility itemsets). Let *minutil*, *minAvg*, *maxAvg*, *minPer* and *maxPer* be positive numbers, provided by the user. An itemset X is a *periodic high-utility itemset* if and only if $minAvg \leq avgper(X) \leq maxAvg$, $minper(X) \geq minPer$, $maxper(X) \leq maxPer$, and $u(X) \geq minutil$.

Table 3. The set of PHUIs in the running example

Itemset	u(X)	$|g(X)|$	Minper(X)	Maxper(X)	Avgper(X)
$\{b\}$	22	3	1	3	1.75
$\{b, e\}$	31	3	1	3	1.75
$\{b, c, e\}$	37	3	1	3	1.75
$\{b, c\}$	28	3	1	3	1.75
$\{a\}$	25	4	1	2	1.4
$\{a, c\}$	34	4	1	2	1.4
$\{c, e\}$	27	4	1	3	1.4

For example, if *minutil* = 20, *minPer* = 1, *maxPer* = 3, *minAvg* = 1, and *maxAvg* = 2, the complete set of PHUIs is shown in Table 3. To develop an efficient algorithm for mining PHUIs, it is important to design efficient pruning strategies. To use the periodicity measures for pruning the search space, the following theorems are presented.

Lemma 2 (Monotonicity of the average periodicity). Let X and Y be itemsets such that $X \subset Y$. It follows that $avgper(Y) \geq avgper(X)$.

Proof. The average periodicities of X and Y are respectively $avgper(X) = |D|/(|g(X)| + 1)$ and $avgper(Y) = |D|/(|g(Y)| + 1)$. Because $X \subset Y$, it follows that $g(Y) \subseteq g(X)$. Hence, $avgper(Y) \geq avgper(X)$. □

Lemma 3 (Monotonicity of the minimum periodicity). Let X and Y be itemsets such that $X \subset Y$. It follows that $minper(Y) \geq minper(X)$.

Proof. Since $X \subset Y$, $g(Y) \subseteq g(X)$. If $g(Y) = g(X)$, then X and Y have the same periods, and thus $minper(Y) = minper(X)$. If $g(Y) \subset g(X)$, then for each transaction $T_x \in g(X) \setminus g(Y)$, the corresponding periods in $ps(X)$ will be replaced by a larger period in $ps(Y)$. Thus, any period in $ps(Y)$ cannot be smaller than a period in $ps(X)$. Hence, $minper(Y) \geq minper(X)$. □

Lemma 4 (Monotonicity of the maximum periodicity). Let X and Y be itemsets such that $X \subset Y$. It follows that $maxper(Y) \geq maxper(X)$ [14].

Theorem 3 (Maximum periodicity pruning). Let X be an itemset appearing in a database D. X and its supersets are not PHUIs if $maxper(X) > maxPer$. Thus, if this condition is met, the search space consisting of X and all its supersets can be discarded.

Proof. By definition, if $maxper(X) > maxPer$, X is not a PHUI. By Lemma 4, supersets of X are also not PHUIs. □

Theorem 4 (Average periodicity pruning). Let X be an itemset appearing in a database D. X is not a PHUI as well as all of its supersets if $avgper(X) > maxAvg$, or equivalently if $|g(X)| < (|D|/maxAvg) - 1$. Thus, if this condition is met, the search space consisting of X and all its supersets can be discarded.

Proof. By definition, if $avgper(X) > maxAvg$, X is not a PHUI. By Lemma 2, supersets of X are also not PHUIs. The pruning condition $avgper(X) > maxAvg$ is rewritten as: $|D|/(|g(X)|+1) > maxAvg$. Thus, $1/(|g(X)|+1) > maxAvg/|D|$, which can be further rewritten as $|g(X)| + 1 < |D|/maxAvg$, and as $|g(X)| < (|D|/maxAvg) - 1$. □

3.2   The Algorithm

The proposed PHM algorithm is a utility-list based algorithm, inspired by the FHM algorithm [4], where the utility-list of each itemset X is annotated with two additional values: $minper(X)$ and $maxper(X)$. The main procedure of PHM (Algorithm 1) takes a transaction database as input, and the $minutil$, $minAvg$, $maxAvg$, $minPer$ and $maxPer$ thresholds. The algorithm first scans the database to calculate $TWU(\{i\})$, $minper(\{i\})$, $maxper(\{i\})$, and $|g(\{i\})|$ for each item $i \in I$. Then, the algorithm calculates the value $\gamma = (|D|/maxAvg) - 1$ to be later used for pruning itemsets using Theorem 4. Then, the algorithm identifies the set I^* of all items having a TWU no less than $minutil$, a maximum periodicity no greater than $maxPer$, and appearing in no less than γ transactions (other items are ignored since they cannot be part of a PHUI by Theorems 1, 3 and 4). The TWU values of items are then used to establish a total order $\succ$ on items, which is the order of ascending TWU values (as suggested in [11]). A database scan is then performed. During this database scan, items in transactions are reordered according to the total order $\succ$, the utility-list of each item $i \in I^*$ is built and a structure named EUCS (Estimated Utility Co-Occurrence Structure) is built [4]. This latter structure is defined as a set of triples of the form $(a, b, c) \in I^* \times I^* \times \mathbb{R}$. A triple (a,b,c) indicates that $TWU(\{a, b\}) = c$. The EUCS can be implemented as a triangular matrix (as shown in Fig. 1 for the running example), or as a hash map of hash maps where only tuples of the form (a, b, c) such that $c \neq 0$ are kept. After the construction of the EUCS, the

---

Algorithm 1. The PHM algorithm

---

input : D: a transaction database,
    $minutil$, $minAvg$, $maxAvg$, $minPer$ and $maxPer$: the thresholds
output: the set of periodic high-utility itemsets

1 Scan D once to calculate $TWU(\{i\})$, $minper(\{i\})$, $maxper(\{i\})$, and $|g(\{i\})|$
   for each item $i \in I$;
2 $\gamma \leftarrow (|D|/maxAvg) - 1$;
3 $I^* \leftarrow$ each item i such that $TWU(i) \geq minutil$, $|g(\{i\})| \geq \gamma$ and
   $maxper(\{i\}) \leq maxPer$;
4 Let $\succ$ be the total order of TWU ascending values on I^*;
5 Scan D to build the utility-list of each item $i \in I^*$ and build the $EUCS$
   structure;
6 **Search** ($\emptyset$, I^*, γ, $minutil$, $minAvg$, $minPer$, $maxPer$, $EUCS$, $|D|$);

Item	a	b	c	d
b	25			
c	61	54		
d	33	45	53	
e	47	54	76	45

Item	a	b	c	d
b	1			
c	4	3		
d	2	2	3	
e	2	3	4	2

Fig. 1. The EUCS   **Fig. 2.** The ESCS

depth-first search exploration of itemsets starts by calling the recursive procedure *Search* with the empty itemset $\emptyset$, the set of single items I^*, γ, $minutil$, $minAvg$, $minPer$, $maxPer$, the EUCS structure, and $|D|$.

The *Search* procedure (Algorithm 2) takes as input an itemset P, extensions of P having the form Pz meaning that Pz was previously obtained by appending an item z to P, γ, $minutil$, $minAvg$, $minPer$, $maxPer$, the EUCS, and $|D|$. The search procedure performs a loop on each extension Px of P. In this loop, the average periodicity of Px is obtained by dividing $|D|$ by the number of elements in the utility list of Px plus one (by Lemma 1). Then, if the average periodicity of Px is in the $[minAvg, maxAvg]$ interval, the sum of the $iutil$ values of the utility-list of Px is no less than $minutil$ (cf. Property 1), the minimum/maximum periodicity of Px is no less/not greater than $minPer/maxPer$ according to the values stored in its utility-list, then Px is a PHUI and it is output. Then, if the sum of $iutil$ and $rutil$ values in the utility-list of Px are no less than $minutil$, the number of elements in the utility list of Px is no less than γ, and $maxper(Px)$ is no greater than $maxPer$, it means that extensions of Px should be explored (by Theorems 1, 3 and 4). This is performed by merging Px with all extensions Py of P such that $y \succ x$ to form extensions of the form Pxy containing $|Px| + 1$ items. The utility-list of Pxy is then constructed by calling the *Construct* procedure (cf. Algorithm 3), to join the utility-lists of P, Px and Py. This latter procedure is mainly the same as in HUI-Miner [11], with the exception that periods are calculated during utility-list construction to obtain $maxPer(Pxy)$ and $minPer(Pxy)$ (not shown). Then, a recursive call to the *Search* procedure with Pxy is done to calculate its utility and explore its extension(s). The *Search* procedure starts from single items, recursively explores the search space of itemsets by appending single items, and only prunes the search space using Theorems 1, 3 and 4. Thus, it can be easily seen that this procedure is correct and complete to discover all PHUIs.

Algorithm 2. The *Search* procedure

---

input : P: an itemset, *ExtensionsOfP*: a set of extensions of P, γ, *minutil*, *minAvg*, *minPer*, *maxPer*, the *EUCS* structure, $|D|$
output: the set of periodic high-utility itemsets

1 **foreach** *itemset Px ∈ ExtensionsOfP* **do**
2      $avgperPx \leftarrow |D|/(|Px.utilitylist| + 1)$;
3   **if** $SUM(Pxy.utilitylist.iutils) \geq minutil \wedge$
        $minAvg \leq avgperPx \leq maxAvg \wedge Px.utilitylist.minp \geq$
        $minPer \wedge Px.utilitylist.maxp \leq maxPer \wedge$ **then** output Px;
4   **if** $SUM(Px.utilitylist.iutils)+SUM(Px.utilitylist.rutils) \geq minutil \wedge$
        $avgperPx \geq \gamma$ and $Px.utilitylist.maxp \leq maxPer$ **then**
5          $ExtensionsOfPx \leftarrow \emptyset$;
6   **foreach** *itemset Py ∈ ExtensionsOfP such that $y \succ x$* **do**
7   **if** $\exists(x, y, c) \in EUCS$ such that $c \geq minutil)$ **then**
8                  $Pxy \leftarrow Px \cup Py$;
9                  $Pxy.utilitylist \leftarrow$ **Construct** (P, Px, Py);
10                 $ExtensionsOfPx \leftarrow ExtensionsOfPx \cup \{Pxy\}$;
11   **end**
12   **end**
13   **Search** $(Px, ExtensionsOfPx, \gamma, minutil, minAvg, minPer, maxPer,$
           $EUCS, |D|)$;
14   **end**
15 **end**

Furthermore, in the implementation of PHM, two additional optimizations are included, which are briefly described next. **Optimization 1. Estimated Average Periodicity Pruning (EAPP).** The PHM algorithm creates a structure called EUCS to store the TWU of all pairs of items occurring in the database, and this structure is used to prune any itemset Pxy containing a pair of items {x,y} having a TWU lower than *minutil* (Line 7 of the search procedure). The strategy EAPP is a novel strategy that uses the same idea but prune itemsets using the average periodicity instead of the utility. During the second database scan, a novel structure called ESCS (Estimated Support Co-occurrence Structure) is created to store $|g(\{x, y\})|$ for each pair of items {x,y} (as shown in Fig. 2). Then, Line 7 of the search procedure is modified to prune itemset Pxy if $|g(\{x, y\})|$ is less than γ by Theorem 4. **Optimization 2. Abandoning List Construction early (ALC).** Another strategy introduced in PHM is to stop

---

Algorithm 3. The Construct procedure

input : P: an itemset, Px: the extension of P with an item x, Py: the
extension of P with an item y

output: the utility-list of Pxy

1 $UtilityListOfPxy \leftarrow \emptyset$;
2 **foreach** *tuple* $ex \in Px.utilitylist$ **do**
3   **if** $\exists ey \in Py.utilitylist$ *and* $ex.tid = exy.tid$ **then**
4   **if** $P.utilitylist \neq \emptyset$ **then**
5   Search element $e \in P.utilitylist$ such that $e.tid = ex.tid$.;
6              $exy \leftarrow (ex.tid, ex.iutil + ey.iutil - e.iutil, ey.rutil)$;
7   **end**
8   **else**
9              $exy \leftarrow (ex.tid, ex.iutil + ey.iutil, ey.rutil)$;
10   **end**
11         $period_{exy} \leftarrow calculatePeriod(exy.tid, UtilityListOfPxy)$;
12         $UpdateMinPerMaxPer(UtilityListOfPxy, period_{exy})$;
13         $UtilityListOfPxy \leftarrow UtilityListOfPxy \cup \{exy\}$;
14   **end**
15 **end**
16 **return** $UtilityListPxy$;

---

constructing the utility-list of an itemset if a specific condition is met, indicating that the itemset cannot be a PHUI. By Theorem 4, an itemset Pxy cannot be a PHUI, if it appears in less than $\gamma = (|D|/maxAvg) - 1$ transactions. The strategy ALC consists of modifying the Construct procedure (Algorithm 3) as follows. The first modification is to initialize a variable max with the value γ in Line 1. The second modification is to the following lines, where the utility-list of Pxy is constructed by checking if each tuple in the utility-lists of Px appears in the utility-list of Py (Line 3). For each tuple not appearing in Py, the variable max is decremented by 1. If max is smaller than γ, the construction of the utility-list of Pxy can be stopped because $|g(Pxy)|$ will not be higher than γ. Thus Pxy is not a PHUI by Theorem 4, and its extensions can also be ignored.

4 Experimental Study

We performed an experimental study to assess the performance of PHM. The experiment was performed on a computer with a sixth generation 64 bit Core i5 processor running Windows 10, and equipped with 12 GB of free RAM. We compared the performance of the proposed PHM algorithm with the state-of-the-art FHM algorithm for mining HUIs. All memory measurements were done using the Java API. The experiment was carried on four real-life datasets commonly used in the HUIM litterature: *retail*, *mushroom*, *chainstore* and *foodmart*. These datasets have varied characteristics and represents the main types of data typically encountered in real-life scenarios (dense, sparse and long transactions). Let

$|I|$, $|D|$ and A represents the number of transactions, distinct items and average transaction length of a dataset. *retail* is a sparse dataset with many different items ($|I| = 16,470$, $|D| = 88,162$, $A = 10,30$). *mushroom* is a dense dataset with long transactions ($|I| = 119$, $|D| = 8,124$, $A = 23$). *chainstore* is a dataset that contains a huge number of transactions ($|I| = 461$, $|D| = 1,112,949$, $A = 7.23$). *foodmart* is a sparse dataset ($|I| = 1,559$, $|D| = 4,141$, $A = 4.4$). The *chainstore* and *foodmart* datasets are real-life customer transaction databases containing real external and internal utility values. The *retail* and *mushroom* datasets contains synthetic utility values, generated randomly [11, 15]. In the experiment, PHM was run on each dataset with fixed *minper* and *minAvg* values, while varying the *minutil* threshold and the values of the *maxAvg* and *maxper* parameters. In these experiments, the values for the periodicity thresholds have been found empirically for each dataset (as they are dataset specific), and were chosen to show the trade-off between the number of periodic patterns found and the execution time. Note that results for varying the *minper* and *minAvg* values are not shown because these parameters have less influence on the patterns found than the other parameters. Thereafter, the notation *PHM V-W-X-Y* represents the PHM algorithm with *minper* = V, *maxper* = W, *minAvg* = X, and *maxAVG* = Y. Figure 3 compares the execution times of PHM for various parameter values and FHM. Figure 4, compares the number of PHUIs found by PHM for various parameter values, and the number of HUIs found by FHM.

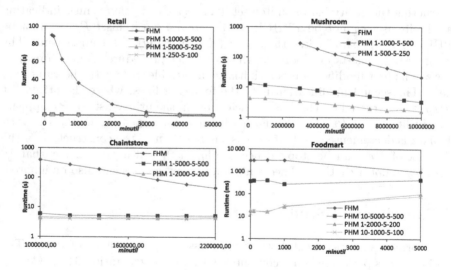

Fig. 3. Execution times

It can first be observed that mining PHUIs using PHM can be much faster than mining HUIs. The reason for the excellent performance of PHM is that it prunes a large part of the search space using its designed pruning strategies based on the maximum and average periodicity measures. For all datasets, it can

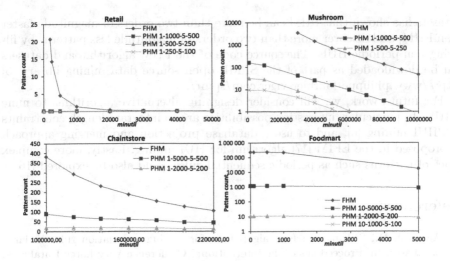

Fig. 4. Number of patterns found

be found that a huge amount of HUIs are non periodic, and thus pruning non periodic patterns leads to a massive performance improvement. For example, for the lowest *minutil*, *maxPer* and *maxAvg* values on these datasets, PHM is respectively up to 214, 127, 100 and 230 times faster than FHM. In general, the more the periodicity thresholds are restrictive, the more the gap between the runtime of FHM and PHM increases. A second observation is that the number of PHUIs can be much less than the number of HUIs (see Fig. 4). For example, on *retail*, 20,714 HUIs are found for *minutil* = 2,000. But only 110 HUIs are PHUIs for PHM 1-1000-5-500, and only 7 for PHM 1-250-5-150. Some of the patterns found are quite interesting as they contain several items. For example, it is found that items with product ids 32, 48 and 39 are periodically bought with an average periodicity of 16.32, a minimum periodicity of 1, and a maximum periodicity of 170. Huge reduction in the number of patterns are also observed on the other datasets. These overall results show that the proposed PHM algorithm is useful as it can filter a huge amount of non periodic HUIs encountered in real datasets, and can run faster. Memory consumption was also compared, although detailed results are not shown as a figure due to space limitations. It was observed that PHM can use up to 10 times less memory than FHM depending on how parameters are set. For example, on *chainstore* and *minutil* = 1,000,000, FHM and PHM 1-5000-5-500 respectively consumes 1,631 MB and 159 MB of memory.

5   Conclusion

This paper explored the problem of mining periodic high-utility itemsets (PHUIs). An efficient algorithm named PHM (Periodic High-utility itemset Miner) was proposed to efficiently discover PHUIs using novel minimum and average periodicity measures. An extensive experimental study with real-life datasets has shown that PHM can be more than two orders of magnitude faster than FHM, and discover more than two orders of magnitude less patterns by filtering non periodic HUIs. The source code of the PHM algorithm and datasets can be downloaded as part of the SPMF open source data mining library [5] http://www.philippe-fournier-viger.com/spmf/.

For future work, we will consider designing alternative algorithms to mine PHUIs. In particular, interesting possibilities are to integrate length constraints in PHUI mining [6], and to use a database projection and merging approach as proposed in the EFIM [17] algorithm for HUI mining. Lastly, more complex types of patterns such as periodic sequential rules could also be explored [16].

References

1. Agrawal, R., Srikant, R.: Fast algorithms for mining association rules in large databases. In: Proceedings of the International Conference Very Large Databases, pp. 487–499 (1994)
2. Amphawan, K., Lenca, P., Surarerks, A.: Mining top-k periodic-frequent pattern from transactional databases without support threshold. In: Proceedings of the 3rd International Conference on Advances in Information Technology, pp. 18–29 (2009)
3. Amphawan, K., Surarerks, A., Lenca, P.: Mining periodic-frequent itemsets with approximate periodicity using interval transaction-ids list tree. In: Proceeding of the 2010 Third International Conference on Knowledge Discovery and Data Mining, pp. 245–248 (2010)
4. Fournier-Viger, P., Wu, C.-W., Zida, S., Tseng, V.S.: FHM: faster high-utility itemset mining using estimated utility co-occurrence pruning. In: Proceedings of the 21st International Symposium on Methodologies for Intelligent Systems, pp. 83–92 (2014)
5. Fournier-Viger, P., Gomariz, A., Gueniche, T., Soltani, A., Wu, C., Tseng, V.S.: SPMF: a Java open-source pattern mining library. J. Mach. Learn. Res. (JMLR) 15, 3389–3393 (2014)
6. Fournier-Viger, P., Lin, C.W., Duong, Q.-H., Dam, T.-L.: FHM+: faster high-utility itemset mining using length upper-bound reduction. In: Proceedings of the 29th International Conference on Industrial, Engineering and Other Applications of Applied Intelligent Systems, p. 12. Springer (2016)
7. Lan, G.C., Hong, T.P., Tseng, V.S.: An efficient projection-based indexing approach for mining high utility itemsets. Knowl. Inform. Syst. 38(1), 85–107 (2014)
8. Kiran, R.U., Reddy, P.K.: Mining rare periodic-frequent patterns using multiple minimum supports. In: Proceedings of the 15th International Conference on Management of Data (2009)
9. Kiran, R.U., Kitsuregawa, M., Reddy, P.K.: Efficient discovery of periodic-frequent patterns in very large databases. J. Syst. Softw. 112, 110–121 (2015)
10. Song, W., Liu, Y., Li, J.: BAHUI: fast and memory efficient mining of high utility itemsets based on bitmap. Int. J. Data Warehous. Min. 10(1), 1–15 (2014)
11. Liu, M., Qu, J.: Mining high utility itemsets without candidate generation. In: Proceedings of the 22nd ACM International Conference Information and Knowledge Management, pp. 55–64 (2012)

12. Liu, Y., Liao, W., Choudhary, A.: A two-phase algorithm for fast discovery of high utility itemsets. In: Proceedings of the 9th Pacific-Asia Conference on Knowledge Discovery and Data Mining, pp. 689–695 (2005)
13. Surana, A., Kiran, R.U., Reddy, P.K.: An efficient approach to mine periodic-frequent patterns in transactional databases. In: Proceedings of the 2011 Quality Issues, Measures of Interestingness and Evaluation of Data Mining Models Workshop, pp. 254–266 (2012)
14. Tanbeer, S.K., Ahmed, C.F., Jeong, B.S., Lee, Y.K.: Discovering periodic-frequent patterns in transactional databases. In: Proceedings of the 13th Pacific-Asia Conference on Knowledge Discovery and Data Mining, pp. 242–253 (2009)
15. Tseng, V.S., Shie, B.-E., Wu, C.-W., Yu, P.S.: Efficient algorithms for mining high utility itemsets from transactional databases. IEEE Trans. Knowl. Data Eng. **25**(8), 1772–1786 (2013)
16. Zida, S., Fournier-Viger, P., Wu, C.-W., Lin, J.C.W., Tseng, V.S.: Efficient mining of high utility sequential rules. In: Proceedings of the 11th International Conference Machine Learning and Data Mining, pp. 1–15 (2015)
17. Zida, S., Fournier-Viger, P., Lin, J.C.-W., Wu, C.-W., Tseng, V.S.: EFIM: a highly efficient algorithm for high-utility itemset mining. In: Proceedings of the 14th Mexican International Conference on Artificial Intelligence, pp. 530–546

Comparative Analysis of Levenberg-Marquardt and Bayesian Regularization Backpropagation Algorithms in Photovoltaic Power Estimation Using Artificial Neural Network

Kian Jazayeri[✉], Moein Jazayeri, and Sener Uysal

Department of Electrical and Electronic Engineering,
Eastern Mediterranean University, Via Mersin 10, Famagusta, Turkey
{kian.jazayeri,moein.jazayeri}@cc.emu.edu.tr,
sener.uysal@emu.edu.tr

Abstract. This paper presents a comparative analysis of Levenberg-Marquardt (LM) and Bayesian Regularization (BR) backpropagation algorithms in development of different Artificial Neural Networks (ANNs) to estimate the output power of a Photovoltaic (PV) module. The proposed ANNs undergo training, validation and testing phases on 10000+ combinations of data including the real-time measurements of irradiance level (W/m^2) and PV output power (W) as well as the calculations of the Sun's position in the sky and the PV module surface temperature (°C). The overall performance of the LM and the BR algorithms are analyzed during the development phases of the ANNs, and also the results of implementation of each ANN in different time intervals with different input types are compared. The comparative study presents the trade-offs of utilizing LM and BR algorithms in order to develop the best ANN architecture for PV output power estimation.

1 Introduction

Solar energy is a renewable and sustainable resource that emerges to meet the energy requirements of today's modern world. The solar energy is converted to direct current (DC) electricity by the Photovoltaic (PV) effect. PV cells are connected to form a PV module (solar panel), which can be connected to other PV modules to construct PV arrays and systems. The importance of developing handling techniques of PV systems is highlighted considering the growing world energy demands and the limitations and threats associated with the traditional energy resources. Artificial Intelligence (AI) techniques are deployed in various applications as an alternative to conventional techniques due to their capabilities in solving complicated practical problems. Artificial Neural Network (ANN) is one of the most popular branches of AI. ANNs are mathematical models that imitate the behavior of biological Neural Systems. An ANN, which is a collection of interconnected computation units, is able to generalize outputs for new inputs after being trained on patterns of training data. ANNs are deployed in many practical applications due to their fault tolerance, flexibility and robustness in

© Springer International Publishing Switzerland 2016
P. Perner (Ed.): ICDM 2016, LNAI 9728, pp. 80–95, 2016.
DOI: 10.1007/978-3-319-41561-1_7

handling noisy data. Some of the ANN application examples in PV systems are given as follows.

Estimation of the daily solar radiation using ANNs is proposed by Elizondo et al. [1] and Williams and Zazueta [2]. Golabl solar radiation prediction using ANN is suggested by Alawi and Hinai [3] and monthly mean daily values of global solar radiation on horizontal surfaces are modeled by Mohandes et al. [4] using Radial Basis Function (RBF) networks. An ANN based total solar radiation time-series simulation model is offered by Mihalakakou et al. [5] and RBF networks are used for estimating total daily solar radiation by Mellit et al. [6].

The right choice of network types and algorithms is essential in order to attain desirable modeling, estimation and prediction outputs using ANNs. Different ANN architectures are employed for PV power estimation purposes by Lo Brano et al. [7]. The ambient temperature, solar irradiance and wind speed data are provided to the ANNs in the mentioned study to estimate the output power of two PV test modules. The authors conclude that the Multi-layer Perceptron (MLP) architecture provides the best performance in terms of the estimation error. A similar study is carried out by Saberian et al. [8] and the minimum temperature, maximum temperature, mean temperature and solar irradiance data are fed to different ANN topologies in order to estimate the output power of a PV module. The authors indicate that the feed-forward MLP with backpropagation training algorithm provides the best performance in PV module power estimation. A comparison of Levenberg-Marquardt (LM) and Bayesian Regularization (BR) backpropagation algorithms for efficient localization in wireless sensor network is presented by Payal et al. [9], a comparison of BR and Cross-Validated Early-Stopping (CVES) backpropagation algorithms for streamflow forecasting is carried out by Wang et al. [10] and a comparative study of backpropagation algorithms in ANN based identification of power system is proposed by Tiwari et al. [11].

In this study, well-detailed and highly accurate data is acquired using appropriate and highly-sensitive measurement equipment to be described in Sect. 2. However no matter how well-detailed and straight-forward-looking the inputs be, the PV power generation relationships are non-linear and cannot be expressed by simple analytical or physical approaches. Especially the effect of the PV module surface temperature highly complicate the relation between the solar irradiance and the PV output power as the PV power generation tendency decreases with increasing PV module surface temperature, which is caused by increasing irradiance. In other words, the solar irradiance has a very complex effect (compliant and opposite effects at the same time) on the PV power generation which cannot be expressed by simple equations or analytical models. The need for using a qualified machine learning technique for PV module output power estimation is highlighted by taking the above into consideration. The scope of this study is to give insight of the competency of the well-known MLP approach with eligible backpropagation algorithm in PV power estimation applications in existence of sufficient inputs. A comparative analysis of LM and BR backpropagation training algorithms in estimation of PV module output power using MLP approach is proposed and the trade-offs of utilizing each algorithm is represented in terms of training error, time, speed, etc. Finally the performances of applying different ANNs trained by LM and BR algorithms for PV power estimation in several ANN implementation intervals with different input types are analyzed.

2 Data Collection

It is aimed to provide the calculated values of the Sun's position in the sky, the angle of incidence and the PV module surface temperature as well as the irradiance level measurements to the Artificial Neural Networks (ANNs) as the input data, and the measurements of the PV module output power as the target data, during the training, testing and validation phases. The only preprocessing applied to the mentioned data is the normalization process which is described in the following sub-sections. Nov. 1st to Nov. 25th, 2015 is selected as the data acquisition interval due to the highly variable meteorological conditions during daylight hours in this period in Cyprus Island where the experiments took place. The ANNs receive highly variable training and testing data in this relatively short time interval. The noise is negligible due to the sensitivity of the measurement equipment utilized for this study. The acquired data for development of the ANNs is described comprehensively in this section.

2.1 The Sun's Position Data

The Sun's position in the sky is defined by the solar altitude and the solar azimuth angles [12]. The solar altitude angle indicates the Sun's elevation from Earth's surface and is expressed as:

$$\gamma_s = \sin^{-1}(\sin\varphi\sin\delta + \cos\varphi\cos\delta\cos\omega) \tag{1}$$

$$\delta = \sin^{-1}\{0.3987\sin(j'80.2° + 1.92(\sin(j' - 2.80°)))\} \tag{2}$$

$$j' = j \times (360/365.25) \tag{3}$$

$$\omega = 15(LST - 12) \tag{4}$$

Where,
φ : latitude of the observation point
δ : solar declination angle
j : Julian day number
ω : hour angle

Because of the irregularity of Earth's orbit as well as human adjustments (time zones and daylight saving application), Local Solar Time (*LST*) is slightly different than Local Time (*LT*) and is represented as:

$$LST = LT + \frac{TC}{60} \tag{5}$$

$$TC = 4(Longitude - LSTM) + EoT \tag{6}$$

$$EoT = 9.87 \sin(2B) - 7.53 \cos(B) - 1.5 \sin(B) \qquad (7)$$

$$B = \frac{360}{365}(j-81) \qquad (8)$$

$$LSTM = 15°.\Delta T_{GMT} \qquad (9)$$

Where,

TC : Time correction factor
EoT : Equation of time
$LSTM$: Local standard time meridian
ΔT_{GMT} : The difference of local time from Greenwich Mean Time

The solar azimuth angle indicates the Sun's deviation from the north axis and is expressed by:

$$\begin{cases} \alpha_s = 180 - \cos^{-1}(\cos \alpha_s) & \text{If } \sin \alpha_s < 0 \\ \alpha_s = 180 + \cos^{-1}(\cos \alpha_s) & \text{If } \sin \alpha_s > 0 \end{cases} \qquad (10)$$

Where,

$$\cos \alpha_s = (\sin \varphi \sin \gamma_s - \sin \delta)/\cos \varphi \cos \gamma_s \qquad (11)$$

$$\sin \alpha_s = \cos \alpha_s \sin \omega / \cos \gamma_s \qquad (12)$$

The mentioned values are normalized between 0 and 1, yielding data close to 1 for the values that have the most impact on the PV module output power and data close to 0 vice versa. The normalized values of the solar altitude and the solar azimuth angles calculated in minutely basis during daylight time interval on Nov. 16th, 2015, as a typical data acquisition period, are shown in Fig. 1.

2.2 The Angle of Incidence

The angle of incidence is the angle between the Sun's beams and a vector perpendicular to the surface of a solar panel and is represented as:

$$\theta = \cos^{-1}\left[\begin{array}{c} \cos(\beta)\cos(Z_s) + \\ \sin(\beta)\sin(Z_s)\cos(\alpha_s - \alpha_m) \end{array}\right] \qquad (13)$$

$$Z_s = 90 - \gamma_s \qquad (14)$$

Where,

β : Tilt angle of the solar panel (45° in this case)
Z_s : Zenith sngle of the Sun
α : Module azimuth angle (in this case: south = 180°)

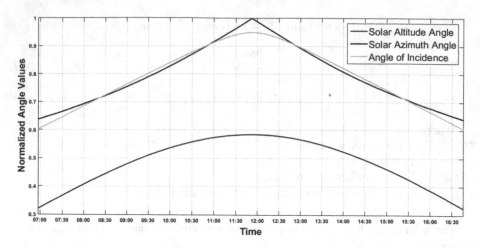

Fig. 1. The normalized values of the solar altitude and azimuth angles and the angle of incidence, calculated on Nov. 16th, 2015.

The normalized values of the angle of incidence, calculated minutely on Nov. 16th, 2015, are shown in Fig. 1. The normalized value takes on 1 when the Sun's beams are perpendicularly received on the solar panel surface and as the Sun's beams deviate from the perpendicular axis, the normalized value decreases down to 0.

2.3   Irradiance Level

The density of the solar radiation power received on a given surface is defined as the irradiance and is measured in Watts per meter square. In this study a south oriented,

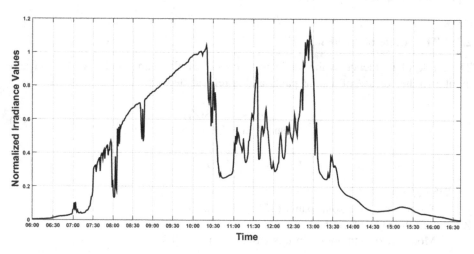

Fig. 2. The normalized irradiance values, measured in (W/m²) on Nov. 16th, 2015.

45° tilted pyranometer is used for field measurements. The measured irradiance values are normalized between 0 and 1. The normalized irradiance values measured minutely on Nov. 16th, 2015 as a typical data acquisition period are shown in Fig. 2.

2.4 PV Module Surface Temperature

The PV Module Surface Temperature has a reverse impact on the PV module performance efficiency since the PV power generation tendency is reduced as the surface temperature increases. According to [13], the PV module surface temperature can be calculated as a function of *NOCT* and the ambient temperature using the following equation:

$$T = T_{ambient} + ((NOCT - 20\,°C)(E_{tot}/800\,\mathrm{Wm^{-2}})) \tag{15}$$

Where,

T = PV Module surface temperature
$T_{ambient}$ = Ambient temperature
$NOCT$ = Nominal operating cell temperature
E_{tot} = Irradiance level (W/m^2)

The minutely temperature ($T_{ambient}$) values in (°C) obtained from the Larnaca International Airport (LCLK) weather station and the measurements of irradiance levels are used for PV module surface temperature calculations. The calculated values are normalized between 0 and 1, yielding values close to 1 for the lower surface temperatures and values close to 0 conversely. The normalized values of the PV module temperature calculated minutely on Nov. 16th, 2015 are shown in Fig. 3.

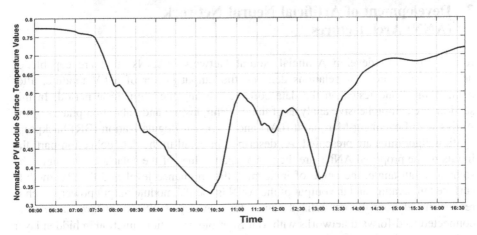

Fig. 3. The normalized PV module surface temperature values, calculated in (°C) on Nov. 16th, 2015.

2.5   Photovoltaic Module Output Power

A south oriented, 45° tilted monocrystalline silicon solar panel (P_{max}: 40 W, V_{OC}: 21.6 V, I_{SC}: 2.56 A) located at 35° 8′ 51″ N, 33° 53′ 58″ E, with 1 meters elevation from the sea level is used for the field measurement purposes. The output power of the PV module directly feeding a constant resistive DC load, is measured (mW) minute by minute and logged after being normalized between 0 and 1. The normalized values of the PV module output power measured in minutely basis on Nov. 16th, 2015, is shown in Fig. 4.

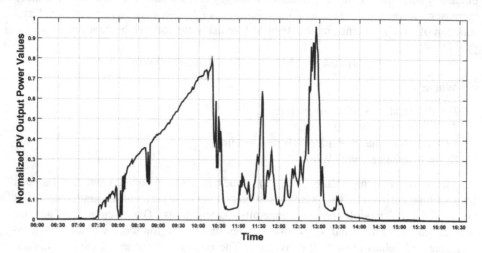

Fig. 4. The normalized PV output power values, measured in (mW) on November 16th, 2015.

3   Development of Artificial Neural Network (ANN) Architectures

It is intended to develop Artificial Neural Networks (ANNs) that are capable of deriving the appropriate relations defining the output power of a PV module, after being trained and tested on the data collected during the acquisition period. In this section, the comprehensive analysis of the training, testing and validation phases of the ANNs with Levenberg-Marquardt (LM) and Bayesian Regularization (BR) backprop-agation algorithms are presented. As described in detail in the data collection part, the inputs to the proposed ANNs are the normalized values of the solar altitude angle, the solar azimuth angle, the angle of incidence, the irradiance level and the PV module surface temperature and the output of the ANN is the PV module output power which is reconstructed from the normalized value. The developed ANNs are three-layer fully connected feed-forward networks with Tan-Sigmoid activation function in hidden layer and linear activation function in output layer. The number of hidden neurons are decided such that the network maintains the required accuracy while the computation time and memory does not exceed certain limits. It is observed the estimation accuracy of the ANN does not show significant improvement for more than 10 neuron sized hidden layers, while the computation time and memory allocation raise significantly with larger hidden layers. Therefore the number of hidden neurons are set to 10. The mentioned ANN architecture is illustrated in Fig. 5. Random small values are assigned to the network weights at the beginning of the training process and the training inputs are fed to the network through the input layer. The training data propagate through the network to reach the output layer. At this point the training error is calculated by comparing the estimated output and the target output which is presented to the network to supervise the learning process. The error is back-propagated in the network to adjust the weights. The training error threshold is set such that the network is kept from being either under-fitted or over-fitted. An under-fitted ANN lacks accuracy in estimation while an over-fitted ANN fails in generalization for new inputs.

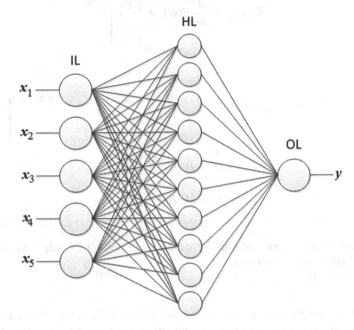

Fig. 5. The proposed ANN architecture (x_1: normalized solar altitude angle, x_2: normalized solar azimuth angle, x_3: normalized angle of incidence, x_4: normalized irradiance, x_5: normalized PV module surface temperature, y: normalized PV module output power, IL: Input Layer, HL: Hidden Layer, OL: Output Layer).

Backpropagation is used to obtain the first and the second derivatives of the error function $E(w)$ with respect to the network weight vector, $w = [w_1\ w_2\ \ldots\ w_N]^T$.

First derivative of the error function with respect to the weight vector is called the Gradient of $E(w)$ and is expressed by Eq. 16. Second derivative of $E(w)$ is the Hessian of $E(w)$ given in Eq. 17. The Levenberg–Marquardt (LM) algorithm developed by Kenneth Levenberg and Donald Marquardt is suitable for ANNs dealing with moderate-sized problems. Wilamowski and Yu [12] introduce the approximation to Hessian matrix indicated in Eq. 18. The update rule of the LM algorithm is presented as Eq. 19. Further mathematical details regarding the LM backpropagation algorithm can be investigated in [14–16].

The BR backpropagation algorithm provides robust estimation for noisy and difficult inputs in the existence of sufficient amount of training data. The algorithm works effectively by eliminating network weights that do not have much impact on the problem solution and shows better performance in avoiding the local minima difficulties. Cross validation is not necessary in BR algorithm, which avoids part of training data from being reserved for validation purposes. Also BR algorithm prevents the ANN from over-training and over-fitting problems. Comprehensive details about BR backpropagation algorithm can be found in [17, 18].

$$\nabla E(w) = \frac{d}{dw}E(w) = \begin{bmatrix} \frac{\partial E}{\partial w_1} \\ \frac{\partial E}{\partial w_2} \\ \cdots \\ \frac{\partial E}{\partial w_N} \end{bmatrix} \tag{16}$$

$$H = \nabla\nabla E(w) = \frac{d^2}{dw^2}E(w) = \begin{bmatrix} \frac{\partial^2 E}{\partial w_1^2} & \frac{\partial^2 E}{\partial w_1 \partial w_2} & \cdots & \frac{\partial^2 E}{\partial w_1 \partial w_N} \\ \frac{\partial^2 E}{\partial w_2 \partial w_1} & \frac{\partial^2 E}{\partial w_2^2} & \cdots & \frac{\partial^2 E}{\partial w_2 \partial w_N} \\ \cdots & & \cdots & \cdots \\ \frac{\partial^2 E}{\partial w_N \partial w_1} & \frac{\partial^2 E}{\partial w_N \partial w_2} & \cdots & \frac{\partial^2 E}{\partial w_N^2} \end{bmatrix} \tag{17}$$

$$H = J^T J + \mu I \tag{18}$$

$$w_{k+1} = w_k - (J_k^T J_k + \mu I)^{-1} J_k e_k \tag{19}$$

Where,
J : Jacobian matrix (matrix of first derivatives with respect to weight vector)
μ : Combination coefficient and,
I : Identity matrix

The training, testing and validation processes of the ANNs are performed on 10695 combinations of data, each combination containing 5 inputs and 1 target output. The normalized values of the solar altitude and azimuth angles, angle of incidence, irradiance and PV module surface temperature are fed to the network through the input layer. The inputs pass through the hidden layer consisting of 10 hidden neurons each having a Tan-Sigmoid transfer function and reach the output layer which contains one neuron with linear transfer function. The estimated output which is in range of 0 to 1 is compared to the learning target output and the error is back-propagated through the network. After several epoch of backpropagation and weight adjustment the training goal is achieved and the network becomes ready to generalize new outputs for unseen inputs.

As mentioned earlier, it is aimed to develop ANNs which are neither under-fitted nor over-fitted. A loose training goal results in a weak network that is not capable of making precise estimations while a very tight training goal will force the network to adjust its weights in order to achieve outputs almost similar to learning targets. Such an over-fitted network provides very accurate results for training inputs but lacks in making generalizations for new and unseen inputs. The Minimum Gradient is set to 1.0e−10 for both ANNs as the training goal in order to achieve accuracy and generalization capabilities at the same time.

In order to create the first ANN, 70 % (7487 paths) of the mentioned data collection is allotted to the training process in which the inputs and the output are presented to the ANN and the weight adjustments between the neurons are done based on the LM back-propagation algorithm. Another 15 % (1604 paths) of the collected data is presented to the ANN as the validation data to determine the generalization abilities of the network. The rest 15 % (1604 paths) of the collected data is given to the network during the testing process, which is carried out independently from the training and validation processes and gives a measure of the network performance. In the testing process, output targets are not presented to the network in order to measure the estimation and generalization abilities of the ANN. The training stops when the validation process show no more generalization.

The same procedure is repeated for construction of the second ANN with BR training backpropagation algorithm. 85 % (9091 paths) and 15 % (1604 paths) of the collected data is presented to this ANN during the training and testing phases respectively. As mentioned before, validation is unnecessary in the BR algorithm which allows further 1604 paths to be added to the training data. The training stops at a pre-set limit of 1000 epochs.

The regression plots and the performance metrics of the ANNs for the training, validation and testing processes are given in Fig. 6 and Table 1 respectively. The performance details of the LM and the BR backpropagation algorithms during the training, testing and validation processes are given in Table 1 and the Mean Absolute

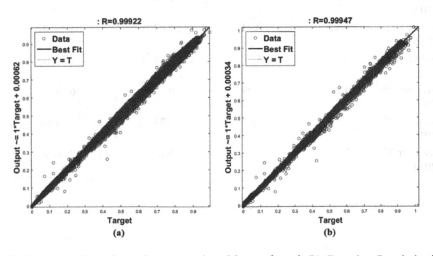

Fig. 6. The regression plots of (a) Levenberg-Marquardt and (b) Bayesian Regularization training backpropagation algorithms.

Table 1. Performance metrics of the Levenberg-Marquardt (LM) and the Bayesian Regularization (BR) training backpropagation algorithms.

Performance metrics	Levenberg-Marquardt (LM) algorithm	Bayesian Regularization (BR) algorithm
Best training performance	1.2549e−04	1.0211e−04
Best validation performance	1.3365e−04	Not applied in BR
Best testing performance	1.0787e−04	1.0418e−04
No. of training epochs	162	1000
Best training epoch	154	1000
Minimum gradient	6.2957e−06	1.2496e−08
Training time (in Seconds)	16.27	114.53

Error (MAE) and the Mean Absolute Percentage Error (MAPE)[1] between the estimated and the measured PV module output power values for ANN implementations from Nov. 26th to Dec 7th, 2015 are given in Table 2.

Table 2. The Mean Absolute Error (MAE) and the Mean Absolute Percentage Error (MAPE) between the estimated and the measured PV module output power values for different ANN implementation periods.

ANN implementation period	Mean absolute error (LM) (mW)	Mean absolute error (BR) (mW)	Mean absolute percentage error (LM)	Mean absolute percentage error (BR)
November 26th, 2015	1.45	1.42	7.05 %	5.87 %
November 27th, 2015	1.12	0.94	5.62 %	5.43 %
November 28th, 2015	1.58	1.04	5.74 %	4.54 %
November 29th, 2015	1.82	1.29	9.74 %	6.28 %
December 2nd, 2015	1.7	0.93	7.33 %	4.77 %
December 7th, 2015	2.16	0.71	3.18 %	2.06 %

---

[1] The noisy data is filtered out in order to maintain reasonable and robust (MAPE) values.

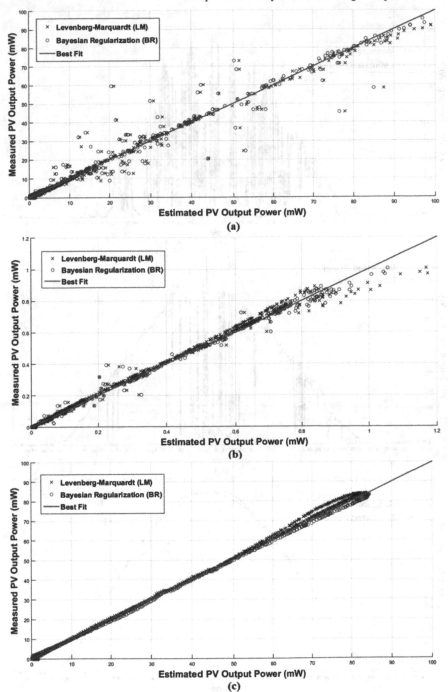

Fig. 7. The measured vs. the estimated PV module output values for ANNs implemented by the LM and the BR algorithms on (a) November 26th, 2015, (b) November 28th, 2015 and (c) December 7th, 2015.

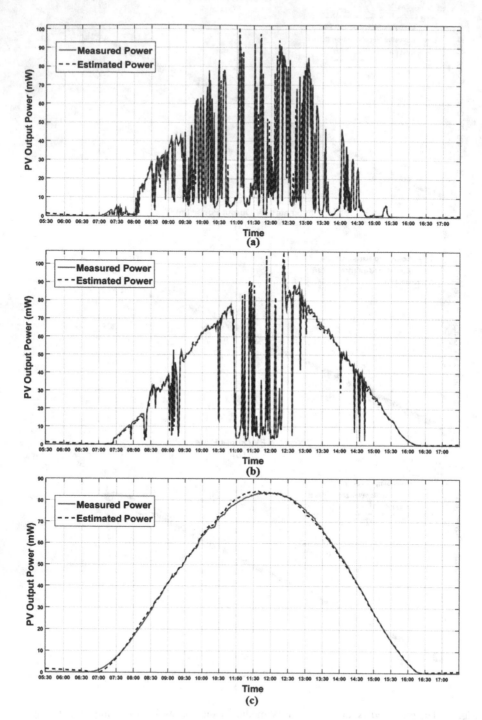

Fig. 8. The measured vs. the estimated PV module output power values for ANN implemented by the BR algorithm on (a) November 26th, 2015, (b) November 28th, 2015 and (c) December 7th, 2015.

4   Implementation of the Developed Artificial Neural Networks (ANNs)

The ANNs created during the development processes described in section-3 are implemented in a period from Nov. 26th, 2015 to Dec. 7th, 2015. During this period the performances of the ANNs for estimation of PV module output power in different meteorological situations varying from highly fluctuating partly cloudy to smooth clear sky conditions are monitored. In order to better express the performance of each ANN, a scatter graph of the measured vs. the estimated PV module output power values for each ANN implementation in three different weather conditions on Nov. 26th, Nov. 28th and Dec. 7th, 2015 are given in Fig. 7. As it is obvious from the figures, the ANN trained by the Bayesian Regularization (BR) backpropagation algorithm shows a better performance than the ANN trained by the Levenberg-Marquardt (LM) backpropagation algorithm especially in higher output values. This performance difference was expected since the BR algorithm is known to work well with noisy and difficult data as described in section-3. Although the estimation performance improvements of the BR algorithm may appear to be relatively small for a single PV module application, the improvement becomes definitely significant when the application is extended to broader PV arrays after taking developmental considerations. On the other hand, the higher performance of the BR algorithm comes with a cost. As it is obvious from Table 2, the training time of the BR algorithm is significantly higher than that of the LM algorithm. The training time of the BR algorithm increases more and more by setting the maximum epoch number to higher limits in order to maintain lower training error. The BR algorithm may not be deployed in time-constrained situations. The PV module output power estimated by the ANN with BR algorithm vs. the measured PV module output power for Nov. 26th, Nov. 28th and Dec. 7th, 2015 are plotted in Fig. 8.

5   Conclusions

The performances of Artificial Neural Networks (ANNs) developed by the Levenberg-Marquardt (LM) and the Bayesian Regularization (BR) training back-propagation algorithms for PV module power estimation are analyzed in this paper. The results show that the BR algorithm provides a better performance than the LM algorithm in PV power estimation, to the cost of higher computation time. The average Mean Absolute Error (MAE) and the Mean Absolute Percentage Error (MAPE) between the estimated and the measured PV module output power values for ANN implementation period from Nov. 26th to Dec. 7th, 2015 are 1.64 (mW) and 6.44 % respectively for the LM algorithm, which are reduced to 1.05 (mW) and 4.83 % by the BR algorithm. The estimation improvement of BR algorithm is highlighted by the fact that the proposed PV power estimation algorithm can be extended to broader PV fleets after taking necessary developmental considerations. On the other hand the training time of the LM algorithm is 16.27 (s), which is increased to 114.53 (s) by the BR algorithm. The training time of the BR algorithm further increases in order to maintain higher accuracy. Consequently it can be concluded that the BR training backpropagation algorithm presents better performance in ANN based PV power estimation purposes and is the right choice where high accuracy is required but this algorithm is significantly time-consuming and may not be utilized where the training speed is of major concern. The LM algorithm is the proper choice in time-constrained situations.

References

1. Elizondo, D., Hoogenboom, G., Mcclendon, R.: Development of a neural network model to predict daily solar radiation. Agric. For. Meteorol. **71**, 115–132 (1994)
2. Williams, D., Zazueta, F.: Solar radiation estimation via neural network. In: Sixth International Conference on Computers in Agriculture, ASAE, Cancun, Mexico, pp. 140–146 (1994)
3. Alawi, S., Hinai, H.: An ANN-based approach for predicting global radiation in locations with no direct measurement instrumentation. Renew. Energy **14**, 199–204 (1998)
4. Mohandes, M., Balghonaim, A., Kassas, M., Rehman, S., Halawani, T.: Use of radial basis functions for estimating monthly mean daily solar radiation. Sol. Energy **68**(2), 161–168 (2000)
5. Mihalakakou, G., Santamouris, M., Asimakopoulos, D.: The total solar radiation time series simulation in Athens, using neural network. Theor. Appl. Climatol. **66**, 185–197 (2000)
6. Mellit, A., Benghanem, M., Hadj Arab, A., Guessoum, A.: Modeling of global solar radiation data from sunshine duration and temperature using the radial basis function networks. In: The IASTED, Grindelwald, Switzerland (2004)
7. Brano, Lo: V., Ciulla, G., Di Falco, M.: Artificial neural networks to predict the power output of a PV panel. Int. J. Photoenergy **2014**, 1–12 (2014)
8. Saberian, A., Hizam, H., Radzi, M., Kadir, M., Mirzaei, M.: Modelling and prediction of photovoltaic power output using artificial neural networks. Int. J. Photoenergy **2014**, 1–10 (2014)
9. Payal A., Rai C., Reddy B.: Comparative analysis of Bayesian Regularization and Levenberg-Marquardt training algorithm for localization in wireless sensor network. In 15th International Conference on Advanced Communication Technology (ICACT), PyeongChang, South Korea, pp. 191–194 (2013)
10. Wang, W., Gelder, P., Vrijling, J.: Comparing Bayesian Regularization and cross-validated early-stopping for streamflow forecasting with ANN models. In: 2nd International Symposium on Methodology in Hydrology, Nanjing, China, pp. 216–221 (2005)
11. Tiwari, S., Naresh, R., Jha, R.: Comparative study of backpropagation algorithms in neural network based identification of power system. Int. J. Comput. Sci. Inform. Tech. **5**(4), 93–107 (2013)
12. Scharmer, K., Greif, J.: The European Solar Radiation Atlas, Vol.1: Fundamentals and Maps, pp. 23–42. Les Presses de l'Ecole des Mines, Paris, France (2000)
13. Luque, A., Hegedus, S.: Handbook of Photovoltaics Science and Engineering, pp. 906–912. Wiley, West Sussex, UK (2003)
14. Wilamowski, B., Yu, H.: Improved computation for Levenberg-Marquardt training. IEEE Trans. Neural Netw. **21**, 930–937 (2010)
15. Levenberg, K.: A method for the solution of certain problems in least-squares. Quart. Appl. Math. **2**, 164–168 (1944)

16. Marquardt, D.: An algorithm for least-squares estimation of non-linear parameters. J. Soc. Ind. Appl. Math. **11**, 431–441 (1963)
17. MacKay, D.: Bayesian interpolation. Neural Comput. **4**(3), 448–472 (1992)
18. Buntine, W., Weigend, A.: Bayesian backpropagation. Complex Syst. **5**, 603–643 (1991)

Data Mining on Divers Alert Network DSL Database: Classification of Divers

Tamer Ozyigit[1]([⊠]), Cuneyt Yavuz[1], Massimo Pieri[2], S. Murat Egi[1],
Bahar Egi[3], Corentin Altepe[3], Danilo Cialoni[2],
and Alessandro Marroni[2]

[1] Computer Engineering Department, Galatasaray University, Istanbul, Turkey
tozyigit@gsu.edu.tr
[2] Divers Alert Network Europe Research Division, Roseto, Italy
[3] Bogazici Underwater Research Center, Istanbul, Turkey

Abstract. Divers Alert Network (DAN) created a database (DB) with a big amount of dive related data which has been collected since 1994 within the scope of Dive Safety Laboratory (DSL) project. The aim of this study is to analyze the DB using data mining techniques. The clustering of divers by their health and demographic information and reveal significant differences in diver groups are the main objectives of this study.

To eliminate time effect of age, divers who participated to only one dive were included in the study. The numbers of one-dive divers is 874. Before applying clustering methods, data cleaning was performed to eliminate the potential mistakes resulting from inconsistencies, inaccuracies and missing information. TwoStep, Gower distances and K-means clustering methods were performed on DB to find the naturally associated clusters. Conventional statistical analyses were performed to understand differences in clusters and between male and female divers.

As the result of these analyses, divers were separated into 3 clusters and distinguishing variables of these clusters were revealed. As TwoStep and Gower Distances are suitable for categorical variables, age and dive activity years were distributed in 3 categories. For K-Means Clustering, original numerical values of these variables was used. The most distinct clusters were formed by TwoStep Clustering. The middle aged male divers with without any health problem are in Cluster 1. Male and female divers with health problems and high rate of cigarette smoking are in the Cluster 2 and old divers with many dive activity years are in the Cluster 3. The search for significant differences in dive-related variables was performed based on the TwoStep Clustering results and separating male and female divers.

Keywords: Data mining · Cluster analysis · Diver groups

1 Introduction

Divers Alert Network (DAN) is collecting data in scope of dive safety research since 1994. The aim of collecting these data is to create a database (DB) consisting of reliable and large number of data in order to support dive related scientific research. Collected

© Springer International Publishing Switzerland 2016
P. Perner (Ed.): ICDM 2016, LNAI 9728, pp. 96–109, 2016.
DOI: 10.1007/978-3-319-41561-1_8

by volunteer divers, this database is the first example of research applications that spread to community in dive related research.

The DB aims at two major goals:

1. An in-depth epidemiological analysis focusing on habits and risks of the diving community;
2. Investigating additional risk factors correlated with the development of circulating bubbles and decompression sickness.

In the beginning, this project is named Safe Dive and the data is collected by taking notes manually. These notes have been replaced by dive computers since 2002 and the project has begun to be called Dive Safety Laboratory (DSL) since then. In 2013, a system which allows sending divers' profiles and other information to DAN was implemented.

An original database was developed, DAN DB, including specific questionnaires for data collection was developed allowing retrospective statistical analysis from 3108 European divers (men 83.3 %, women 16.7 %) who made 50151 dives over five years. A specific file format called DAN DL7 was developed in order to allow to receive all the personal and dive date in the same format. The original format DAN DL7 was developed by Petar J. Denoble, M.D., in the beginning of the 2000s, revised by DAN Europe team in the 2015 for the future use with the new web based data collection.

All dives were considered started when the divers descend over one meter depth and finished when they reached the same deep without any return at more depth within five minute. In our samples we had previously excluded all the dives less than five meters depth and 10 min of diving time. New recently developed hypothesis indicating that is necessary to investigate and identify further risk factors that may correlate with an increase in the incidence of bubbles formation and DCS.

To obtain useful information about divers, dives and dive safety, the use of data mining techniques was necessary. The objective of this paper is to reveal demographic/physical profiles of divers and clustering these profiles using clustering techniques to investigate if there are different groups consisting of distinct properties. The result of this study will be used to relate the diver groups with risk scoring of different dive types and dive profiles.

Building such databases supports multivariate statistical studies and data mining applications in diving. DAN has been publishing annual reports on dive safety and accidents since 1988 [1]. Ozyigit et al. applied clustering of decompression sickness using dive accident data provided by DAN America [2]. Another study is the use of multi criteria decision making systems for selecting personal for manned underwater operations [3].

For the first step of study, divers are separated into different groups using 3 different cluster analysis methods and the distinguishing variables of one-dive diver clusters are revealed in cluster tables. Finally the study focused on dive-related data of one-dive divers to uncover significant differences between diver groups and male-female divers.

2   Method

Originally, DAN Europe stores the DSL data in MSSQL Database. In order to keep diver identities confidential, the relevant data was removed and we obtained the data in tabular form consisting of multiple tables arranged by diver, dive and dive event variables. There is an ID variable (DiverID and DiveID respectively) for each diver and dive. Similarly, there is an ID variable (EventID) for dive events which consist of dive sets performed in 48 h intervals.

These tables were transferred to MySQL 5.5 developed by Oracle Corp. (500 Oracle Parkway, Redwood Shores, CA 94065) in order to perform data merging and extract required variables to be used in cluster analysis.

Data is separated into two groups. The first one consists of demographical and anamnesis information of divers and the latter consists of pre-dive and post-dive information collected from Daily Dive Log forms prepared filled by divers and digital data collected from dive computers.

Data repetition is caused by the fact that a diver may have more than one dive in one event and forms being refilled for every dive. Keeping records in this fashion created several issues for arranging data for the analysis. Most important of those was the necessity to eliminate this repetition effect.

Since there are more than one dive and event records for a diver in different dates and diver data is collected for each dive, changing of repeated data about diver in time caused another problem. A diver may participate to an event at the age of 30 and another one later at the age of 35. Considering divers' age is also an important variable for clustering, it is required to eliminate time effect of variable values so divers who participated to only one dive are included in analysis.

13 tables are included in database where diver, dive and dive event data are recorded. These tables are dive, event, record header, diver identification and demographics, dive header at start, dive profile, dive log details, dive safety report, diver additional anamnesis, dive additional data, dive additional problems, dive additional medications. Because data used for analysis is separated into these tables, it was required to aggregate selected variables into one table. Database diagram is shown in Fig. 1.

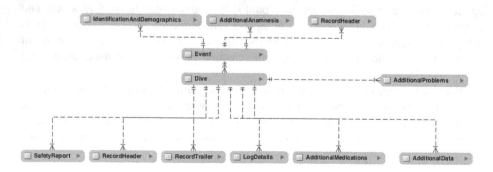

Fig. 1. Database diagram

Primarily these tables and their variables were examined and variables which will be used for cluster analysis were selected. The criteria for data selection is to choose divers' personal data (e.g. dive activity years) that may have effect on dive-related data. Variables which have a large number of missing data (e.g. height and weight) were eliminated since they reduce the number of data and analysis' reliability even if they are very important.

Diver data used for analysis were age, sex, dive activity years, allergy, asthma, back pain, back surgery, cigarette smoking, diabetes, ear-sinus problems, hearth problems, muscle pain, nervous system disorder, vascular diseases and sea sickness. Age and dive activity years were numeric variables and others were binary (0-1) variables. As TwoStep and Gower Distances are suitable for categorical data, age and dive activity years were distributed in 3 categories to be included into analysis. For K-Means Clustering, original numerical values of these variables was used for clustering.

Two-Step clustering is based on forming preliminary groups and re-clustering of those groups, then obtaining a cluster tree [4, 5]. K-Means uses local search approach to separate points into k groups. Firstly, k preliminary cluster is selected arbitrarily and fewer clusters are formed based on those preliminary clusters' centroids [6]. K-means has a very wide usage area including biology and computer graphics [7]. Gower Distances method developed by Gower is a distance measure used on binary variables [8]. Diver clusters obtained from clustering analysis are given in the next section.

For data cleaning and performing clustering with Gower Distances method, R, a language and environment for statistical computing and graphics developed by John Chambers and colleagues was used. The TwoStep and K-Means Clustering were performed using IBM Statistical Package for the Social Sciences (SPSS) version 20 released by IBM Corporation, New York. Chi-square, ANOVA and student t tests were performed using SPSS.

Past and present anamnesis information of divers was transformed into one variable which denotes if that diver has that disease or not. The analysis were performed using two tables consisting of divers who participated to only one dive and to one event which were obtained basing on DiverID variable.

In order to be able to perform clustering with categorical variables, age and activity year variables were separated into 3 categories by examining the box plots shown in Fig. 2.

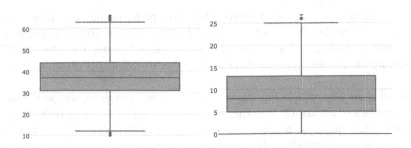

Fig. 2. Box plots for age (left) and activity years (right)

Observing the box plots above, the optimum intervals age and dive activity years categories were built by minimum to 1st quarter, 1st quarter to 3rd quarter and 3rd quarter to maximum. These categories are shown in Tables 1 and 2.

Table 1. Age categories

Interval	Age category
12 < 30	Young
30 < 45	Middle aged
45 < 70	Old

Table 2. Activity years diving categories

Interval	Age category
0 < 5	Few
5 < 13	Average
>13	Many

The study on dive-related data was based on investigating differences between clusters and male and female divers. Examined variables were "alcohol before dive", "exercise before dive", "state of rest before dive", "minimum water temperature", "thermal comfort", "diving platform", "maximum depth", "workload", "breathing gas", "apparatus", "equipment malfunctions" and any "symptoms". For categorical dive-related variables, chi-square test for dependency on diver clusters and male/female divers were applied. For numerical variables one-way ANOVA test for multiple groups and student t tests for pairwise comparisons were applied to find significant discrepancies for each combination of cluster and for male-female divers.

3   Results

Before cleaning of data, there were 3108 divers and 50151 dives performed by these divers. Total event count was 20052. There were 16.13 dives and 6.45 events per diver. Dive count per event was 2.5.

After cleaning, the number of divers who participated to only one dive was 874 (120 female, 754 male). Divers who are aged between 12 and 70 are included in the analysis and average age for divers who participated to only one dive is 37.52. Box plots for number of dives per diver and number of events per diver are shown in Fig. 3.

Observing the box plots in Fig. 3, we can say that most of the divers participated to one or few dives or events. On the other hand, we can see by the outliers that there are divers who participated to a great number of dives and events which dramatically increase the average number of dives and events per diver.

To clear the data, inconsistent and inaccurate values were omitted (e.g. divers who aged 150, whose dive activity years is more than 70 years and whose gender is

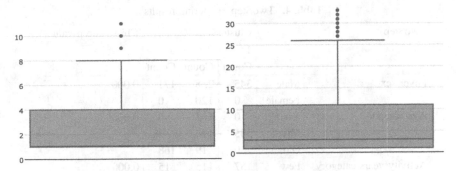

Fig. 3. Box plots for dive per diver (left) and event per diver (right)

undefined in database). Before omitting there were 930 divers who participated only one dive. After filtering this number became 874. Number of filtered divers is shown in Table 4. Finally, clustering analysis was performed on cleaned data which has divers aged between 12 and 70. The number of filtered divers is given in Table 3.

Table 3. Number of filtered divers by variables.

	One dive divers
Age (<12 and >70)	53
Activity years (>70)	0
Sex (undefined)	3

Clustering is performed using Two-Step and Gower Distances methods which are effective with categorical data. To provide a different perspective to the analysis K-means method which is more effective on numerical data was carried out as well.

Divers are separated into 3 clusters using 3 clustering analysis methods. The distinguishing variables for each group obtained from TwoStep Clustering are shown in bold in Table 4.

When we observe the Table 4, we can say that the first cluster is formed by middle aged male divers who without any health problem.

The second cluster is formed by male and female divers (all female divers are in the Cluster 2) with high rate of cigarette smoking. Especially allergy, ear/sinus problems and sea sickness have high rates of occurrence in this group. But other health problems have also higher rates than other 2 clusters. We can name this group as the group with health problems or unhealthy divers.

The Cluster 3 if formed by older male divers. The diving activity in years is greater than other groups and most of the heart problems are observed in this group.

The chi-square tests show highly significant specific distribution of variables into groups except diabetes, nervous system disorder and vascular disease which are rarely seen. This means the one-dive diver clusters are formed by significantly different groups of divers.

Table 4. TwoStep clustering results

TwoStep		Clusters			Chi-square p value
		1	2	3	
		Count	Count	Count	
Diver sex	Male	**345**	238	171	0.000
	Female	0	120	0	
Age category	Young	0	165	0	0.000
	Middle	**345**	174	3	
	Old	0	19	**168**	
Activity years category	Few	57	115	15	0.000
	Average	175	197	64	
	Many	113	46	**92**	
Allergy	Present	0	**48**	3	0.000
Asthma	Present	0	**7**	0	0.006
Back pain	Present	0	**12**	3	0.003
Back surgery	Present	0	2	3	0.045
Cigarette smoking	Present	0	**105**	20	0.000
Diabetes	Present	0	0	1	0.128
Ear/sinus problem	Present	0	**36**	0	0.000
Hearth problem	Present	0	2	**11**	0.000
Muscle pain	Present	0	9	3	0.015
Nervous system disorder	Present	1	1	0	0.783
Vascular disease	Present	0	3	0	0.114
Sea sickness	Present	0	**18**	3	0.000

The results of Gower Distances clustering are given in Table 5. When we look at the chi-square p values, we can say that the distinguishing variables having effect on cluster variables are sex, age category and dive activity years. Observing the clusters, we see that dive activity years is the only influential variable on cluster formations.

Although ear/sinus problem and sea sickness have higher percentages (9 % and 5 % respectively) in Cluster 2, similar to the Cluster 2 obtained from TwoStep method, the chi-square significances and a quick observation of the Table 5 show that variable occurrence frequencies are distributed randomly into 3 clusters.

As mentioned before, in K-Means clustering, the numerical variables are taken into account rather than categorical variable. We used the original numerical data for age and dive activity years when performing clustering using this method and as expected, the distinguishing variables were these two.

Observing the Table 6, we can say that younger divers with few dive activity years are in the Cluster 1, older divers with many diving activity years are in the Cluster 2 and middle aged divers with average dive activity years are in the Cluster 3.

Table 5. Gower distances clustering results

Gower distances		Clusters			Chi-square p value
		1	2	3	
		Count	Count	Count	
Diver sex	Male	366	149	239	0.000
	Female	70	38	12	
Age category	Young	92	65	8	0.000
	Middle	271	104	147	
	Old	73	18	96	
Activity years category	Few	0	**187**	0	0.000
	Average	**436**	0	0	
	Many	0	0	**251**	
Allergy	Present	29	13	9	0.195
Asthma	Present	4	2	1	0.685
Back pain	Present	8	3	4	0.964
Back surgery	Present	1	3	1	0.104
Cigarette smoking	Present	62	37	26	0.020
Diabetes	Present	0	1	0	0.159
Ear/sinus problem	Present	16	16	4	0.001
Hearth problem	Present	4	3	6	0.304
Muscle pain	Present	5	4	3	0.596
Nervous system disorder	Present	0	1	1	0.353
Vascular disease	Present	2	1	0	0.539
Sea sickness	Present	11	9	1	0.011

Another remark is that, there are few female divers in Cluster 3, the group of old and experienced divers.

The categorical variable frequencies are random except heart problems. Heart problems have a high rate of occurrence in the Cluster 3 which was expected as older divers are in this cluster.

One-way ANOVA tests were applied to numerical variables (age and dive activity years) to investigate whether the means of all groups are equal or significantly different. Table 6 shows that the diver groups are significantly different in these variables. However the other (categorical) variables have high p values of chi-square tests indicating these variables are not dependent to K-Means clusters.

As the most distinct clusters were formed by TwoStep Clustering, analysis on dive-related data were performed on the clusters obtained by this method. The percentages and averages for diver clusters and male-female diver are given in Table 7.

Older (male) divers have the higher percentage in "alcohol before dive". However the chi-square p value is 0.3471 which means that there is no significant dependency between clusters and alcohol before dive. To analyze the significant differences between groups deeply, chi-square tests for cluster combinations were performed as well.

Table 6. K-means clustering results

K-means clustering		Clusters						Chi-square p value
		1		2		3		
		Count	Mean	Count	Mean	Count	Mean	
Diver sex	Male	321		84		349		0.000
	Female	77		2		41		
Diver age years			**30**		**50**		**43**	0.000*
Diver activity years			**6**		**27**		**10**	0.000*
Allergy	Present	30		2		19		0.096
Asthma	Present	6		0		1		0.098
Back pain	Present	7		2		6		0.875
Back surgery	Present	1		1		3		0.469
Cigarette smoking	Present	73		8		44		0.007
Diabetes	Present	0		0		1		0.537
Sinus problem	Present	24		1		11		0.027
Hearth problem	Present	0		6		7		0.000
Muscle pain	Present	4		2		6		0.591
Nervous system disorder	Present	1		0		1		0.896
Vascular disease	Present	2		0		1		0.713
Sea Sickness	Present	10		0		11		0.297

*One-way ANOVA test p values

For Cluster 1 and Cluster 2, the p value is 0.3580. For Cluster 1 and Cluster 3, p value is 0.5156. The most significant difference was found between Cluster 2 and Cluster 3 with the p value of 0.1597 but still less than 0.05.

For "exercise before dive", the chi-square p value for all 3 clusters is 0.0004, showing that exercise level before dive differs significantly according to clusters. Most active group before dive is middle aged healthy divers. The percentage of light exercise is high in Cluster 3, group of old and experienced divers. The Cluster 2, group of male and female divers with health problems has the highest percentage of no exercise.

After group combinations analysis, p value of chi-square test between Cluster 1 and Cluster 2 is 0.01498, Cluster 1 and Cluster 3 is 0.0760, Cluster 2 and Cluster 3 is 0,0760. The most significant difference is between Cluster 1 and Cluster 2 (healthy and unhealthy divers).

For "state of rest before dive", the chi-square p value for all 3 clusters is 0.0006, showing that this variable is significantly dependent to clusters. The p value of chi-square test between Cluster 1 and 2 is 0.0069, between Cluster 1 and Cluster 3 is 0.1670 which is not highly significant and between Cluster 2 and Cluster 3 is 0.0002. The biggest difference is between Cluster 2 and Cluster 3. We can say that the Cluster 2 differs from other clusters in state of rest before dive.

Table 7. Dive-related variable percentages and averages

Dive-related variable		Clust. 1	Clust. 2	Clust. 3	Male	Female
Alcohol before dive	Yes %	37.97	34.64	40.94	38.46	29.17
Exercise before dive	None %	31.88	43.30	25.73	34.48	40.83
	Light %	47.25	41.06	58.48	46.95	46.67
	Moderate %	16.81	13.13	11.11	16.46	12.50
	Heavy %	4.06	2.51	4.68	4.11	0.00
State of rest before dive	Rested %	85.51	80.73	90.64	85.81	76.67
	Tired %	14.46	18.99	7.02	12.73	22.50
	Exhausted %	2.03	0.28	2.34	1.46	0.83
Avg. min water temp.	°C	13.65	13.22	13.85	13.25	15.19
Thermal comfort	Hot %	1.16	0.56	0.58	0.80	0.83
	Pleasant %	86.67	76.54	81.29	81.70	80.00
	Cold %	11.59	21.51	16.37	16.31	18.33
	Very cold %	0.58	1.40	1.75	1.19	0.83
Platform	Beach/Shore %	31.30	36.59	25.73	32.10	31.17
	Small boat. %	10.72	14.53	6.43	10.34	18.33
	Charter boat. %	51.01	43.02	58.48	50.80	39.17
	Live-aboard %	0.29	0.28	0.00	0.27	0.00
	Other %	6.67	5.59	9.36	6.50	8.33
Average max. depth	Meters	31.13	29.73	35.34	31.73	29.14
Workload	Resting %	46.67	38.83	39.77	42.97	36.67
	Light %	40.00	44.41	51.46	43.77	45.83
	Moderate %	12.17	13.97	8.19	12.07	12.50
	Severe %	1.16	2.51	0.00	1.06	4.17
	Exhausting %	0.00	0.28	0.58	0.13	0.83
Breathing gas	Air %	85.51	87.99	73.10	85.55	87.50
	Nitrox %	5.22	6.15	11.70	7.03	5.83
	Trimix %	4.06	2.23	8.77	4.77	0.83
	Other %	5.22	3.63	6.43	4.64	5.83
Apparatus	Scuba open %	95.94	98.32	95.32	96.29	100.00
	Rebreather %	1.74	0.84	2.34	1.72	0.00
	Other %	2.32	0.84	2.34	1.93	0.00
Equipment malfunction	BC %	1.16	0.84	2.34	1.46	0.00
	Breathing Ap. %	1.74	2.23	0.58	1.72	1.67
	Depth gauge %	0.00	0.56	0.00	0.13	0.83
	Dive computer %	0.29	0.00	0.00	0.13	0.00
	Face mask %	1.74	3.91	2.34	2.12	6.67
	Fins %	0.00	1.12	0.00	0.13	2.50
	Thermal protect. %	0.58	3.35	0.58	1.59	2.50
	Weight belt %	1.45	1.68	0.58	1.33	1.67
	Other %	2.03	0.56	0.00	1.06	0.83
Any symptom	Yes %	7.83	3.07	14.62	7.16	7.50

We didn't observe significant differences between groups in minimum "water temperature". The one-way ANOVA test p value is 0.554. The averages of clusters are given in Fig. 7. The p value of student t test between Cluster 1 and Cluster 2 is 0.3975, between Cluster 1 and Cluster 3 is 0.7499 and between Cluster 2 and Cluster 3 is 0.3429.

For "thermal comfort" during dive, the chi-square p value is 0.0183 which means the thermal comfort during dive depend on the clusters. The Cluster 2, the group of male and female divers with health problems complains more about cold water. The chi-square p value performed to Cluster 1 and Cluster 2 is 0.0022. For Cluster 1 and Cluster 3 p value of chi-square is 0.2223 and for Cluster 2 and Cluster 3 is 0.5785. This means that the biggest difference in thermal comfort is between Cluster 1 and Cluster 2, groups of healthy and unhealthy divers.

For diving "platform", the chi-square p value is 0.0095 showing that the clusters differ in diving platform. Comparing clusters 1 and 2, the p value of chi-square test is 0.1861 and for 1 and 3 is 0.1812 which are not very significant. The p value of chi-square test between Cluster 2 and Cluster 3 is 0.0007 showing that these two diver groups are very distinct in diving platform.

The one-way ANOVA test p value for "average maximum depth" of clusters is 0.0034 meaning that there are significant differences between clusters. The student t test p value for Cluster 1 and Cluster 2 is 0.2620, for Cluster 1 and Cluster 3 is 0.0215 and for Cluster 2 and Cluster 3 is 0.0009. These p values show that Cluster 3, older and experienced divers dive significantly deeper than other groups.

For "workload", the chi-square p value for all clusters is 0.0350 meaning that there are significant differences between diver groups. Chi-square p value for Cluster 1 and Cluster 2 is 0.1661, for Cluster 1 and Cluster 3 is 0.035 and for Cluster 2 and Cluster 3 is 0.0602. The Cluster 3 differs significantly from other two clusters in workload.

For "breathing gas", the p value of chi-square test is 0.0004, indicating significant differences between clusters. The experienced divers use more nitrox, Trimix and other gases than other groups. Chi square p values of pairwise tests justify this as the value for Cluster 1 and Cluster 2 is 0.353352, for Cluster 1 and Cluster 3 is 0.0037 and for Cluster 2 and Cluster 3 is 0.0001.

For "breathing apparatus", we didn't find significant dependency on clusters. The chi-square p value is 0.3006. The divers use open circuit scuba with high percentage. The pairwise chi-square tests p value for Cluster 1 and Cluster 2 is 0.1589, for Cluster 1 and Cluster 3 is 0.8971 and for Cluster 2 and Cluster 3 is 0.1321.

For "equipment malfunction", the highest percentages are in the Cluster 2 especially in face mask and thermal protection. The chi-square p value for all clusters is 0.0106 which indicates significant differences between clusters. The p value of chi-square test between Cluster 1 and Cluster 2 is 0.0151, between Cluster 1 and cluster 3 is 0.4141 and between Cluster 2 and Cluster 3 is 0.083588. The most important difference is between Cluster 1 and Cluster 2, groups of healthy and unhealthy divers. We can say that the Cluster 2 is the most distinct group from other clusters.

For "any symptom" The Cluster 3, group of old and experienced (and deep) diver has the highest percentage. The chi-square p value is 0.000008, indicating very highly significant dependency of this variable to clusters. The test p value between Cluster 1 and 2 is 0.0053, between Cluster 1 and Cluster 3 is 0.0158 and between Cluster 2 and Cluster 3 is 0.000001.

All 120 female divers are in the Cluster 2 but there also 238 male divers in this cluster which makes it a mixed cluster. In order to reveal differences between male and female divers, their dive-related data was examined in the same way as diver clusters.

The percentage of alcohol before dive for male divers is 38.46 % and for female divers is 29.17 %. The p value of chi-square test is 0.05037.

None of the female divers perform heavy exercise before dive. The percentage of "No Exercise" is higher in females while males have higher percentage of moderate exercise. However the p value of chi-square test is 0.0939 indicating that significance of difference between males and females in terms of exercise before dive is not very high.

For "state of rest before dive", the p value of chi-square test is 0.0155 indicating a significant difference between male and female divers in state of rest before dive. We can say that female divers feel more tired than male divers before dive.

When it comes to "water temperature", we observe that female divers prefer diving in warmer water than male divers. The student t test p value is 0.0028 meaning that the difference between male and female divers is significant. On the other hand, there is no big difference in "thermal comfort". The p value of chi-square test is 0.9380.

For "diving platform" The biggest difference is in small boat platform (50.80 % for males and 39.17 % for females). The p value of chi-square test is 0.04790.

Male divers tend to dive slightly deeper than female divers (average 31.73 m for male divers and 29.14 m for female divers). The p value of student t test is 0.0325.

There is significant difference in "workload" between males and females. Severe and exhausting workloads for female divers' are higher. The p value of chi-square test is 0.03997 meaning that workload depends on sex.

Breathing gas does not depend significantly on sex as p value of chi-square test is 0.2113. However 36 male divers and only 1 female diver used Trimix. When we focus on Trimix specifically, the difference is significant as p value is 0.0464.

All of the female divers used open circuit scuba while 3.71 % of male divers used other apparatus. The p value of chi-square test is 0.1001 which indicates a weak significance for dependency of diving apparatus to sex.

Male divers experienced equipment malfunction problem in 9.68 % of dives while this rate is 16.67 %. We can see that the face mask problems rate is very high in female divers compared to male divers. The p value of chi-square test is 0.0022 meaning that there is a significant difference between males and females.

There is not significant difference in symptom occurrence between male and female divers (7.15 % for male divers and 7.50 % for female divers). The p value of chi-square test is 0.8942.

4   Conclusion

The loss of data to get rid of time effect was very important. We were able to use the data of 874 one-dive divers. However the total number of divers was 3108. Fortunately we will be able to use most of dive data (after cleaning missing and inaccurate data) when analyzing type of dives and relating them to results of dives (problems, signs/symptoms and bubble counts) for the next step of the project.

Although not mentioned in paper, we performed similar analysis on divers who participated to one event. The number of one-event divers was 1699. We observed that clusters for divers who participated to only one dive and one event are created similarly. Still we used the clusters on one-dive divers for the analysis on dive related variables to get rid of time effect as diver's condition may change daily during a 48 h event.

TwoStep and Gower Distances methods clusters data based on binary variables, whereas K-means creates clusters based on numerical variables (age and dive activity years).

TwoStep clustering was the most suitable method for clustering given the type of diver's data especially with age and dive activity years transformed to categorical variables.

Obtained groups have meaningful characteristics. One group consists of middle aged male divers who have no health problems. Another group consists of male and female divers who have higher rates than other two clusters in following problems: cigarette smoking, allergy, asthma, back pain, ear/sinus problems and sea sickness. We named this group as divers who have health problems. Another group consists of old and experienced divers. Hearth problems in this group have higher rate than other groups.

For dive-related data, the group of unhealthy divers differs from others in exercise before dive (highest rate of no exercise), state of rest before dive (highest rate of tired), equipment malfunctions (especially in face mask and thermal protection problems) and any symptoms (lowest rate). The group of older and experienced divers differs from other groups in maximum depth (deeper than other groups), workload (highest rate in resting), breathing gas (highest rates of nitrox and Trimix) and any symptom (highest rate). The group of middle aged male divers with no health problems has significantly higher rates than the unhealthy divers group in exercise before dive and thermal comfort, lower rate in equipment malfunctions.

Male and female divers are significantly different in state of rest before dive, minimum water temperature, maximum depth and equipment malfunctions.

To avoid problems on prospective data collection studies mentioned in method section, data should be recorded in single table and for each diver, keeping one record especially when recording demographical data.

Important data such as weight and height should be collected more caringly for all divers. Furthermore, current database mostly consists of data about health information. For much more comprehensive and effective data mining studies, collecting data about socio-economic status about divers is also important (having their own diving equipment, diving outside country, etc.)

Diving experience of divers is also an essential data for this study. The variable named DiveActivityYears indicates since when the diver is diving. However this variable doesn't provide accurate information about experience of divers. Therefore, variables expressing number of dives a diver has participated in his/her lifetime should be included in the database. Likewise, certificate degree information and diving courses taken by divers also give information about divers' experience hence these variables should also be included in the database for providing a much better analysis.

Physiological data about pre-dive and post-dive is highly important in terms of dive health research. To be able to carry out detailed studies about dive profile and bubble measurement, collecting these data should be considered crucial. Besides, a more detailed physiological statistics like water loss measurements, vascular measurements can be obtained even if they require devices which may be hard to keep in dive sites.

Further analysis can be carried out for observing the data of the "aging" divers who participate to multiple events over the life time of the DB.

The characteristics of diver clusters can be compared with the similar statistics of general population to analyze if divers differentiate from non-diver population in terms of variables like health, smoking, drug use, alcohol use.

Another future research direction is investigating multi-variable relationship between diver profile, dive profile - bubble measurement relation. By this way, risky diver groups, risky dive profiles and the possible results of combinations of dive and dive profiles can be examined.

Acknowledgement. This project has been financed by Galatasaray University, Scientific Research Project Commission - Project No. 15.401.001.

References

1. Divers Alert Network, 1988–2011 Annual Diving Reports. https://www.diversalertnetwork. org/medical/report/. Accessed 15 Oct 2015
2. Ozyigit, T., Egi, S.M., Denoble, P., Balestra, C., Aydin, S., Vann, V., Marroni, A.: Decompression illness medically reported by hyperbaric treatment facilities: cluster analysis of 1929 cases. Aviat Space Environ. Med. **81**(1), 1–5 (2010)
3. Ozyigit, T., Egi, S.M.: Commercial diver selection using multiple-criteria decision-making methods. Undersea Hyperb. Med. **41**(6), 565–572 (2014)
4. Chang, H.L., Yeh, T.H.: Motorcyclist accident involvement by age, gender and risky behaviors in Taipei. Taiwan. Transp. Res. Part F **10**, 109–122 (2007)
5. Chiu, T., Fang, D., Chen, J., Wang, Y., Jeris, C.: A robust and scalable clustering algorithm for mixed type of attributes in large database environment. In: Proceedings of the 7th ACM SIGKDD International Conference on Knowledge Discovery and Data Mining, pp. 263–268, San Francisco (2001)
6. Lloyd, S.P.: Least squares quantization in PCM. IEEE Trans. Inf. Theory **28**(2), 129–137 (1982)
7. Aggarwal, C.C., Yu, P.S.: Data mining techniques for associations, clustering and classification. In: Zhong, N., Zhou, L. (eds.) PAKDD 1999. LNCS (LNAI), vol. 1574, pp. 13–23. Springer, Heidelberg (1999)
8. Gower, J.C.: A general coefficient of similarity and some of its properties. Biometrics **27**(4), 857–871 (1971)

Multiple Health Phases Based Remaining Useful Lifetime Prediction on Bearings

Junjie Chen[1]([✉]), Xiaofeng Wang[2], Wenjing Zhou[2], Lei Zhang[3], and Fei Liu[1]

[1] Institute of Automations, Jiangnan University, Wuxi 214000, China
chenjunjie3711@sina.cn, fliu@jiangnan.edu.cn
[2] Corporate Technology, Siemens Ltd., Beijing 100102, China
{xiaofengwang,zhou.wenjing}@siemens.com
[3] College of Internet of Things, Nanjing University of Posts and Telecommunications,
Nanjing 210003, China
lei.z@njupt.edu.cn

Abstract. Bearings are key components for all industrial machinery systems. The health status of bearings has great impact on the performance of rotating machineries. Remaining useful lifetime (RUL) estimation on bearings can effectively improve the reliability and availability of industrial machineries. In this paper, a multiple health phases based method is proposed for RUL estimation with application to bearings. Bags of word is brought into the method to model the time-frequency domain features of bearing vibration signals. Besides that, a gaussian mixture model is utilized to model the lifetime of various bearings to build accurate lifetime prediction model. Finally, the experiments demonstrate that the proposed method achieves a good performance comparing with other existing methods.

Keywords: Machine bearings · Remaining useful lifetime · Bag of words · Multi-variate gaussian distribution · Gaussian mixture model

1 Introduction

Bearings are key components for all industrial machinery systems. They generally work in tough industrial environment, and many various faults may arise frequently. Any kind of fault may cause the breakdown of the whole industrial systems, which could lead to a huge economic or safety damage. If the remaining useful life(RUL) of bearings could be predicated accurately, then the potential damage could be avoided by advanced maintenance implementations, to achieve a maximum lifetime and minimum maintenance costs of industrial machinery systems. Therefore, RUL estimation on bearings has attracted more and more attention in recent years [1,2].

Currently, RUL estimation methods can be put into three categories, which are physical model based methods, domain expert based methods, and data-driven methods. In this paper, a data-driven method for RUL prediction is proposed. There are a rich amount of data-driven RUL prediction methods, such

© Springer International Publishing Switzerland 2016
P. Perner (Ed.): ICDM 2016, LNAI 9728, pp. 110–124, 2016.
DOI: 10.1007/978-3-319-41561-1_9

as ENN (Elman neural network) [5], ANN (artificial neural network) [6]. Most of these methods focus on learning the trend feature of RUL from historical the life time trajectory of bearings [5,6]. However, the physical feature of bearings is time changing during their life time in real world, which gets few attention in past research work. As the physical features changes of bearing can directly reflect the health condition of bearing. Therefore, the changing physical features of bearings during their life time is modelled in the proposed method. This is achieved by separating the lift time of bearings into different health phases. Bag of words to represent the time-frequency feature of the vibration signals of bearings. Multi-variate gaussian distribution is used to model the time durations of different health phases. The basic idea is based on the intuition that a bearing that takes a short time from normal health condition to abnormal condition is likely to break down more quickly than the one that takes a long time from normal to abnormal.

The main contributions of this paper can be summarized as:

- A bag-of-word model is proposed to represent the salient physical features implied in STFT transformed vibration data, and a health phase identification model(classifier) is setup based on these salient features.
- Multi-variate gaussian distribution is used to model the health changing trend of RUL of bearing.

2 Related Work

The key point of RUL estimation is to extract appropriate features and build a prediction model. There are three kinds of main features: time domain, frequency domain and time-frequency domain features. Time-frequency feature is used more and more widely because it combines the advantages of the two others. Rosero et al. [3] and Wang and Chen [4] compared the results of some time-frequency domain features like STFT(Short-Time Fourier Transform), WVD(Wigner Ville Distribution), WPD(Wavelet Packet Decomposition) in detecting bearing faults, finally they found STFT method showed weaker effectiveness than others. But STFT has low calculation and definitude physical meaning, it is necessary to find a method for being used cooperatively. With many successful applications of bag-of-words in text and image analysis. We import bag-of-words model in the field of fault diagnosis for the first time. Certainly, the rationality and operability of the model is critical. For this purpose, a lot of methods have been proposed. Chen proposed a fusion method using Elman neural network to modify residual series of grey model, which is used to predict gear remaining useful life [5]. Ali et al. also used an artificial neural network to predict the RUL [6]. Many other methods based on artificial neural network have also been proposed. These methods are instable for prediction because it's hard to determine the neural network structure and train them. Reference [7,8] introduced some Gauss process regression-based prediction methods, which can obtain the distribution of the remaining useful life. But the disadvantage is the poor rationality about the historical and current data.

It's a common way to estimate the RUL of components based on their current health state [9]. And SVM shows superior performance in identifying different degradation state compared with the other classifiers [10]. Wuest [11] use SVM to identify the state drivers in manufacturing programmes and have shown the effectiveness. Patil also predict the remaining useful life by multistage SVM [12].

3 Multiple Health Phases Based RUL Estimation

3.1 Overview of Multiple Health Phases Based RUL Estimation

The general process of the proposed method for RUL estimation on bearings is demonstrated in Fig. 1. Firstly, a series of segments are obtained with the same size. Multiple feature vectors are extracted from each of the local segments using Short-time Fourier Transform (STFT). Secondly, bag of words model is proposed to represent the time-frequency features of bearing vibration signals. Similar to the bag of words model in [13], all local segments from the training sets are clustered by k-means clustering and all the cluster centers form a code book. A vibration signal could then be featured by a histogram of the words. Thirdly, the life time of bearings are divided into several phases. In other word, the life time of bearings can be represented by a series of duration time vectors, which describes how long bearings stay in different phases. These vectors are described by a Gaussian Mixture Model (GMM). The weight parameters in GMM are defined as Pearson correlation coefficient between two signals in frequency domain. At last, the RUL of bearings is estimated by predication algorithms.

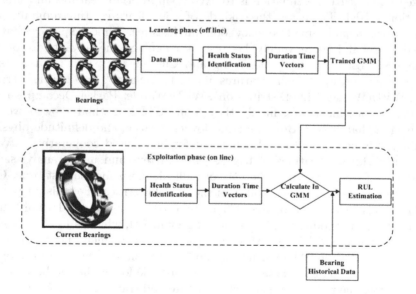

Fig. 1. RUL estmation steps

3.2   Identification of the Bearing Health Phases

Bearing Health Phases Classification. The simplest way to identify bearings health phases is using ISO standard, but it has strong limitations. Because ISO standard is empirical from all kinds of working conditions and it can't match some specific platforms. So the method which has a combination of absolute standard and relative standard is preferred.

Absolute Standard. For a specific platform, the absolute standard is considered. According to the reference of ISO80186-3 (see Fig. 2). Benchmark data sets are chosen and divided into 4 stages:

1. New machine condition (Green);
2. Unlimited long-term operation allowable (Yellow);
3. Short-term operation allowable (Orange);
4. Vibration causes damage (Red).

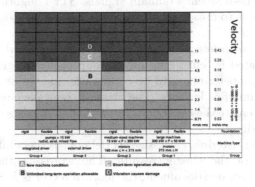

Fig. 2. ISO80186-3 standard for the assessment of vibration (Color figure online)

Relative Standard. By using absolute standard, we have four segments of health phases of the benchmark data set. Then corresponding time-frequency features can be extracted from these segments. The features contain the particular information of supply frequency, environmental noise, bearings fault frequency, etc. Then classifiers are used to identify bearings phases of the test data set, which could reach better matching than absolute method.

Feature Extraction Using Bag-of-Words. In this section, we introduce the bag of words model based representation for vibration signals classification(see Fig. 3). Such representation neglects the order of local segments within a vibration signal and express the vibration signal as a histogram of codewords.

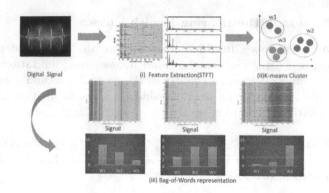

Fig. 3. The flowchart of the proposed bag-of-words approach

Local Segments Extraction. By sliding a window with a fixed length continuously, local segments are extracted from each window. These local segments are then transformed into time-frequency domain signals. STFT algorithm is performed for each local segment to obtain the feature vectors.

Bag-of-Words Representation. For the analysis on images and videos, codebook is commonly generated by clustering on a group of local segments, ie., all the clustering centers are determined to be codewords. Images or videos are represented as a histogram of codewords. The classical k-means clustering algorithm is generally used to create the codebook [15], although it exists some other methods, such as mean-sift [16].

With k-means clustering, a k-size codebook is generated, which contains the centroid of the cluster results. The words in this codebook are $w(1), w(2), \ldots w(k)$. For one document, the feature vector is $F = (f_1, f_2, \ldots, f_k)$.

$$f(i) = \frac{[n(i)/\sum_1^k n(i)]}{[m(i)/\sum_1^k m(i)]} \tag{1}$$

where $i = 1 \ldots k$, $n(i)$ is the amount of the ith word in this document; $m(i)$ is the amount of this word in all documents;

Classifier. Some classifiers such as Artificial Neural Networks (ANN) [17], Probabilistic Neural Networks (PNN) [18], and Decision Tree classifiers [19] are widely used for phases classification. In this paper, we use the Support Vector Machine (SVM) as a classifier, with the feature vector (Eq. 1) of train data set as classification algorithm input.

It should be noted that the proposed feature extraction based on bag of words model, is not limited with SVM classifier. This mechanism can also be applied in other classifiers such as the ANN, PNN and some others to improve the classification performances.

3.3   Modeling of the Bearing Life Time

According to the aforementioned method, the life time of bearing is modeled as a time series consisting of four health phases. The four health phases are denoted as phase I, phase II, phase III and phase IV. Obviously, for a bearing, there are strong correlations between time durations of health phases. Thus, RUL of a bearing can be predicted based on the time durations of previous health phases of bearings. For instance, a bearing in health phase I for 2 years may have 1 year remaining life time. Then, for a bearing, its life time can be represented by a health duration vector:

$$T^i = (t_1^i, t_2^i, t_3^i, \ldots,) \tag{2}$$

T^i is regarded as a training sample for learning prediction model, where i is the training sample index. $t_1^i, t_2^i, t_3^i, \ldots$, are the time durations that bearing has in corresponding health phases. The sum of components of T^i will be the total life time of the bearing.

To model the correlation between time durations of different health phases, multi-variate gaussian model are adopted to describe the probabilistic distribution of health duration vectors. Meanwhile, as there are bearings with different qualities or working in various conditions, Gaussian Mixture Model (GMM) is used to model the general life time of the bearings.

GMM Introduction. A GMM is a linear combination of multiple Gaussian distributions [20]. Its definition is given as follows.

$$f(x) = \sum_{k=1}^{K} \pi_k \mathcal{N}(x|\Theta_k) \tag{3}$$

where each component function $\mathcal{N}(x|\Theta_k)$ is a Gaussian distribution. $\Theta_k = (\mu_k, \Sigma_k)$ is the model parameters, μ_k is mean and Σ_k is covariance matrix. The multivariate Gaussian distribution of a D-dimensional vector x is in form [21]:

$$\mathcal{N}(x|\Theta) = \frac{1}{(2\pi)^{D/2}} \frac{1}{|\Sigma|^{1/2}} \exp\left\{-\frac{1}{2}(x-\mu)^T \Sigma^{-1}(x-\mu)\right\} \tag{4}$$

where μ represents a D-dimensional mean vector, σ denotes a $D \times D$ covariance matrix, and $|\Sigma|$ is the determinant of the matrix σ. The prior distribution π_k denotes the probability that observation x_n belongs to the kth class ω_k. It can be seen π_k is independent of observation x_n. Furthermore, π_k should satisfy the following constrains:

$$0 \le \pi_k \le 1, \sum_{k=1}^{K} \pi_k = 1; k = 1, \ldots, K \tag{5}$$

GMM Based Bearing Life Time Model. The health duration vectors are clustered with clustering methods, such as k-means. For each cluster, a multivariate gaussian model is used to model the distribution of health duration vector in that cluster. Then, the linear combination of gaussian models completely describes the bearing life time in probabilistic perspective. The model parameters (μ_k, Σ_k) can be inferred by maximum likelihood method as:

$$\mu_k = \bar{x} = \frac{1}{n_k} \sum_{i=1}^{n_k} x_{k,i} \tag{6}$$

$$\Sigma_k^2 = \frac{1}{n_k} \sum_{i=1}^{n_k} (x_{k,i} - \bar{x})^2 \tag{7}$$

To simplify the model computation, the weight parameters π_k in GMM are estimated as:

$$\pi_k = \frac{|\{T^i | T^i \in C_k\}|}{|\{T | T \in C\}|} \tag{8}$$

C_k is the kth health duration vector cluster, $C = C_1 \cup C_2 \cup \ldots$ is the union of all the clusters.

3.4   RUL Estimation

When health phase is identified, the RUL of bearing is predicted as shown in Fig. 1. RUL estimation is based on the maximization of conditional probability of health duration. For bearings, the conditional probability of health duration is defined as

$$P(T_{remain} | T_{past}) = \frac{P(T_{past}, T_{remain})}{P(T_{past})} \tag{9}$$

T_{past} represents the time durations of previous health phases that bearing had passed, while T_{remain} represents the time durations of health phases that bearing will have. $T = (T_{past}, T_{remain})$ describes a possible life time of a bearing through the health phase perspective. For example, $T_{past}^i = (1000, 300)$ describes that a bearing is in health phase I for 1000 h and in health phase II for 300 h. Then $P(T_{remain}^i = (150, 30) | T_{past}^i)$ is the conditional probability that bearing will be in health phase III for 150 h and in phase IV for 30 h, given past time durations T_{past}^i.

Given a past health duration vector T_{past}, the remaining health duration can be estimated by maximizing the conditional probability of health duration with Expectation-maximization (EM) algorithm.

$$\underset{T_{remain}}{\operatorname{argmax}} \sum_{k=1}^{K} \pi_k \mathcal{N}(T_{remain} | \Theta_k, T_{past}) \tag{10}$$

Because (T_{past}, T_{remain}) is a multi-variate gaussian distribution, with mean μ and covariance matrix Σ:

$$\mu = \begin{bmatrix} \mu_{past} \\ \mu_{remain} \end{bmatrix} \tag{11}$$

$$\Sigma = \begin{bmatrix} \Sigma_{past,past} & \Sigma_{past,remain} \\ \Sigma_{remain,past} & \Sigma_{remain,remain} \end{bmatrix} \tag{12}$$

then conditional probability $P(T_{remain}|T_{past})$ is also a guassian distirbution with mean $\mu_{remain|past}$ and $\Sigma_{remain|past}$:

$$\mu_{remain|past} = \mu_{remain} + \Sigma_{remain,past}\Sigma_{past,past}^{-1}(T_{past} - \mu_{past}) \tag{13}$$

$$\Sigma_{remain|past} = \Sigma_{past,past} - \Sigma_{remain,past}^{T}\Sigma_{remain,remain}^{-1}\Sigma_{remain,past} \tag{14}$$

The following are the examples of RUL estimation based on observations of different health durations of bearings.

1. If the health duration vector is $T_{past} = (0,0,0)$, this means the bearing is in health phase I. Then, the most possible remaining lifetime is the $T_{remain} = (t_1', t_2', t_3')$ which maximizes the following gaussian mixture model.

$$\underset{t_1',t_2',t_3'}{\operatorname{argmax}} \sum_{k=1}^{K} \pi_k \mathcal{N}(T|\Theta_k) \tag{15}$$

Here t_i' represents the time duration of health phase i. So the bearing's RUL is $t_1' + t_2' + t_3' - t_{used}$, where t_{used} is the time that has been passed after bearing was launched.

2. If the current bearing is in yellow alert phase which corresponds to health phase II, then the health duration vector is $T_{past} = (t_1, 0, 0)$. The most possible $T_{remain} = (t_2', t_3')$ is the one that maximize conditional probability of health duration as:

$$\underset{t_2',t_3'}{\operatorname{argmax}} \sum_{k=1}^{K} \pi_k \mathcal{N}(T|\Theta_k, (t_1)) \tag{16}$$

The remaining useful life is $t_1 + t_2' + t_3' - t_{used}$.

3. If the current bearing is in orange alert phase with means bearing is in health phase III, then the health duration vector is $T_{past} = (t_1, t_2, 0)$. Thus $T_{remain} = (t_3')$ is the one that satisfies

$$\underset{t_3'}{\operatorname{argmax}} \sum_{k=1}^{K} \pi_k \mathcal{N}(T|\Theta_k, (t_1, t_2)) \tag{17}$$

The RUL is $t_1 + t_2 + t_3' - t_{used}$.

4. If the health phase is at red alert which indicates bearing is in health phase IV. This means bearing is damaged and the remaining useful lifetime is 0.

4   Application and Experimental Results

In this section, vibration signals of bearings during the whole lifetime, which are acquired from an accelerated degradation test bed, are used to verify the effectiveness of the proposed method.

4.1   Introduction to the Vibration Data

An experimental system named PRONOSTIA [14] is shown in Fig. 4. This system is designed to obtain accelerated degradation data of bearings under different operating conditions. In order to do bearing accelerated degradation testing in a short time, a radial force equal to the maximum dynamic load of the bearing is applied to the bearings on the tested. Vibration signals are captured by the vibration sensors fixed on the bearing. The sampling frequency is 25.6 kHz. Each sample time is 0.1s, i.e. Each sample contains 2,560 data points, and the sampling interval is 10s. The vibration signals are transmitted into a PC through NI CDAQ cards. The end of the bearing remaining useful life is the time when the amplitude of the vibration signal exceeds 20g. Every cases is assigned with a unique identifier in the form of $Bearingi - j$, with i denoting the index of the operating conditions and j denoting the case index. The detail of the condition is shown in Table 1, and this paper choose some training sets and test sets (see Table 2).

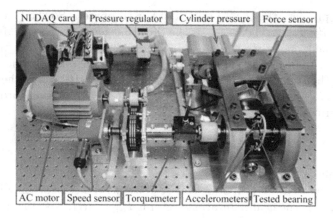

Fig. 4. Overview of the experimental system

4.2   Identification of the Bearing Health Phases

Bearing1-1 is chosen as the benchmark by absolute standard. The figure of Bearing1-1 RMS is divided into 4 phases (see Fig. 5). The raw signals of four benchmark phases are also shown in

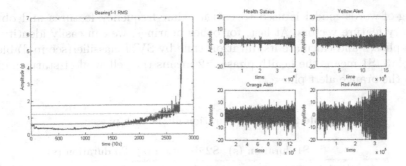

Fig. 5. Identification of Bearing1-1 health phases by absolute standard

The spectrograph representing time-frequency features is shown in Fig. 6. With the degradation of this bearing status, the color becomes deeper and deeper. i.e., the energy at each frequency band gets more and more powerful.

Table 1. Bearing operating conditions

Symbol	Description	Cond. 1	Cond. 2
L	Load (N)	4000	4200
N	Speed (rev/min)	1800	1650

Table 2. Learning and test sets

	Bearing cases	
Training sets	Bearing1-1	Bearing2-1
	Bearing1-2	Bearing2-1
	Bearing1-3	Bearing2-3
	Bearing1-4	Bearing2-4
	Bearing1-5	Bearing2-6
Test sets	Bearing1-6	Bearing1-7
	Bearing2-5	Bearing2-7

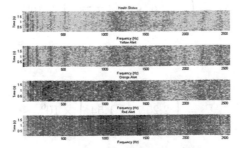

Fig. 6. The spectrograph of benchmark phases

Bag-of-words model is used to extract time-frequency features and obtain necessary feature vectors. At last, for other bearings, we can easily identify the health phases and calculate its duration time by SVM classifier(see in Table 3). In Table 3, S1 means the health phase, S2 means the yellow alert status and S3 means the orange alert phase.

Table 3. Bearings duration life in each phases

	S1 duration (s)	S2 duration (s)	S3 duration (s)
Bearing1-1	20496	3906	3484
Bearing1-2	8254	74	379
Bearing1-3	17805	3129	2816
Bearing1-4	10848	1113	2316
Bearing1-5	24109	520	0
Bearing2-1	8805	230	74
Bearing2-2	7844	78	55
Bearing2-3	19461	63	27
Bearing2-4	7430	39	39
Bearing2-6	6859	78	63

4.3   Modelling of GMM

Now there are 10 feature vectors after identifying the phases of train bearings. To be brief, this paper divides these features into 2 groups by k-means clustering, the result of clustering is shown in Table 4. Each group obeys Gaussian distribution. According to (6) and (7), $\mu_1, \Sigma_1, \mu_2, \Sigma_2$ can be calculated:

$$\mu_1 = (2.05, 0.19, 0.16) \times 10^3 \tag{18}$$

$$\Sigma_1 = \begin{pmatrix} 16.0 & -4.79 & -5.57 \\ -4.79 & 8.11 & 7.78 \\ -5.57 & 7.78 & 7.55 \end{pmatrix} \times 10^6 \tag{19}$$

$$\mu_2 = (8.34, 0.27, 0.49) \times 10^3 \tag{20}$$

$$\Sigma_2 = \begin{pmatrix} 8.15 & 2.25 & 4.71 \\ 2.25 & 7.32 & 1.54 \\ 4.71 & 1.54 & 3.41 \end{pmatrix} \times 10^6 \tag{21}$$

Table 4. The result of clustering

	Bearing cases	Whole life (s)
Group1	Bearing1-1	27886
	Bearing1-3	23750
	Bearing1-5	24629
	Bearing2-3	19551
Group2	Bearing1-2	8707
	Bearing1-4	14277
	Bearing2-1	9109
	Bearing2-2	7977
	Bearing2-4	7508
	Bearing2-6	7000

4.4 RUL Prediction

To demonstrate the method in this paper better, we have done some comparative experiments which identify bearing phases with different features. The results of the RUL prediction over 10 testing sets can be found in Table 5.

The $Error^{1-4}$ of the RUL estimation are calculated with different features. Error [22] and Error [23] are references from other research.

The error is defined by

$$error = \frac{estimatedRUL - trueRUL}{trueRUL} * 100\,\% \qquad (22)$$

In [23], two sets (Bearing1-3, Bearing2-7) are estimated to already be in failure state at the time of RUL prediction, so these cases are marked as N/A.

Table 5. The estimation of RUL and error comparison

Test set	True. RUL(s)	Est. RUL(s)	$Error^1$	$Error^2$	$Error^3$	$Error^4$	Error [22]	Error [23]
Bearing1-3	5726	3898	45.83 %	31.93 %	23.73 %	27.14 %	91.44 %	N/A
Bearing1-4	2886	2726	−2.57 %	5.56 %	97.02 %	97.02 %	97.05 %	319.62 %
Bearing1-5	1609	976	84.46 %	39.35 %	−202.14 %	−127.67 %	69.56 %	30.57 %
Bearing1-6	1457	125	91.42 %	91.42 %	91.42 %	89.86 %	66.43 %	154.8 %
Bearing1-7	7570	4460	96.95 %	41.08 %	54.23 %	41.07 %	93.52 %	55.01 %
Bearing2-3	7527	11687	63.77 %	−55.26 %	33.31 %	−51.63 %	94.55 %	12.24 %
Bearing2-4	1390	2800	85.67 %	−101.34 %	80.89 %	−101.40 %	70.50 %	93.62 %
Bearing2-5	3089	2308	−306.06 %	25.30 %	−46.14 %	25.28 %	86.73 %	19.06 %
Bearing2-6	1289	2875	−117.27 %	−123.03 %	84.24 %	−123.03 %	69.21 %	238.5 %
Bearing2-7	578	390	−568.91 %	32.54 %	−601.35 %	−568.91 %	29.31 %	N/A

[a] The method with STFT, SVM, GMM
[b] The method with STFT, BOW, SVM, GMM
[c] The method with WPD, SVM, GMM
[d] The method with WPD, BOW, SVM, GMM

Table 6. The comparison of error

	$Error^a$	$Error^b$	$Error^c$	$Error^d$	Error [22]	Error [23]
Overall mean	146.29 %	54.68 %	131.45 %	125.30 %	76.72 %	115.35 %
Test mean	218.64 %	72.01 %	207.81 %	205.72 %	64.37 %	101.91 %

^{a}The method with STFT, SVM, GMM
^{b}The method with STFT, BOW, SVM, GMM
^{c}The method with WPD, SVM, GMM
^{d}The method with WPD, BOW, SVM, GMM

From Table 5, we can obtain the over mean value and test mean of error (see Table 6). Some conclusions can be drawn by the comparison of error:

1. In extracting time-frequency feature vectors, WPD (Wavelet Packet Decomposition) is better than STFT (Short Time Fourier Transform).
2. As a method for further feature extraction, BOW (Bag-of-Words) could improve the accuracy in some degree.
3. The best combination is STFT + BOW + SVM + GMM, which has the minimum error of RUL estimation.

5 Conclusion

A new method is proposed for RUL prediction on bearings in this paper. A bag of words model is proposed to identify feature vectors of bearing status. Gaussian mixture model is utilized to describe the different phases of bearing life time, and EM algorithm is used to predict RUL for bearings. With comparison with two other existing methods, the experimental results prove the effectiveness of the proposed method. The proposed method can also be applied to RUL estimation on other machinery components, such as gearboxes.

Acknowledgements. This work is supported by Natural Science Foundation of Jiangsu Province under Grant BK20150854 and Scientific Funding of Nanjing University of Posts and Telecommunications under Grant NY214025.

References

1. Qian, Y., Yan, R., Hu, S.: Bearing degradation evaluation using recurrence quantification analysis and Kalman filter. IEEE Trans. Instrum. Measur. **63**(11), 2599–2610 (2014)
2. Li, N., Lei, Y., Lin, J., et al.: An improved exponential model for predicting remaining useful life of rolling element bearings. IEEE Trans. Ind. Electron. **62**(12), 7762–7773 (2015)

3. Rosero, J., Cusido, J., Espinosa, A.G., et al.: Broken bearings fault detection for a permanent magnet synchronous motor under non-constant working conditions by means of a joint time frequency analysis. In: Industrial Electronics (2007)
4. Wang, H., Chen, P.: Fuzzy diagnosis method for rotating machinery in variable rotating speed. IEEE Sens. J. 11(1), 23–34 (2011)
5. Chen, X., Liu, M.: Gear remaining useful life prediction based on grey neural network. AA 3, 4 (2015)
6. Ali, J.B., Chebel-Morello, B., Saidi, L., et al.: Accurate bearing remaining useful life prediction based on Weibull distribution and artificial neural network. Mech. Syst. Sig. Process. 56, 150–172 (2015)
7. Hong, S., Zhou, Z., Lu, C., et al.: 1547. Bearing remaining life prediction using Gaussian process regression with composite kernel functions. J. Vibroengineering, 17(2) (2015)
8. Boskoski, P., Gasperin, M., Petelin, D., et al.: Bearing fault prognostics using Rnyi entropy based features and Gaussian process models. Mech. Syst. Sig. Process. 52, 327–337 (2015)
9. Chiacho, J., Chiachio, M., Sankararaman, S., et al.: Condition-based prediction of time-dependent reliability in composites. Reliab. Eng. Syst. Saf. 142, 134–147 (2015)
10. Huang, H.Z., Wang, H.K., Li, Y.F., et al.: Support vector machine based estimation of remaining useful life: current research status and future trends. J. Mech. Sci. Technol. 29(1), 151–163 (2015)
11. Wuest, T.: Application of SVM to identify relevant state drivers. In: Identifying Product and Process State Drivers in Manufacturing Systems Using Supervised Machine Learning, pp. 153–188. Springer International Publishing, Berlin (2015)
12. Patil, M.A., Tagade, P., Hariharan, K.S., et al.: A novel multistage Support Vector Machine based approach for Li ion battery remaining useful life estimation. Appl. Energy 159, 285–297 (2015). MLA
13. Niebles, J.C., Wang, H., Fei-Fei, L.: Unsupervised learning of human action categories using spatial-temporal words. Int. J. Comput. Vis. 79(3), 299–318 (2008)
14. Nectoux, P., Gouriveau, R., Medjaher, K., et al.: PRONOSTIA: an experimental platform for bearings accelerated degradation tests. In: IEEE International Conference on Prognostics and Health Management, PHM 2012, pp. 1–8. IEEE Catalog Number: CPF12PHM-CDR (2012)
15. Fei-Fei, L., Perona P.: A bayesian hierarchical model for learning natural scene categories. In: IEEE Computer Society Conference on Computer Vision and Pattern Recognition, CVPR 2005, vol. 2, pp. 524–531. IEEE (2005)
16. Jurie, F., Triggs B.: Creating efficient codebooks for visual recognition. In: Tenth IEEE International Conference on Computer Vision, ICCV 2005, vol. 1, pp. 604–610. IEEE (2005)
17. Guo, L., Rivero, D., Dorado, J., et al.: Automatic epileptic seizure detection in EEGs based on line length feature and artificial neural networks. J. Neurosci. Methods 191(1), 101–109 (2010)
18. Güler, İ., Übeylı, E.D.: ECG beat classifier designed by combined neural network model. Pattern Recogn. 38(2), 199–208 (2005)
19. Safavian, S.R., Landgrebe, D.: A survey of decision tree classifier methodology. IEEE Trans. Syst. Man Cybern. 21(3), 660–674 (1991)
20. Peel, D., MacLahlan, G.: Finite Mixture Models. Wiley, New York (2000)
21. Bishop, C.M.: Pattern Recognition and Machine Learning. Springer, Berlin (2006)

22. Wang, T.: Bearing life prediction based on vibration signals: a case study and lessons learned. In: 2012 IEEE Conference on Prognostics and Health Management (PHM), pp. 1–7. IEEE (2012)
23. Singleton, R.K., Strangas, E.G., Aviyente S.: Time-frequency complexity based remaining useful life (RUL) estimation for bearing faults. In: 9th IEEE International Symposium on Diagnostics for Electric Machines, Power Electronics and Drives (SDEMPED), pp. 600–606. IEEE (2013)

Innovations in News Media: Crisis Classification System

David Kaczynski, Lisa Gandy[✉], and Gongzhu Hu

Central Michigan University, Mt Pleasant, MI 48858, USA
gandy1l@cmich.edu

Abstract. Research in crisis management is a relatively new area of study, originating in the 1980s. Researchers have created several different models that separate organizational crises into discrete stages, such as pre-crisis, crisis and post-crisis. In this article we discuss a natural language based crisis detection system which classifies news articles relating to crises into the appropriate crisis stage. We use news articles from the New York Times as a source of training data, and use this data along with state of the art data mining and machine learning algorithms as the core of the system. In the future, our system may be expanded to identify and evaluate crisis management strategies, suggest crisis management strategies for the current state of a crisis, or provide stakeholders with summaries of crises in news media.

Keywords: Data mining · Machine learning · Crisis management

1 Introduction

A major research area in crisis communication is how to prevent or repair damage to the reputation of organizations before, during, or after the occurrence of crises. The creation of various crisis communication theories have been facilitated through case studies and by measuring the financial and/or reputation repair results of applied strategies.

In this article, we explore the use of data mining techniques in identifying characteristics of news articles relating to organizations in crisis. While there are many types of crises, staged crisis models, and crisis communication strategies, we begin our research by attempting to classify organizational data as pertaining to a specific stage of a crisis. After using data mining and classification techniques we create a system which can take any news article and classify it as pre-crisis, crisis or post-crisis.

We use news media as a source of data for training in our system. The methodology that we apply can be repeated in order to add more articles to the data set in the future and retrain the model. This may increase the accuracy for classifying crises that are not yet in the training data. Our methods may also be repeated for the purpose of adding more labels in order to classify more specific characteristics of the crises or of the crises response strategies being employed.

© Springer International Publishing Switzerland 2016
P. Perner (Ed.): ICDM 2016, LNAI 9728, pp. 125–138, 2016.
DOI: 10.1007/978-3-319-41561-1_10

There are several contributions of this paper. Though incident management systems do exist [6,19,30], our incident management system is built on a major theory of crisis communication, the Coombs model [8]. This model divides crises into three stages: pre-crisis, crisis and post crisis. Our system focuses on incident identification using news articles, which are widely prevalent due to the Internet, whereas other systems use numeric data. Our system requires no data preparation, as it can classify the text of any article, whereas again other systems often require the use of numeric data which must be prepared before use. In addition, our system has been evaluated on many sets of crises, whereas other systems often focus on one type of crisis, such as crime [6], traffic accidents [20], or weather [30]. In summary, our work focuses on a natural language based real time crisis identification system which requires no data preparation by the user and has already been evaluated on several existing crises.

In the background section we provide the preliminary information for the research. The section about our system walks through the proposed data mining process in detail including experimental results and the evaluation of the results. We then discussed the implementtion of our system as a Web application, followed by a briefly discuss current research in crisis management and data mining. We conclude with a discussion of a proposed system based on our methodology and discuss future work.

2   Related Work

Researchers have studied crisis management for many years and quite a number of systems were developed. In this section, we discuss some previous work related to crisis management that focused on text mining.

2.1   Crisis Detection Systems

Current crisis detection systems focus on using data mining techniques to either identify or predict a type of crisis. Peng et al. [30], introduced a three stage system which ultimately uses Multicriteria Decision Making to determine the risk of an incident occurring. This system was demonstrated using agrometeorological disaster data. The system was in the conceptual stage and has not been implemented. Chen et al. [6] proposed a general framework for crime detection. Their system was well designed and worked well for the give data set, however it differs from our work as it focuses on the detection of new crimes using numeric data, whereas our system focuses on detecting corporate crises using text data in the form of news articles.

Similar to the work in [6,30], Harms et al. [15] used association rule mining to find patterns in numeric data to assess climate conditions both locally and globally. French and Niculae [11] explored problems existing in current predictive models used in emergency management and suggest alternatives. Kararsova et al. [19] also used association rule mining to determine the relationships spatially between fire and rescue incident locations. Papamichail and French [29] created a system that supports decision making in terms of emergencies in a nuclear power plant. Berndt et al. [4] demonstrated that data warehousing can be used to explore novel situations regarding bioterrorism as well as suggestions for investigations after an incident occurs. Berndt et al. have also demonstrated that data warehousing can be used in supporting quality assurance related to medical care [5].

2.2   Natural Language Processing

The Internet era leaves us with a huge number of articles in text format. Most of these articles lack metadata, and therefore classification of text is an increasingly useful sub-domain of general classification. Automatic text classification is useful in many fields, such as classifying product reviews [9], spam filtering [1], guiding financial investments [39], and, among others. In regards to the use of machine learning in text classification, support vector machines [17] as well as Naive Bayes [28] have long been used successfully to this end. In addition, the J48 decision tree algorithm has also proved to be useful in regards to text classification [45]. Jain et al. [16] demonstrated that text classification is even more successful when classifiers are combined using simple or weighted voting.

Besides text classification, text and sentiment classification are another two important subfields of natural language processing. Entity extraction seeks to extract people, places, and other important entities from text in an unsupervised manner. Etzioni et al. [10] created the KnowItAll system which not only extracts entities but finds relations between the entities. Takeuchi and Collier [42] used support vector machines to extract medical terminology from medical journal abstracts. Miller et al. [25] also used entity extraction to find important people and places mentioned in broadcast news. In regards to sentiment classification, Pang et al. [28] demonstrated the successful classification of movie reviews into positive and negative through the use of support vector machines. Go et al. [12] successfully classified tweets using tweets with emoticons as training data. Wan [44] used English corpora and machine translation to classify the sentiment of product reviews written in Chinese.

3   NLP-based Crisis Identification System

Our NLP (Natural Language Processing) based crisis identification system is trained using a collection of 80 articles from the New York Times. We will first give an overview of the system architecture, and then discuss the process of data collection and move on to the inner workings of the system itself.

3.1   System Architecture

The overall architecture of our news media article classification system is shown in Fig. 1. In the configuration file `build.sbt`, shown as STB in the figure, we specify all external dependencies by group ID, artifact ID, and version. When the application is executed, if a dependent library is not found locally (MySQL in the figure), the framework will search for and download the library remotely from the Maven Central Repository, which is a repository of build artifacts [2]. This allows the code to be more portable between machines by reducing the hassle in collecting external libraries.

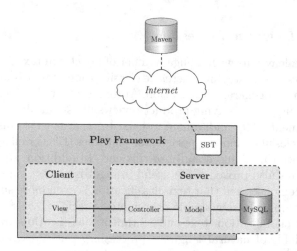

Fig. 1. System architecture — model of the Play Framework integrating the various technical components

We use a simple Model-View-Controller design pattern for the separation of responsibilities using the integrating tool Play Framework [32]. The role of the Model is to translate the data in the database into Scala or Java data types. We define an `Article` class in the Model that stores all of the article meta data as `String` values. The View is strictly responsible for displaying and download-ing results, such as word counts, classification results, and spreadsheets. The Controller is where the vast majority of the data mining process takes place.

Play Framework is a web application framework that is an integrated, easy-to-deploy web server, supports for Java's imperative and Scala's functional pro-gramming paradigms, supports native XML and JSON parsing, and provides a prototype-friendly templating library for creating HTML5 content quickly. These features are useful for the work presented in this paper.

3.2   Data Collection

In order to create a system which can identify new crises we must collect data for training. The New York Times Online Archive is utilized to facilitate the collection of crisis-related news media. The web interface for the archive allows users to search for articles by specifying a search term, date range, result type (article or blog), author, or section [43]. A total of 80 articles were collected about organizational crisis. The articles were downloaded as .html files and labeled manually as one of the three stages of the three-stage crisis model (see Table 1). Due to the nature of news media data, crises were only selected as training data if the organizations had news media exposure during all three stages in regard to a single crisis.

Table 1. Distribution of articles in training set

Class	Number of articles
Pre-crisis	36
Crisis	18
Post-crisis	26
Total	80

The articles collected cover eight organizations that experienced a crisis between 2003 and 2013. An article is labeled with the pre-crisis class as long as the main topic of the article is not about a crisis that has occurred or is occurring. An article is labeled as crisis class if the news article is the first exposure of the crisis to the public through the news media. An article may also be in the crisis class if the organization has not yet contained the crisis or if the crisis is still expanding. An article is labeled as post-crisis if the impact of the crisis has been fully realized or if the organization is reacting to a crisis using a post-crisis communication strategy.

Per article, we collect the publication date of the article, the name of the organization involved, the text of the article, and its url.

Listed below is a brief summary of each crisis:

(1) *Space Shuttle Columbia:* On February 2, 2003, the space shuttle Columbia disintegrated on re-entry into the earth's atmosphere, killing the seven astronauts on board [38].

(2) *BP Oil Spill:* On the evening of April 20, 2010, an explosion on the BP-owned oil rig Deepwater Horizon left 11 workers missing and later declared dead [36].

(3) *Apple-FoxCon:* Apple experienced negative publicity over its relationship with the Chinese electronic manufacturing company FoxCon when a factory explosion in May of 2011 killed two and injured over a dozen others [3].

(4) *News of the World:* was revealed in July of 2011 that employees of News of the World, a tabloid owned by Rupert Murdoch's News Corporation, had hacked the phones of British soldiers and of victims of violent crimes [23].

(5) *JP Morgan Chase Losses:* In May of 2012, the multi-billion dollar financial institution JP Morgan Chase announced trading losses of over $2 billion USD [34].

(6) *Lance Armstrong Doping:* When the Unites States Anti-Doping Agency made public its evidence in a doping case against Lance Armstrong in the Fall of 2012, there were negative impacts for both his public image and for the charity that he co-founded, the Livestrong Foundation [24].
(7) *Exxon pipeline:* On March 29, 2013, Exxon Mobil's Pegasus pipeline ruptured near a small town in central Arkansas [40].
(8) *Target Data Breach:* From November 27 to December 15, 2013, the second largest breach of a retailer in history occurred, acquiring successfully the personal information of customers of the American retail chain Target [31].

3.3 Data Preparation

Text Preparation. In early experiments with the crisis data, we found that the remove of named entities from the text greatly increased the accuracy of the crisis identification model. We found that the named entities led to overfitting for some articles at the expense of others. Therefore the first step in data preparation is the identification and removal of named entities. We used Stanford's Named Entity Recognizer (NER), for entity recognition. The parser is configured to identify seven classes of named entities: time, location, organization, person, money, percent, and date. All of the entities recognized by the parser are then removed.

After all of the entities are removed the term frequency/inverse document frequency for all words are calculated using the Weka machine learning tool. This vector of tf-idf scores becomes the bag of words feature used in later classification. Bag of words refers to the method of treating each word as a feature without concern for each word's relation to other words in terms of ordering. The settings used to create the word vector in Weka are given below in Table 2.

Table 2. Parameters of the Weka `StringToWordVector` filter

Parameter	Value
IDFTransform	True
TFTransform	True
attributeIndices	(*The column of the article text*)
normalizeDocLength	Normalize all data
stemmer	LovinsStemmer
useStoplist	True
wordsToKeep	1000000

Sentiment Classification. There are two sets of metrics that are being collected for sentiment scores. The first is a scalar metric of positive or negative sentiment that ranges from −5 to 5. The second set of metrics is a vector of six sentiment categories: positive, negative, litigious, uncertain, modal weak, and modal strong. Because all of the sentiment analysis is executed programmatically in our own code, we use the version of the articles that have been stemmed using the Porter stemmer Java library. We also apply the Porter stemmer to the dictionaries that contain the sentiment data for words. This allows us to look-up words from the articles in the dictionaries by their stems and to not have to worry about small variations in highly similar words.

The first, scalar metric is calculated using Algorithm 1. We start by looking up every word of an article in the AFINN-111 dictionary map [26]. The word is the key in the map, and the value is the sentiment of the word, ranging from negative five to positive five. The sentiment of an article is calculated by summating the sentiments of the words. In order to normalize the sentiment of articles, we divide the total by the total number of words (N) in the article that were contained in the dictionary map. The result is the sentiment of the article from a scale of -5 to 5.

Algorithm 1. Calculate Scalar Sentiment

    Input: A – an article
    Input: D – AFINN dictionary map
    Output: S – sentiment measure of A, a scale value from -5 to 5
1 **begin**
2     $S \leftarrow 0$
3     $N \leftarrow 0$
4   **foreach** $word \in A$ **do**
5   **if** $word \in D.keys$ **then**
6         $S \leftarrow S + D[word]$
7         $N \leftarrow N + 1$
8   **end**
9     $S \leftarrow S \div N$
10   **return** S
11 **end**

The second, vector metric is calculated using Algorithm 2. We begin by looking up every word of an article in each of six word lists, one for each category of sentiment. The number of words in each category is totaled for an article. We start by looking up every word of an article in each of six word lists, one for each category of sentiment. The number of words in each category is totaled for an article. Then, the number of words in each category is divided by the number of words in any category. The result is the percentage of words that belong to each category of sentiment. Note that some words belong to multiple categories of sentiment, so the total of the percentages can exceed 100 %. For example, "always" is both modal-strong and positive, and "apparently" is both modal-weak and uncertain [22].

Algorithm 2. Calculate Categorical Sentiment as Vector

Input: A – an article
Input: D – list of dictionaries, one for each category
Output: V – vector of sentiment measures of A
1   **begin**
2      $V \leftarrow \{0, 0, 0, 0, 0, 0\}$
3      $N \leftarrow 0$
4   **foreach** $word \in A$ **do**
5   **for** $i = 0$ *to 5* **do**
6   **if** $word \in D[i]$ **then**
7                  $V[i] \leftarrow V[i] + 1$
8                  $N \leftarrow N + 1$
9   **end**
10   **end**
11   **for** $i = 0$ *to 5* **do**
12          $V[i] \leftarrow V[i] \div N$
13   **end**
14   **return** V
15   **end**

3.4   Feature Selection

In this step we select the features which are most useful in classification as a form of feature selection. At this point we have 4,424 features, 4,417 of these features are tf-idf scores for individual stemmed words, and the remaining seven features consist of six scalar sentiment scores and one continuous sentiment score. We use the J48 decision tree implementation provided in the Weka [13] machine learning toolkit for feature selection. In a decision tree, the nodes with the lowest depth offer the most information gain. Since all of the features of our data are continuous variables, the nodes in the resulting decision tree contain a threshold for which a feature provides information. The final decision tree contains the features which provide the most useful information in the classification of articles, and features that provide redundant or no information are omitted from the decision tree.

In the case of our data set, twelve word stems provided the most information in regards to classifying news articles: accis, advic, breach, commander, deb, den, emerg, encrypt, handl, hear, review, and widespread. In Fig. 2, we can see the values for which the TF-IDF scores were considered useful for these features, and in Table 3 we can see the words which each stem represents.

3.5   Classification: Naive Bayes Model

In this section, we measure the performance of applying Weka's Naive Bayes classifier to the features extracted from the New York Times articles relating to crises. We compare the results of applying the Naive Bayes classifier on all 4,424 features to the results of applying the Naive Bayes classifier on the twelve text features selected from the J48 decision tree analysis, and the results of applying the Naive Bayes classifier to the twelve selected text features plus the sentiment features. The three models are evaluated by the true positive rate, false positive rate, precision, recall, and receiver operating characteristic (ROC Curve). The weighted averages are based on the distribution of the data set that belongs to each class. The target class is pre-crisis, crisis, or post-crisis and is judged on a per document basis rather than a topic basis.

```
J48 pruned tree
------------------

advic <= 0
|   deb <= 0
|   |   encrypt <= 0
|   |   |   den <= 1.506716
|   |   |   |   commander <= 1.609757
|   |   |   |   |   emerg <= 1.295889
|   |   |   |   |   |   breach <= 0
|   |   |   |   |   |   |   accis <= 0
|   |   |   |   |   |   |   |   hear <= 1.115454
|   |   |   |   |   |   |   |   |   widespread <= 0.701946
|   |   |   |   |   |   |   |   |   |   handl <= 0
|   |   |   |   |   |   |   |   |   |   |   review <= 0: pre crisis (36.0)
|   |   |   |   |   |   |   |   |   |   |   review > 0: post crisis (2.0)
|   |   |   |   |   |   |   |   |   |   handl > 0: post crisis (2.0)
|   |   |   |   |   |   |   |   |   widespread > 0.701946: post crisis (4.0)
|   |   |   |   |   |   |   |   hear > 1.115454: post crisis (5.0)
|   |   |   |   |   |   |   accis > 0: post crisis (9.0)
|   |   |   |   |   |   breach > 0: post crisis (3.0/1.0)
|   |   |   |   |   emerg > 1.295889: crisis event (3.0/1.0)
|   |   |   |   commander > 1.609757: crisis event (3.0/1.0)
|   |   |   den > 1.506716: crisis event (3.0)
|   |   encrypt > 0: crisis event (4.0)
|   deb > 0: crisis event (3.0)
advic > 0: crisis event (3.0)

Number of Leaves  :     13

Size of the tree :      25
```

Fig. 2. Results of J48 decision tree analysis

Table 3. Reverse look-up of word stems

Stem	Potential matches
accis	accident, accidental, accidentally
advic	advice
breach	breach, breached, breaches, breaching
commander	commander, commanders
deb	debatable, debate, debated, debates, debating
den	den, denature, denial, denials, denied, denier, deniers, denies, dens, deny, denying
emerg	emerge, emerged, emergence, emergencies, emergency, emergent, emerges, emerging
encrypt	encrypt, encrypted, encryption, encrypts
handl	handle, handled, handles, handling
hear	hear, hearing, hearings, hears
review	review, reviewed, reviewing, reviews
widespread	widespread

The confusion matrix for all three Naive Bayes models, as well as the precision recall and F1 measures are given in Table 4(a), (b) and (c).

Table 4. Confusion matrices for the three Naive Bayes models

(a) All features

| | | Predicted | | |
		A	B	C
Actual	A	32	2	2
	B	5	8	5
	C	7	5	14

(b) Selected features with sentiment

| | | Predicted | | |
		A	B	C
Actual	A	35	1	0
	B	1	15	2
	C	0	2	24

(c) Selected features only

| | | Predicted | | |
		A	B	C
Actual	A	35	1	0
	B	1	16	1
	C	1	13	12

When we take a closer look at the detailed accuracy of the classification models, we can see the metrics supporting the initial impression of the confusion matrices. The standard measures of the classification accuracy are gicven in Table 5. The measures for pre-crisis, crisis, and post-crisis are given in Table 5(a), (b) and (c), and the weight averages are given in Table 5(d).

Table 5. Classification accuracy measures

(a) Pre-crisis article classification

Attribute	All Features	Selected Features
TP Rate	88.9%	97.2%
FP Rate	27.3%	2.3%
Precision	72.7%	97.2%
Recall	88.9%	97.2%
ROC Area	80.6%	97.8%

(b) Crisis article classification

Attribute	All Features	Selected Features
TP Rate	44.4%	83.3%
FP Rate	11.3%	4.8%
Precision	53.3%	83.3%
Recall	44.4%	83.3%
ROC Area	69.0%	90.3%

(c) Post-crisis article classification

Attribute	All Features	Selected Features
TP Rate	53.8%	92.3%
FP Rate	13.0%	3.7%
Precision	66.7%	92.3%
Recall	53.8%	92.3%
ROC Area	72.1%	92.4%

(d) Weighted averages

Attribute	All Features	Selected Features
TP Rate	67.5%	92.5%
FP Rate	19.0%	3.3%
Precision	66.4%	92.5%
Recall	67.5%	92.5%
ROC Area	75.2%	94.4%

In the model with all of the features present, the weighted average percentage of correctly classified instances is 67.5 %, just over two thirds (see Table 5(d)). While this is about twice the accuracy of randomly guessing when given three classes, there is still plenty of room for improvement. The impression of the model with all 4,424 features worsens when we look at the true positive rate for the individual classes. Because there are more pre-crisis articles than any other class, the higher true positive rate for pre-crisis articles has the most influence on the weighted average. In the model with selected features, not only is the weighted average of true positive rate much higher at 92.5 % (Table 5(d)), but the true positive rate varies much less between the three classes (97.2 %, 83.3 %, and 92.5 %, respectively). This accuracy was calculated using 5-fold cross validation All other metrics in the model with the selected features improve as well: the false positive rates lower, precision and recall increase, and the ROC area increases for all three classes as shown in Fig. 3(a), (b) and (c).

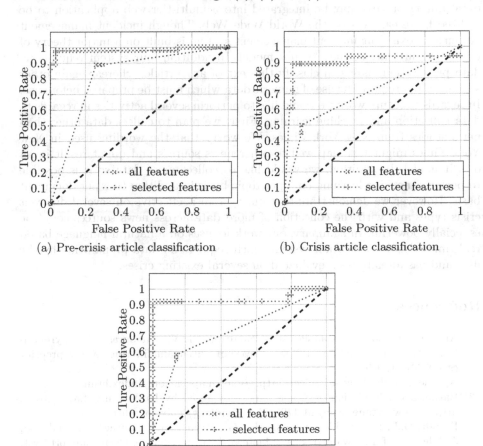

(a) Pre-crisis article classification

(b) Crisis article classification

(c) Post-crisis article classification

Fig. 3. ROC curves

4   Conclusion

In this paper, we set out to use data mining and machine learning techniques to create a reusable methodology that can be applied towards the empirical research of crisis communication. Because research in crisis communication has been dominated by case studies, our initial effort in this area has been to create a corpus of news media data that is labeled with characteristics of crisis communication theory. We aggregated and labeled news media data from the New York Times Article Archive, and we were successfully able to train a classification model to recognize the occurrence of specific terms in order to identify characteristics of organizational crisis. We also explored how our corpus and data mining process may be integrated into a multi-tier web application to be utilized by researchers on the World Wide Web. Though incident management systems do exist our incident management system is built on a major theory of crisis communication, the Coombs model. In addition, our system requires no data preparation, as it can classify the text of any article, whereas again other systems often require the use of numeric data which must be prepared before use. In regards to future work we would like to perform several activities in regards to data collection and machine learning. First, we plan to collect data other than news stories from New York Times. We worry that the wording used in New York Times might not map well to other news sources and might result in an overfit model. In addition, we would like to collect more data so it is possible to perform machine learning on a per topic basis in addition to a per document basis. However we demonstrate that our system is effective on several different crisis types and with the collection of more data across news sources could be especially powerful. In summary, our work focuses on a natural language based real time crisis identification system which requires no data preparation by the user and has already been evaluated on several existing crises.

References

1. Androutsopoulos, I., Koutsias, J., Chandrinos, K.V., Paliouras, G., Spyropoulos, C.D.: An evaluation of Naive Bayesian anti-spam filtering. arXiv preprint cs/0006013 (2000)
2. Apache: Apache Maven Project. http://maven.apache.org/index.html
3. Barboza, D.: Explosion at Apple Supplier Caused by Dust, China Says (2011). http://www.nytimes.com/2011/05/25/technology/25foxconn.html
4. Berndt, D.J., Fisher, J.W., Craighead, J.G., Hevner, A.R., Luther, S., Studnicki, J.: The role of data warehousing in bioterrorism surveillance. Decis. Support Syst. 43(4), 1383–1403 (2007)
5. Berndt, D.J., Fisher, J.W., Hevner, A.R., Studnicki, J.: Healthcare data warehousing and quality assurance. Computer 34(12), 56–65 (2001)
6. Chen, H., Chung, W., Xu, J.J., Wang, G., Qin, Y., Chau, M.: Crime data mining: a general framework and some examples. Computer 37(4), 50–56 (2004)
7. Coombs, W.T.: Ongoing Crisis Communication: Planning, Managing, and Responding. Sage Publications, Thousand Oaks (1999)

8. Coombs, W.T.: Protecting organization reputations during a crisis: the development and application of situational crisis communication theory. Corp. Reput. Rev. **10**(3), 163–176 (2007)

9. Dave, K., Lawrence, S., Pennock, D.M.: Mining the peanut gallery: opinion extraction and semantic classification of product reviews. In: Proceedings of 12th International Conference on World Wide Web, pp. 519–528. ACM (2003)

10. Etzioni, O., Cafarella, M., Downey, D., Popescu, A.M., Shaked, T., Soderland, S., Weld, D.S., Yates, A.: Unsupervised named-entity extraction from the web: an experimental study. Artif. Intell. **165**(1), 91–134 (2005)

11. French, S., Niculae, C.: Believe in the model: mishandle the emergency. J. Homeland Secur. Emerg. Manag. **2**(1), 1–16 (2005)

12. Go, A., Bhayani, R., Huang, L.: Twitter sentiment classification using distant supervision. CS224N Project Report, pp. 1–12, Stanford (2009)

13. Hall, M., Frank, E., Holmes, G., Pfahringer, B., Reutemann, P., Witten, I.H.: The WEKA data mining software: an update. SIGKDD Explor. Newslett. **11**(1), 10–18 (2009)

14. Han, J., Kamber, M., Pei, J.: Data Mining: Concepts and Techniques, 3rd edn. Morgan Kaufmann, Waltham (2012)

15. Harms, S.K., Deogun, J., Saquer, J., Tadesse, T.: Discovering representative episodal association rules from event sequences using frequent closed episode sets and event constraints. In: Proceedings IEEE International Conference on Data Mining (ICDM 2001), pp. 603–606. IEEE (2001)

16. Jain, G., Ginwala, A., Aslandogan, Y.A.: An approach to text classification using dimensionality reduction and combination of classifiers. In: Proceedings of the 2004 IEEE International Conference on Information Reuse and Integration, IRI 2004, pp. 564–569. IEEE (2004)

17. Joachims, T.: Text categorization with support vector machines: learning with many relevant features. In: Nédellec, C., Rouveirol, C. (eds.) ECML 1998. LNCS, vol. 1398, pp. 137–142. Springer, Heidelberg (1998)

18. Jones, K.S.: A statistical interpretation of term specificity and its application in retrieval. J. Doc. **28**(1), 11–21 (1972)

19. Karasová, V., Krisp, J.M., Virrantaus, K.: Application of spatial association rules for improvement of a risk model for fire and rescue services. In: Proceedings of ScanGIS 2005 (2005)

20. Lee, D.H., Jeng, S.T., Chandrasekar, P.: Applying data mining techniques for traffic incident analysis. J. Inst. Eng. **44**(2), 90 (2004). Singapore

21. Liang, X.: Government crisis communication on the microblog: a theory framework and the case of Shanghai metro rear-end collision. In: Proceedings of 6th International Conference on Theory and Practice of Electronic Governance, ICEGOV 2012, pp. 248–257. ACM (2012)

22. Loughran, T., McDonald, B.: When is a liability not a liability? Textual analysis, dictionaries, and 10-Ks. J. Financ. **66**(1), 35–65 (2011)

23. Lyall, S.: British Tabloid Hacked Missing Girl's Voice Mail, Lawyer Says (2011). http://www.nytimes.com/2011/07/05/world/europe/05britain.html

24. MacLaggan, C.: Exclusive: Livestrong Cancer Charity Drops Lance Armstrong Name from Title (2012). http://tinyurl.com/zrx9slu

25. Miller, D., Schwartz, R., Weischedel, R., Stone, R.: Named entity extraction from broadcast news. In: Proceedings of DARPA Broadcast News Workshop, pp. 37–40 (1999)

26. Nielsen, F.: AFINN (2011). http://www2.imm.dtu.dk/pubdb/p.php?6010

27. Nielsen, F.: A new ANEW: evaluation of a word list for sentiment analysis in microblogs. arXiv preprint (2011). arXiv:1103.2903
28. Pang, B., Lee, L., Vaithyanathan, S.: Thumbs up? Sentiment classification using machine learning techniques. In: Proceedings of ACL-02 Conference on Empirical Methods in Natural Language Processing, vol. 10, pp. 79–86. Association for Computational Linguistics (2002)
29. Papamichail, K.N., French, S.: Design and evaluation of an intelligent decision support system for nuclear emergencies. Decis. Support Syst. **41**(1), 84–111 (2005)
30. Peng, Y., Zhang, Y., Tang, Y., Li, S.: An incident information management framework based on data integration, data mining, and multi-criteria decision making. Decis. Support Syst. **51**(2), 316–327 (2011)
31. Perlroth, N.: Target's Nightmare Goes On: Encrypted PIN Data Stolen (2013). http://bits.blogs.nytimes.com/2013/12/27/targets-nightmare-goes-on-encrypted-pin-data-stolen
32. Framework, P.: Play 2.1 documentation. http://www.playframework.com
33. Porter, M.F.: An algorithm for suffix stripping. Prog.: Electron. Libr. Inf. Syst. **14**(3), 130–137 (1980)
34. Reuters: Banks Suffer After-Hours On Word of Chase Losses (2012). http://www.nytimes.com/2012/05/11/business/daily-stock-market-activity.html
35. Rish, I.: An empirical study of the Naive Bayes classifier. In: IJCAI Workshop on Empirical Methods in Artificial Intelligence, vol. 3, no. 22, pp. 41–46 (2001)
36. Robertson, C.: Search Continues After Oil Rig Blast (2010). http://www.nytimes.com/2010/04/22/us/22rig.html
37. Safavian, S.R., Landgrebe, D.: A survey of decision tree classifier methodology. IEEE Trans. Syst. Man Cybern. **21**(3), 660–674 (1991)
38. Sanger, D.E.: Loss of the Shuttle: the Overview; Shuttle Breaks Up, 7 Dead (2003). http://www.nytimes.com/2003/02/02/us/loss-of-the-shuttle-the-overview-shuttle-breaks-up-7-dead.html
39. Schumaker, R.P., Chen, H.: Textual analysis of stock market prediction using breaking financial news: the Azfin text system. ACM Trans. Inf. Syst. (TOIS) **27**(2), 12:1–12:19 (2009)
40. Schwirtz, M.: Oil Pipeline Ruptures in Arkansas (2013). http://www.nytimes.com/2013/03/31/us/oil-pipeline-ruptures-in-arkansas.html
41. Seeger, M.W., Sellnow, T.L., Ulmer, R.R.: Communication and Organizational Crisis. Praeger Publishers, Westport (2003)
42. Takeuchi, K., Collier, N.: Bio-medical entity extraction using support vector machines. Artif. Intell. Med. **33**(2), 125–137 (2005)
43. The New York Times: New York Times Article Archive (2014). http://www.nytimes.com/ref/membercenter/nytarchive.html
44. Wan, X.: Using bilingual knowledge and ensemble techniques for unsupervised chinese sentiment analysis. In: Proceedings of Conference on Empirical Methods in Natural Language Processing, pp. 553–561. Association for Computational Linguistics (2008)
45. Youn, S., McLeod, D.: A comparative study for email classification. In: Elleithy, K. (ed.) Advances and Innovations in Systems, Computing Sciences and Software Engineering, pp. 387–391. Springer, Heidelberg (2007)

Development of Issue Sets from Social Big Data: A Case Study of Green Energy and Low-Carbon

Chun-Che Huang[1(✉)], Yu-Jie Fang[1], Shian-Hua Lin[2],
Wen-Yau Liang[3], and Shu-Rong Wu[1]

[1] Department of Information Management, National Chi Nan University,
Puli Township, Nantou County 545, Taiwan, ROC
cchuang@ncnu.edu.tw, tram716@gmail.com,
m920075@gmail.com
[2] Department of Computer Science and Information Engineering,
National Chi Nan University, Puli Township 545, Taiwan, ROC
shlin@csie.ncnu.edu.tw
[3] Department of Information Management,
National Changhua University of Education, Changhua 500, Taiwan, ROC
wyliang@cc.ncue.edu.tw

Abstract. "Energy" has been one element of the development of human civilization, also a power for national industry, construction and economic development. The green energy has become the cornerstone in sustainable development to secure such energy supply but may accommodate opinions from controversial perspectives when this subject is discussed. This study develops an interactive big data system, which aims at aggregating data from Facebook, PTT, news, and provides an interactive interface for energy domain experts. The "interaction" characterizes the seamless integration between users and the system to construct the controversial issue sets of energy, which could be identified and established autonomously in this study. The approach using tags of the link in two controversial issues can help end-users effectively query on demand. The energy relevant issues can be fully aware and provided to the decision makers from the positive and negative viewpoints.

Keywords: Social big data · Green energy · Controversial issues · Text mining

1 Introduction

Due to digital technologies, more and more human activities leave imprints whose collection, storage and aggregation can be readily automated. In particular, the use of social media results in the creation of datasets, which may be obtained from platform providers or collected independently with relatively little effort as compared with traditional sociological methods. The text message of social stream could be related to specific social connections between information senders/authors and recipients/readers. From social media data sources, the data sources can be characterized by their different formats and contents, their very large size, and the online or streamed generation of

© Springer International Publishing Switzerland 2016
P. Perner (Ed.): ICDM 2016, LNAI 9728, pp. 139–153, 2016.
DOI: 10.1007/978-3-319-41561-1_11

information, defined as *social big* data [1]. Social media plays an important role to reflect public opinions. However, in the domain of energy, little research of social big data is studied to recognize the uncovered social responses, specifically controversial issues effectively.

Controversial issues may be disseminated from opinion leaders in social big data, resolved with some solutions. From developmental perspective of resilience, as [2] argue, implies the presence of latent resources that can be activated, combined, and recombined in new situations as challenges arise. It is crucial to identify the controversial issue relevant to people, resource and solution. The concepts of controversial issue, scholar defined issues as "When face to the same events, different individual, organization or group, based on different positions, opinions or adjustment of resource, will bring about competition." It could be seen, if we does not have any conflict, competition and difference view, the issues will never form, even bring other problem. According to [3], becoming an issue depends on originators and catalyst. Once the two interact to each other, the issue will be emerge. However, issues could be forming by occasional emergency, such as disaster and rescue or the war by International conflicts. In addition, other scholar defined hot issues as the period within the issues frequently be discussed [4].

In this study, to crawl social big data, an interactive social big data (**SBD**) system for social response is designed and implemented. The system aggregates data from Web sources (Facebook, PTT, news) and provides an interactive interface for domain experts to browser information. To recognize the semantic and sentiment of social response, **5W** (Who, What/event, When, Where and What/Which) and **4S** (Sentiment: positive, negative, neutral and ironic) metadata attributes are defined and designed into the interactive interface. SBD extracts 5W object-relations and integrates with the domain concept hierarchy to form a knowledge base, where the proposed axioms identify controversial issues.

The remaining parts of the paper include literature survey in Sect. 2. The solution approach is proposed in Sect. 3. The case of green energy and low-carbon is studied to show the superiority of the proposed approach. Section 5 concludes this study.

2 Literature Review

2.1 Social Big Data and Text Mining

Social streaming contents become ubiquitous with our life such as blog, social network and web forums. The text message of social stream could be related to specific social connections between information senders/authors and recipients/readers. Social text stream influences the way of people's daily communication, not only individual users but also groups run social connections through social text streams. The media data sources could be characterized by their different formats and contents, their very large size, and the online or streamed generation of information, defined as *social big* data [1]. Social big data based on the analysis of vast amounts of data that could come from multiple distributed sources but with a strong focus on social media. Hence, social big data analysis [5, 6] is inherently interdisciplinary and spans areas such as data mining, machine learning, statistics, graph mining, information retrieval, linguistics, natural language processing, the semantic Web, ontologies, and big data computing, among others.

Applications of social big data are numerous. For example, in health care [7], finance [8], marketing [9], and so on. In order to analyze the social big data properly, the traditional analytic techniques and methods (data analysis) require adapting and integrating them to the new big data paradigms emerged for massive data processing. Different big data frameworks such as Apache Hadoop [10] and Spark [11] have been arising to allow the efficient application of data mining methods and machine learning algorithms in different domains. Based on these big data frameworks, several libraries such as Mahout [12] and SparkMLib [13] have been designed to develop new efficient versions of classical algorithms. In addition, a considerable number of studies have been made on formal text stream data. The social streaming data are substantially different from formal web streaming data. For instance, several studies [14–16] focus on analyzing the semantics of text streaming data and social network data. Different text mining approaches are proposed to classify large amount of documents into different issues or identify entities [14]. In addition, many applications are solved by event detection algorithms such as newswires [17] and blogs [18]. Several text-based semantic analyses apply TDT (Topic Detection and Tracking) [19] to mine topics from general text data. Social text streaming data contains rich social connection information, context sensitivity and temporal information. Therefore, events can be detected via social text stream [17].

In general, there are two categories of existing event detection algorithms: document pivot methods and feature-pivot methods. One is to utilize the semantics distance between documents to cluster documents [17]. The other one is to discover events by grouping words together [20]. In recent years, there has been renewal of interest in social text streaming data [17]. Some researches detect events from twitter and aim to harvest collective intelligence [19, 20]. However, traditional documents and searching techniques not only make knowledge difficult to discover, but also make it brittle. Consequently, it could only be applied in very limited application [21]. Although many scholars studying text mining focus on the structured knowledge generation in the defined formats like rules or procedures, only a few attempt to emphasize semi-structured or unstructured knowledge. To capture each document, the diversified requirements are required in expressing with a series of dimensions or abstractions, e.g., 5W1H in the Zachman Framework [22] to relate to the context. This format of 5W1H (what, where, who, when, and why) represents a suitable solution approach to externalize knowledge in organizations because it can capture the nature of each dimension (perspective) and integrate the target knowledge. Applying the Zachman Framework, knowledge in organizations could be transformed systematically as semi-structured knowledge documents [23]. In current social big data, lack of appropriated labeled data for semi-structured knowledge documents obstruct to apply text mining to deal with the issue identification problem.

2.2   Issue Identification

The controversial issue identification is the first job in issue management, which is an important concept in modern public relation theory and practice, indicated by the United States public relations scholar Chase [24]. In social media, user's online interactive behavior with others often makes some user generated contents popular. The modeling and prediction of the popularity of online content are an important research issue and can facilitate many key application domains [25]. From the issue identification perspective: It is a task that aims to group together the materials that discuss the same event (URL, IAD). To become as hot issues, there are two kinds of reasons (Cao et al. 2015): One kind of issues becomes "HOT" because their contents are very popular and general, so that everybody can produce related contents or follow them. The other kind of issues becomes "HOT" because they have significant distribution on the social network, where their contents are so specific that only can be produced and propagated by a limited user group. Therefore, the popularity, generality and contribution to social network makes the issue identification is crucial.

In social media, an event is considered as a life form with stages of birth, growth, decay and death. To track life cycles of events, we use the concept of energy function. Like the endogenous fitness of an artificial life agent, the value of energy function indicates the liveliness of a news event in its life span [26]. The life cycle of an event is analogous to living beings. With abundant nourishment (i.e., related documents for the event), the life cycle is prolonged; conversely, an event or living fades away when nourishment is exhausted. Improper tracking algorithms often unnecessarily prolong or shorten the life cycle of detected events [27]. To develop an issue set is important in issue management. However, large articles in big data world make issue identification not easy and time consumption. An approach to establish issue set automatically is a challenged. Particular emotion, either positive/negative opinion provision is useful to decision makers in social media. From the survey of previous literature, it is concluded:

1. There has been an increasing interest in analyzing social big data. However, detecting social issues, e.g., controversial issues in the era of information explosion requires state-of-the-art analytics techniques [28].
2. In some social big data, e.g., the case of Twitter, trend keyword lists are provided for the user's convenience. However, it is still not easy to determine the details based on a few simple keywords. The keywords usually relate to the hot issues at any time so many documents will contain pertinent details, such as news on the Internet. Thus, to provide detailed information about a controversial issue, it is necessary to identify relationships, e.g., 5W among social big data [28].

Next, the solution approach, including an interactive big data system and method to identify controversial issues are proposed to develop controversial issue sets aiming to identify issues automatically with both the positive and negative opinions from social big data.

3 Solution Approach

The proposed interactive big data system first is based on "search engine techniques," collecting and retrieving data in social media. Base on the Facebook (FB) API technique, crawl data of public social network activities are captured with extracted metadata according to the FB API format. As for metadata extraction of news article from news sites, the semi-structured metadata extraction technique is implemented to integrate manual efforts with labeling programs so that the system can extract metadata from dozens of websites efficiently and effectively. According to the big data techniques, an interactive issue map is proposed.

In this study, an controversial issue is defined and characterized as follows:

(1) The core of the problem.
   An issue which is generated after a certain period of dispute and discussion is the core of the problem that should be clearly referred to in social big data.
(2) Having positive and negative comments and opinions.
(3) Structural
   An issue sets in this study are constructed structurally and could be extended and related to other issues; as a result, the issue sets are labeled and defined according to the principle of 5W1P1N. 5W stands for WHO, WHERE, WHEN, WHAT AND HOW. When it comes to dispute, no matter whether the issue is right or wrong, the cons and pros of opinion are accepted and categorized as 4V, which stands for Like Viewpoint, Dislike Viewpoint, Neutral Viewpoint, and ironic Viewpoint. The public opinion is divided into 1P1N which means Positive Viewpoint and Negative Viewpoint.
(4) Hierarchical
   Issues are labelled under the principle of 5 W. They are structural and a hierarchical structure can be produced to represent the issues. The issue sets are constructed hierarchically and could be extended and related to sub/super issues.

3.1 Semi-structured Metadata from Social Networks

In order to collect information to identify controversial issues, 5W and 4S (Sentiments) are proposed for metadata.

- Which: While the issues are generated constantly in social big data, it is crucial to collect data and classify keywords for taxonomy. It is anticipated that, before controversial issues occurred in Internet, the conceptual keywords are detected and formed as tag clouds [29, 30]. Among the tag clouds, the experts can easily discover desired issues just in time.
- When: The system automatically collects temporal data from FB, PTT, news to form new issues given by an evolutional process.
- What: As the issue is warmed up, the contents, opinions, sentiments are generated gradually. The system can extract things of people, subjects, time, places, and things and generate metadata in databases for quick searches.

- Who: As the issue is focused progressively, users try to express their opinions, comments, feelings and groups of users are formed. The system should analyze the social big data and classify the users as Opinion Leaders, Fans or Followers, and Readers. Consequently, the people clusters are formed.
- Where: As an issue is popular, various activities are initiated, rather than still in cyber BA [31]. People involve in these activities in physical location and large of data are collected instantly. The relationships of 5W become matured. At this time, milestone or checkpoint of the issue is required to recognize.

In addition, four sentiments are used to interpret participant's responses. Positive: According to the particular issue, the attitude trends to support with a positive perspective. Negative: According to the particular issue, the attitude trends to withdraw with a negative perspective. Neutral: According to the particular issue, the attitude trends to be neutral without any particular trend. Satire: According to the particular issue, the sematic is hard to be identified by program since some positive words may imply "disagree" emotion.

At the initial stage, FB is focused on this study since it is one of large social media in Taiwan. Groups and IDs in FB are collected using Google Search according to a particular keyword. Mining from Facebook ID's, according to each issue, controversial issues and opinion leaders/main followers are identified and linked to 5W relationships.

According to each Facebook ID (fbid), the fbid's public attributes are extracted through a browser-based crawler, named human-like FB robot (HLFBR). Attributes of an fbid definitely have impact on the desire to the FB user's social activity. Through sharing positive and negative perspectives, controversial issues will be identified based on atypical events/challenges occurrence of 4S (four Sentiments) perspectives. Consequently, fbid can be classified by four roles: **Latent Opinion Leader (LOL)** with many FB personal fans, friends or followers. **Authors** post new issue, message, and news. **Repliers** respond according to any post from authors. **Followers** frequently respond to their Authors and LOL, specifically to negative opinions. The followers may be supporters, or observers occasionally respond to LOL.

The system can extract 5W: WHO (fbid), WHEN (post time), WHERE (tag location), WHAT (recognize objects or things from the content analyzed with Text Mining → Keywords → Cloud Tags → Things). In this process, the following parameters are acquired:

Post *Likes:* the larger of "LIKES", the hotter issue is. The fbid posts Likes can be related to its attitude attributes.

Post *Comments:* the larger of "Comments", the more popular issue is. The content of comments also can be analyzed and abstracted and summarized as graph information to compute the quality of fbid.

Post *Shares:* it refers to transmit rate. If there is content in "Share", the content is also be analyzed to understand its attitude attributes.

Post *Clicks:* it refers to the number of readers. Unfortunately, it is not provided by FB. This study will refer to the parameters of "links" in Google Search to estimate the number of readers.

3.2   Social Big Data System

To resolve the crawling problem, a Social Big Data System (SBDS) is configured in Fig. 1, where each module is described from left to right:

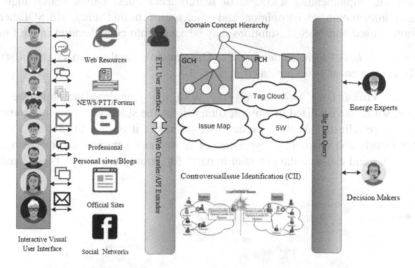

Fig. 1. The social big data flow and architecture

Web Resources: Web resources at the left side will be extracted and analyzed according to different types of natures of resources from the perspectives of 5W.

- Important websites: News, Ptt/Forums sites provide abundant web pages for extract hot or emergent social events, in which the data set also provide plenty resources to recognize people caring about the event.
- Who poses opinions according to which issue?
- Professional Personal Web Sites/Blogs: This resource types are good to identify opinion leaders. Most opinion leaders have their own blogs or sites.
- Official Sites: One of the important sources that can capture "Who, What, When, Where, Why" from government documentaries.
- Social Network: Different from official websites, social network may contain "5W" information in one platform, but they could be hided and not easy to be discovered.

Extract-Transform-Load (ETL): ETL is an important interface of Big Data System (BDS) corresponding to different types of Web Resources. Use metadata built in ETL to map sources and target data. In this study, metadata are captured and mapped to Dublin Core, widely used in digital libraries as the common format for cross-domain metadata exchange. The process is named as *M2DC mapping*, in this paper.

Domain Concept Hierarchies (DCH): After M2DC mapping, the BDS analyzes possible conceptual structure, such as Tag Cloud, Issue Map, and "5W" keyword (object) network. In this way, "Hypothesis" is proposed to explore the truth and decision rules. Through DHC, a knowledge Framework is developed.

Big Data Query (BDQ): Aforementioned modules may locate at different servers containing various databases. Incorporating the outcomes of BDQ and indices, the DBs are built via implementing a couple of search techniques; Quick search inquiries necessary information, social opinion, and other issues. In this study, MS SQL Server is definitely used and NoSQL solutions may be taken into consideration in the future.

Controversial Issue Identification (CII): This module analyze social big data to identify controversial issue presented in Sect. 3.4.

Interactive Visual User Interface (IVUI): With the graphic visual interface user interface, which is visualized in multiple dimensions, the solution approach parameters are easy to be adjusted. With the visual user interface, it is easy for experts to view relevant reports and charts (e.g. hyper-tree or issue map) summarized for the controversial issues and evaluate the potential impacts (diagram presented in Figs. 2 and 3).

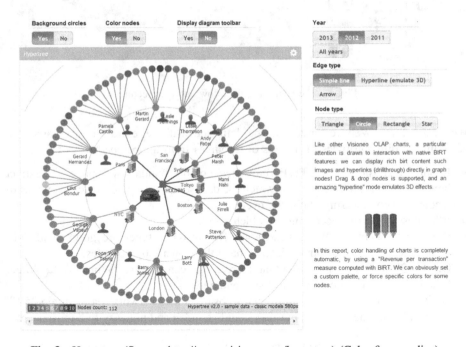

Fig. 2. Hypertree (Source: http://www.visioneo.org/hypertree) (Color figure online)

3.3   The Interaction Between the SBD System and the User

Based on the real time analysis of ETL and 5 W4S metadata, the interactive system crawls social text streams generates daily reports for resilience data analyzer with collection of keywords in DCH, attitude attributes of 4S, and 5W metadata. With the keywords collection and Cloud Keyword Base (CKB), the sematic is analyzed for

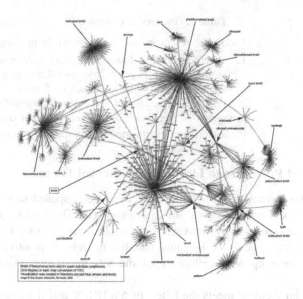

Fig. 3. Issue map (Source: Wandora GNU GPL v3 open source)

phrases and issues are formed. The issues are validated with resilience experts to remove the error issue (a checkpoint). The relevant issues are sent to R Matrix analytics, where opinion leaders, latent/hot controversial issues can be identified with the parameters of weighting, TFT-IDF, frequency. SBD and UI integration flow is shown in Fig. 4.

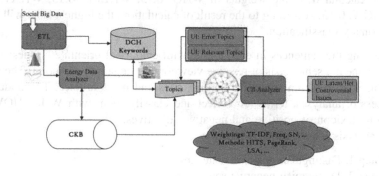

Fig. 4. Interaction between SBD system and user

In summary, the social big data system will generate the required data and information for the issue identification. The interactive process is presented in Table 1, where Input, Process, and Output are shown.

Table 1. Interactive process

Input	Social web resources (news, Ptt, search from Google, SN, ...)
Process/UI/resilience data analyzer	1. Latent keywords → Latent issue (UI): predict and update true/false, mark P/N perspective → Dictionary as references → Keywords → Issues → Scores 2. Timeline analysis: Timeline keyword/Content/fbid histogram → Trends → Checkpoints (relevant issues)
Output	Parameters for identification of controversial issues

3.4   Axioms and Procedure to Identify Controversial Issues

In this study, two axioms and a solution procedure are proposed to present the data structure of the labels and then yield the final issue sets. The exploring procedure is initiated at the bottom level by processing of articles, categorizing labels, detecting issues, analyzing the models of labelling and the models of positive and negative viewpoints, and ends up with building up issue sets under two axioms and the 4 processes.

Axiom 1: If an incident meets the labels of 5W1P1 N and if there are more than n articles accumulated in consecutive d days, then it is defined as an issue and included in an issue set.

(1) The categorization of labels in an article:
Based on 4W (WHO, WHERE, WHAT, HOW), the frequency of 4W is calculated. The n most frequently used terms are extracted to compare with the title.

- If only one label matches, then this label will be the primary classification.
- If there are more than two matching labels after the comparison, the labels will be calculated by the weights of WHO: 0.15, WHERE: 0.15, WHAT: 0.55, HOW: 0.15; according to the result of calculation, the highest label will be the primary classification.

(2) Analyzing the sentences in each article with emotional orientated issues:
Determine if a sentence is of positive viewpoint or negative viewpoint. The source of this study is from social networking sites and news articles. To conduct the process of analyses, terms are collect and classify them with WHO, HOW and polarity lexicon of positive and negative adjectives.
The analysis is processed as follows:

Step 1. Distinguish positive sentences
Step 2. Distinguish negative sentences

Step 3.

If positive word occurs once in a sentence,
  it will be set as +1; it occurs n times, then it will be $(+1) * n$.
If negative word occurs once in a sentence,
  it will be set as -1; it occurs m times, then it will be $(-1) * m$.
By adding up the above two points, the formula $z = (+1)*n+ (-1)*m$ is obtained.

  If $z \leq$ negative *filter value*, the sentence is classified as negative.
  If $z = 0$, the sentence is of neutral opinion or of no opinion
  If z is between *filter value and negative filter value*, the sentence cannot be distinguished as either positive or negative; expertise judgment is required.
  If $z >$ *filter value*, the sentence is classified as positive.

  Note that the filter value may be defined according to different domains of study and different users.

(3) To determine if a sentence is of positive or negative opinion, to see if it correlates with the positive or negative viewpoint in a similar article with similar viewpoints and then produce annotations.

  Axiom 2: Hierarchical analysis of Issues

(1) {Issue A} $\subset$ Independent Issues
(2) {Issue B, Issue C} $\subset$ Common Issues
(3) {Issue E} $\subset$ {Issue D} $\subset$ Subordinate Issues
(4) {Independent Issue, Common Issues, Subordinate Issues} $\subset$ ANY Issues

  The standard procedure is proposed to automatically find out as well as build up issues and then classify them as either positive viewpoint or negative viewpoint in issue sets. The procedure is as follows:

  Step 1. Apply search engines and Web technology. Web Crawler program can automatically crawl articles on the website every day through Google search and the hyperlinks of web pages.
  Step 2. Use label modules process sentence phrasing, segmentation of words, and the comparison and labeling of terms, and the classification of articles.
  Step 3. Use the emotional modules of positive and negative viewpoints to analyze the positive and negative sentences in articles
  Step 4. Conducting the process of comparing issues.

  Compare the sentiment in articles with the labels in issue sets. In the process of comparison, axiom 2 is used to analyze the relationship among issues and distinguish if they are common, subordinate or independent with one another.

4 Case Study

Energy has been one of the most important elements in the development of civilization and has been the major power for industry and economic growth. Fossil fuels have been the main source of energy for over one hundred years; however, the storage of fossil fuels is decreasing. To make the situation even worse, the overuse of fossil fuels has resulted in environmental pollution and global warming.

As a result, European countries, the U.S. and Japan have been devoted to the development of green energy since 1970s. Recently, due to the rise of oil price, green energy has become more and more popular; techniques in the utilization wind power, solar power, photoelectricity, and biomass have improved quite a lot. More and more countries are promoting new energy policy not only ensuring the safety of using energy but also making advancement in the technology of energy. However, both sides of the coin have to be taken into account, and the advantages and disadvantages of green energy have to be taken into consideration. Green energy may not cause the destruction of ecology, while fossil fuels cause the destruction of ecology that arouses people's fear. Dispute concerning energy occurs and the public have positive and negative opinions while discussing this issue. Solar power, photoelectricity, and electricity generated by the wind may come from the nature, and they are non-carbon, and clean, but they cannot be cost-effective and is not a stable source of energy with limited utilization. Natural gas releases less pollution into the air but it is costly; coal is rich in storage and easy to get, but it emits carbon dioxide that is bad for the environment and people's health; nuclear power is cheap and clean, but it is dangerous and the disposal of nuclear waste is a serious problem.

The purpose of this case is to use the proposed SBD system to do phrasing and word segmenting for news/articles from social networking websites. By comparing and analyzing the sentiment, articles can be classified. In addition, through the utilization of positive and negative labelling models, polarity lexicons and sentiment are analyzed. At the same time, current issues and discussions concerning these issues can be automatically found out and dealt with. As a result, the positive and negative viewpoints can be fully manifested and thus understood by decision makers.

For example, the case collects the Chinese news related to carbon capture and storage (CCS) from social media between May 2006 and March 2015. With the proposed solution approach, the results are obtained in Table 2.

Next, an experiment is made to compare the correctness of issues produced by the previous study and the proposed approach. The previous study used keyword frequency and TDT method to identify issue sets. The results show the proposed approach can more precisely recommend the issue. The previous study may produce the issue which are not satisfied the definition of issue. For example, some event should not be included since they have less positive/negative options. The event is not controversial and required no further disputation or action to clarify the arguments. Table 3 presents the comparison qualitatively.

Table 2. The results of examplified issues in the issue sets

Issue	CCS					
WHEN	8/25/2010	11/7/2012	6/25/2013	6/26/2013	6/28/2013	...
WHERE	Yunlin	Changhua	Mailiao	Taixi	Miaoli	...
WHO	Industrial technology research institute	Taiwan cement	Corporation, Taiwan	Environmental protection administration	International energy agency	...
HOW	CO2	Environmental assessment specification	Energy policy	Induced seismicity	Gas leak	
WHAT	Carbon capture and storage	Carbon storage	Carbon capture	Ocean storage		
Positive	Storing carbon dioxide underground either into deep saline formations or depleted oil or gas fields is believed to be the best way to return carbon dioxide to the Earth. Developed countries such as Norway, Australia, Canada and the United States have already had quite advanced technologies concerning carbon dioxide storage; Taiwan is able to follow their steps in no time					
	Due to the fact that the widespread carbon dioxide emissions may cause global warming and climate change, many countries aims at the technology of Carbon Capture and Storage (CCS), which is the process of capturing waste carbon dioxide and storing it into deep geological formations					
Negative	CPC Corporation, Taiwan will conduct a CCS pilot project at Miaoli. However, some scientists and environmental groups cast doubt on this project. They have expressed concern about the active fault lines across Taiwan and the frequent occurrence of earthquakes in Taiwan. Is it safe to have such a large-scale CCS project here?					
	It may be quite risky in Taiwan					

Table 3. The comparison of previous study and proposed approach

	Previous study	The proposed approach
Focus	Using keywords only Frequencies of keywords are considered only The less frequency, the larger weight to classify data	Use tag to classify issues Use tag to correlate issues Consider 5W and P/N opinions
Advantage	Simple and agile	Can clarify the distribution of each phrase and 5W in articles High correctness of classification Low vague issues
Weakness	Cannot produce the distribution and importance level of terms High incorrectness of classification	Time consumption Issue with low emotional opinion may be ignored

5 Conclusion

This study presents a social big data system that integrates search engine techniques, article labelling models, phrasing and word segmenting modules for extracting semantical concepts from news and articles on social networking websites. By comparing and analyzing the sentiment, articles or replied comments can be classified into 4S types. Also, through the utilization of positive and negative labelling models, polarity lexicons and sentiment lexicons are analyzed. The positive and negative viewpoints can be fully manifested and thus understood for decision makers. The following problems are required to further explore:

– The approaches applied in English articles were studies before. The integration of English and Chinese characteristics into the same system is challenged.
– Comments from Facebook and PTT contain a lot of noisy data that dramatically influences the performance of the system.
– The computation of weight depends on users or domain experts. Sensitivity analysis is required to make sure the correctness of weight usage.

References

1. Bello-Orgaz, G., Jung, J.J., Camacho, D.: Social big data: recent achievements and new challenges. Inf. Fusion **28**, 45–59 (2016)
2. Sutcliffe, K.M., Vogus, T.J.: Organizing for resilience. Posit. Organ. Scholarsh.: Found. New discipl. **94**, 110 (2003)
3. Cobb, R.W.: Participation in American politics: The Dynamics of Agenda-Building. Johns Hopkins University Press, Baltimore (1983)
4. Bonsón, E., Torres, L., Royo, S., Flores, F.: Local e-government 2.0: social media and corporate transparency in municipalities. Gov. Inf. Q. **29**(2), 123–132 (2012)
5. Cambria, E., Rajagopal, D., Olsher, D., Das, D.: Big social data analysis. In: Akerkar, R. (ed.) Big Data Computing, pp. 401–414. Chapman and Hall/CRC publication (2013)
6. Manovich, L.: Trending: the promises and the challenges of big social data. In: Debates in the Digital Humanities, pp. 460–447 (2011)
7. Young, S.D.: Behavioral insights on big data: using social media for predicting biomedical outcomes. Trends Microbiol. **22**(11), 601–602 (2014)
8. Nguyen, T.H., Shirai, K., Velcin, J.: Sentiment analysis on social media for stock movement prediction. Expert Syst. Appl. **42**(24), 9603–9961 (2015)
9. Jang, H.J., Sim, J., Lee, Y., Kwon, O.: Deep sentiment analysis: mining the causality between personality-value-attitude for analyzing business ads in social media. Expert Syst. Appl. **40**(18), 7492–7503 (2013)
10. White, T.: Hadoop: The Definitive Guide. O'Reilly Media, Sebastopol (2009)
11. Zaharia, M., Chowdhury, M., Franklin, M.J., Shenker, S., Stoica, I.: Spark: cluster computing with working sets. In: Proceedings of 2nd USENIX Conference on Hot Topics in Cloud Computing, vol. 10, p. 10 (2010)
12. Anil, R., Dunning, T., Friedman, E.: Mahout in Action, pp. 1–2. Manning Publications Co., New York (2011)

13. Meng, X., Bradley, J., Yavuz, B., Sparks, E., Venkataraman, S., Liu, D., Tal-Walkar, A.: MLlib: Machine Learning in Apache Spark (2015). arXiv preprint arXiv:1505.06807
14. Bekkerman, R.: Automatic categorization of email into folders: benchmark experiments on Enron and SRI corpora (2004)
15. Mei, Q., Liu, C., Su, H., Zhai, C.: A probabilistic approach to spatiotemporal theme pattern mining on weblogs. In: Proceedings of 15th International Conference on World Wide Web, pp. 533–542 (2006)
16. Allan, J., Carbonell, J.G., Doddington, G., Yamron, J., Yang, Y.: Topic detection and tracking pilot study final report (1998)
17. Zhao, Q., Mitra, P.: Event detection and visualization for social text streams. In: ICWSM (2007)
18. Krause, A., Leskovec, J., Guestrin, C.: Data association for topic intensity tracking. In: Proceedings of 23rd International Conference on Machine Learning. pp. 497–504. ACM (2006)
19. Weng, J., Lee, B.S.: Event detection in Twitter. In: ICWSM, vol. 11, pp. 401–408 (2011)
20. Li, R., Lei, K.H., Khadiwala, R., Chang, K.C.C.: Tedas: a Twitter-based event detection and analysis system. In: 2012 IEEE 28th International Conference on Data Engineering (ICDE). pp. 1273–1276. IEEE (2012)
21. Miller, J.A., Potter, W.D., Kochut, K.J.: Knowledge, data, and models: taking an objective orientation on integrating these three. IEEE Potentials **11**(4), 13–17 (1992)
22. Inmon, W.H., Zachman, J.A., Geiger, J.G.: Data Stores, Data Ware-Housing, and the Zachman Framework: Managing Enterprise Knowledge. McGraw-Hill Companies, Inc., New York (1997)
23. Huang, C.C., Kuo, C.M.: Transformation and searching of semi-structured knowledge in organizations. J. Knowl. Manag. **7**(4), 106–123 (2003)
24. Chase, W.H.: Public issue management: the new science. Publ. Relat. J. **33**(10), 25–26 (1977)
25. Kong, Q., Mao, W., Zeng, D., Wang, L.: Predicting popularity of forum threads for public events security. In: 2014 IEEE Joint Intelligence and Security Informatics Conference (JISIC), pp. 99–106 (2014)
26. Chen, C.C., Chen, Y.-T., Sun, Y., Chen, M.-C.: Life cycle modeling of news events using aging theory. In: Lavrač, N., Gamberger, D., Todorovski, L., Blockeel, H. (eds.) ECML 2003. LNCS (LNAI), vol. 2837, pp. 47–59. Springer, Heidelberg (2003)
27. Chen, C.C., Chen, Y.T., Chen, M.C.: An aging theory for event life-cycle modeling. IEEE Trans. Syst. Man Cybern. Part A Syst. Hum. **37**(2), 237–248 (2007)
28. Kim, H.G., Lee, S., Kyeong, S.: Discovering hot topics using Twitter streaming data social topic detection and geographic clustering. In: 2013 IEEE/ACM International Conference on Advances in Social Networks Analysis and Mining (ASONAM), pp. 1215–1220. IEEE (2013)
29. Kuo, B.Y., Hentrich, T., Good, B.M., Wilkinson, M.D.: Tag clouds for summarizing web search results. In: Proceedings of 16th International Conference on World Wide Web, pp. 1203–1204. ACM (2007)
30. Koutrika, G., Zadeh, Z.M., Garcia-Molina, H.: Data clouds: summarizing keyword search results over structured data. In: Proceedings of 12th International Conference on Extending Database Technology: Advances in Database Technology, pp. 391–402. ACM (2009)
31. Nonaka, I., Konno, N.: The concept of "B, A": building a foundation for knowledge creation. Knowl. Manag.: Crit. Perspect. Bus. Manag. **2**(3), 53 (2005)

Early Prediction of Extreme Rainfall Events: A Deep Learning Approach

Sulagna Gope[1(✉)], Sudeshna Sarkar[1], Pabitra Mitra[1], and Subimal Ghosh[2]

[1] Department of Computer Science and Engineering, Indian Institute of Technology, Kharagpur, India
sulagna.student12@gmail.com, {sudeshna,pabitra}@cse.iitkgp.ernet.in
[2] Department of Civil Engineering, Indian Institute of Technology, Bombay, India
subimal@civil.iitb.ac.in

Abstract. Prediction of heavy rainfall is an extremely important problem in the field of meteorology as it has a great impact on the life and economy of people. Every year many people in different parts of the world suffer from the severe consequences of heavy rainfall like flood, spread of diseases, etc. We have proposed a model based on deep neural network to predict extreme rainfall from the previous climatic parameters. Our model comprising of a stacked auto-encoder has been tested for Mumbai and Kolkata, India, and found to be capable of predicting heavy rainfall events over both these regions. The model is able to predict extreme rainfall events 6 to 48 h before their occurrence. However it also predicts several false positives. We compare our results with other methods and find our method doing much better than the other methods used in literature. Predicting heavy rainfall 1 to 2 days earlier is a difficult task and such an early prediction can help in avoiding a lot of damages. This is where we find that our model can give a promising solution. Compared to the conventional methods used, our method reduces the number of false alarms; on further analysis of our results we find that in many cases false alarm has been raised when there has been rainfall in the surrounding regions. Thus our model generates warning for heavy rain in surrounding regions as well.

Keywords: Machine learning · Deep learning · Stacked auto-encoder

1 Introduction

Early prediction of heavy rainfall has always been a challenge in the field of weather forecasting. Early rainfall alert helps in relocating the population which could be affected, operating the flood control systems effectively, preparing the disaster mitigation team, etc. which minimizes the social and economic losses. This problem has become even more challenging with changing climatic patterns. According to [7] extreme rainfall events are expected to increase in changing climate. Therefore a proper scientific understanding of the rainfall extremes has become very important for correct prediction. Every year some metropolitan

© Springer International Publishing Switzerland 2016
P. Perner (Ed.): ICDM 2016, LNAI 9728, pp. 154–167, 2016.
DOI: 10.1007/978-3-319-41561-1_12

cities in India specially Mumbai and Kolkata experience very heavy rainfall during monsoon which brings life to a standstill in these places. Both these regions are urbanised and have high population of people living here, this makes it extremely difficult to take preparatory measures like relocation, rainfall alert broadcasting, etc. for high rainfall in a short notice like in 6 h or even less.

Currently, weather prediction is mainly based on numerical weather prediction (NWP) models. This in turn requires more detailed study of the physical processes responsible for heavy rainfall and simulate them, which is a computationally heavy process. Instead of doing that, we intend to use a data-centric approach and apply machine learning and data mining techniques to understand and predict rainfall.

2   Literature Survey

NWP models make use of a number of differential equations based on the laws of physics, fluid motion, atmospheric science, etc. The present weather conditions are fed to the models to get the prediction. Though these models perform well in predicting other weather conditions, they have not been efficient in predicting heavy rainfall events well in advance [14,17]. Statistical and probabilistic models have also been used to overcome the shortcomings of the NWP models. Numerous works on precipitation prediction in India are available [22,23], most of which have tried to relate extreme rainfall with anomalous weather behaviour. Though these models could predict rainfall in general, they again failed to predict extreme rainfall events exclusively, in advance.

Later on [21] used a clustering technique to identify the atmospheric parameters and the regions undergoing significant changes during extreme events. These parameters and regions act as fingerprints of extreme events which can be used for further classification into extremes or non-extremes. Nayak and Ghosh [19] modified the above method by using anomaly frequency method (AFM) of feature extraction and support vector machines (SVM) for classification. Munir's method could reduce the huge number of false alarms generated by the fingerprinting technique, but still the false alarms remained high. Several other machine learning techniques are available in literature for weather prediction, which make use of artificial neural network, support vector machines and bayesian networks. Though deep learning has become extremely popular in some fields like image processing, natural language processing, speech, etc., it has not yet been used much in the field of weather forecasting. There are only a very few work on application of deep learning in meteorology. Liu et al. [18] developed a deep neural network model to predict temperature, dew point, mean sea level pressure and wind speed in the next few hours. A deep hybrid model using deep neural network and probabilistic graphical model was proposed for forecasting weather parameters like temperature, wind, etc [9]. Recently a recurrent neural network model namely convolutional LSTM has been proposed [25] to address the precipitation nowcasting problem using radar echo data set. However none of these methods have dealt with extreme rainfall prediction problem in particular, which is an anomalous weather event, and has a direct impact on the lives of people. Here we have made an attempt to predict extreme rainfall events much ahead of time, compared to the state of art methods. We have used a stacked auto-encoder model for feature learning and reduction. The reduced features have been used for classification.

The AFM-SVM Method: Munir et al. used the anomaly frequency method (AFM) for extracting features from the entire set of weather parameters. In this method a positive and negative anomaly threshold has been defined as follows: $\delta^+ = \bar{X} + 1.25\bar{SD}$ and $\delta^- = \bar{X} - 1.25\bar{SD}$, where $\bar{X}$ denotes the climatological mean of a weather variable and $\bar{SD}$ denotes the climatological standard deviation of the variable in a particular grid at a specific time instant. Weather parameters whose values exceed the positive threshold are said to have positive anomalous behaviour and parameters whose values are below the negative anomaly threshold are said to have negative anomalous behaviour. The features which consistently show anomalous behaviour during extreme rainfall events are extracted from the entire set of features, considering the spatial and temporal extent of the features and the remaining weather features are ignored. The extracted features are then used for classification. A two-phase SVM is trained to predict night and day extremes separately. This method performs much better than the fingerprinting technique [21] but still generates a large number of false alarms.

3 Objective

The available methods for heavy rainfall prediction are able to predict only 6 h prior to the event. We want to predict these extreme (heavy) rainfall events much earlier that is about 1 or 2 days before with greater precision. This will ensure that least damage is caused by heavy rainfall events. All the weather variables data that are needed for prediction are collected for the entire Indian sub-continent whose latitude ranges from 5 degrees to 40 degrees north and longitude ranges from 65 degrees to 100 degrees east. The region has been shown in Fig. 1 which is sub-divided into grids. In total there are 21 weather variables used, for over the entire region shown in figure, collected on a daily and six-hourly basis. Munir's method showed that heavy rainfall can occur due to some anomalous weather features prevalent in regions that are far away from the region of interest. We have thus used the weather features over the entire Indian subcontinent. This will help in capturing the non-homogeneity in land-sea interaction, weather system and topography, over entire India, which affects rainfall in all parts of the country. The weather data has been obtained from the National Centers for Environmental Prediction/National Center for Atmospheric Research (NCEP/NCAR) reanalysis data. The region is divided into 225 grids. There are a total of $21 * 225 = 4725$ variables available for each day (if daily data is taken). This is further increased if the 24 h and 48 h prior combined features are used, making it a total of 9450 features. Such huge set of features if used for training a machine learning model may lead to overfitting. This calls for feature

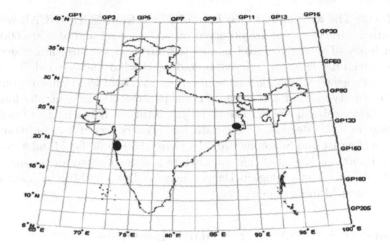

Fig. 1. Indian subcontinent and the regions of interest

reduction or feature extraction from the original set of features, before going into classification. Instead of relying solely on the anomalous features and extracting them greedily, we have used deep learning approach of learning and subsequently reducing the feature set. Weather condition is a combination of many complex interwind structures which need to be decoded to fully understand the weather system and the processes governing it. We believe that the different complex structures comprising the weather system can be learned and understood if they are unveiled in multiple layers just like images. We have thus used a deep neural network model for the purpose. Though deep learning has not yet been used much in weather prediction, we have tried to explore it's ability in this domain. Here we have used a deep neural network architecture, namely the stacked auto-encoder for feature reduction. The reduced features are then used for further classification.

4   Problem Statement

In this work we have addressed the problem of predicting the occurrence of heavy rainfall events in Mumbai and Kolkata in monsoon season (months of June, July, August and September), when there is maximum number of heavy rainfall events in these regions. The prediction is to be done based on the weather conditions over the region and it's surroundings in the past 48 to 6 h. Using historical data of some the important weather variables and rainfall over the mentioned regions, our model is trained. It is then tested with a new set of feature values of the present time and used to predict whether heavy rainfall will take place or not in the next 6–48 h.

Data Used: The weather data has been obtained from NCEP/NCAR website. The weather variables used are observed at surface level and at 850-, 600- and 400-hPa levels. The surface level parameters are temperature, mean sea level pressure, precipitable water, relative humidity, U-wind and V-wind. The 850-, 600-, 400-hPa level variables are air temperature, vertical wind velocity (omega), relative humidity, u-wind and v-wind. These parameters are generally found to influence heavy rainfall events. The reanalysis data has a spatial grid resolution of 2.5 degrees * 2.5 degrees. These data has been collected for the months of June, July, August and September for the years 1969–2008 for Mumbai and the years 1980–2000 for Kolkata. The rainfall data is also collected for the same time period from India Meteorological Department (IMD). The Mumbai data is the same as used in Munir's paper.

5   Stacked Autoencoder Feature Learning

Hierarchical feature learning techniques help in extracting non-linear characteristics from the input in multiple layers. Deep belief network (DBN) was introduced by Hinton and Salakhutdinov [12], which solved the vanishing gradient problem of very deep neural architecture. After that a number of variants of deep neural network were proposed by others. Bengio et al. [6] and Ranzato et al. [20] trained the DNN by layerwise unsupervised pre-training followed by finetuning over the entire network. This approach is also known as the greedy layerwise unsupervised pre-training. In this study, we want to learn the weather attributes which are mainly in the form of real numbered values and thus have chosen stacked auto-encoder architecture of deep learning for the purpose.

A simple auto-encoder is an unsupervised one layered neural network where the input $X = x_1, x_2, x_3, ..., x_n$ is a n dimensional feature vector. The output is given by

$$h_{W,b}(X) = f(W^T X) = f(\sum_{i=1}^{n} W_i x_i + b) \tag{1}$$

where $f : \mathbb{R} \mapsto \mathbb{R}$ is a non-linear transformation function and W and b are the weights and bias of the network respectively. The objective is to make $h_{W,b}(X) \approx X$, that is to learn the feature set and regenerate it. The hidden layer gets to learn a compressed representation of the input, such that the original input can be regenerated from it. The loss function of an autoencoder with a single hidden layer is given by,

$$J(W, b) = [\frac{1}{m} \sum_{i=1}^{m} \frac{1}{2} \|h_{W,b}(x_i) - x_i\|^2)] + \frac{\lambda}{2} \sum_{i}^{n} \sum_{j}^{nhid} (W_{j,i})^2 \tag{2}$$

where m is the number of training examples, nhid is the number of units in hidden layer, considering only one hidden layer. The second term in Eq. (2) is the regularization term and λ is the weight decay parameter. The autoencoder tries to minimize Eq. (2) by gradient descent. An autoencoder can also have hidden layer whose size is greater than the size of input layer. In that case a sparsity constraint is imposed on the hidden units. The autoencoder is still able to discover interesting patterns in the input set. A hidden unit is said to be active or firing if it's output is close to 1 and inactive if it's output is close to 0.

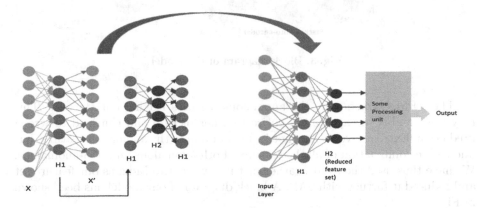

Fig. 2. Layer-wise pre-training in SAE followed by stacking of hidden layers for further processing.

A denoising autoencoder (DAE) is an improvement of the autoencoder, which is designed to learn more robust features and prevent the autoencoder from simply learning the identity. A stacked autoencoder is formed by stacking multiple such denoising autoencoders one on top of the other to form a deep neural network. Each layer is trained separately to adjust the parameters of that layer. After all the layers are pre-trained in this fashion, they are combined together. Finally finetuning is done over the entire network so that the output of the SAE becomes a good approximation of the input. An example of layer-wise pre-training followed by hidden layer stacking is shown in Fig. 2. The reduced feature set obtained after stacking the hidden layers can be used as input features for further processing.

6   Outline of Proposed Approach

6.1   Proposed Model

Our prediction model has two phases: Feature Learning and feature compression followed by classification. SAE has been used to get a compressed representation of the feature set. For classification we have used support vector machines (SVM) and also neural network. The configuration of the SAE used has been discussed in the later sections. The classifier parameters have to be adjusted to get the best performance. The inherent problem of biased data set has also been dealt with effectively to get the best performance. For this purpose a cost-sensitive SVM has been used.

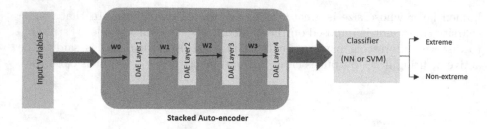

Fig. 3. Block diagram of the model

The AFM extracts only the anomalous features leaving out the rest. This may not be able to capture the entire weather picture and can lead to wrong prediction. Extreme events may not just occur due to the anomalous features but due to the combination of all the features, both anomalous and non-anomalous. We have thus used the total weather picture over entire India as our feature set and reduced it further with SAE. A block diagram of our model has been shown in Fig. 3.

6.2   Methodology for Extreme Rainfall Prediction

We have focused and tested our model for prediction mainly in two cities in India, namely Mumbai and Kolkata. Unlike data in other domains, weather data has some peculiar characteristics that should be kept in mind while dealing with them. For instance, there is a season-to-season and also year-to-year variation in the weather parameters. Sometimes there are significant changes in the weather patterns of consecutive months. Due to this reason we have focused our study on the monsoon months when there is maximum amount of heavy rainfall events.

7   Experimental Results

Our model has been used on the weather parameter data obtained for over the entire Indian subcontinent, to predict rainfall in Mumbai and Kolkata. Prediction is performed 6, 24, 36 and 48 h before the extreme event. The entire dataset has been divided into training and test periods. For Mumbai the training set consists of the rainfall events from 1969 to 1999 and the test set consists of rainfall events from 2000 to 2008 and for Kolkata the training period ranges from 1980–1995 and the test period ranges from 1996–2000. Total rainfall amount (throughout the day) exceeding 75 mm has been considered to be an extreme rainfall event for Mumbai and total rainfall amount exceeding 60 mm has been considered to be an extreme event for Kolkata. The thresholds considered are slightly on the higher side due to the fact that both Mumbai and Kolkata receive high intensity rainfall but in short durations.

7.1 Stacked Auto-encoder Configuration

There are 21 weather parameters (as mentioned above) used in this study. In literature these parameters have been found to be important factors influencing rainfall. We have the values of all the parameters for each of the grid points at different time steps. This results in a huge feature set, which is difficult to process and use. We have tried to reduce the feature space size by using a stacked auto-encoder, which reduces the feature space in such a way that the original feature set can be rebuilt from the reduced set. Thereby it does not eliminate any feature but captures all in a compact form.

Here we have built a 4 layer SAE model. The entire set of features is fed as input to the model. The first hidden layer has 2500 units, the second layer has 1000 units, the third layer has 500 units and the innermost layer has 200 units. The number of units in each layer has been set by trial and error method. For example we found that 2500 units are good enough for learning in the first layer compared to 3000 units. In order to reduce the size of feature space further, we have gradually learned the features in multiple layers and have obtained a compact representation of the feature set. Batch learning has been followed during layer-wise pre-training. The sparsity parameter value has been adjusted to tune the model. Dropout is used to prevent over-fitting. All these parameters are adjusted by trials such that the outputs of the network generated are good approximations of the input. The innermost hidden layer activation (output) which becomes the reduced feature set can now be used to carry out further classification.

7.2 Evaluation Methodology

The AFM-SVM method predicts rainfall using 6 to 48 h prior weather data. Thus the results can be obtained only 6 h prior to the event. This time may not be enough for evacuation, relocation or preparation of disaster mitigation team. Here we intend to predict atleast 24 h before or 48 h before. We have also done prediction in 6–48 h lead time just like the AFM-SVM method. The weather parameters are fed to the stacked autoencoder and the reduced feature space is obtained for further classification into extreme and non-extreme events. However there are a very few extreme rainfall cases compared to normal or no rainfall cases, that is the data set is biased. This naturally gives rise to a large number of false positives and false negatives during classification. This problem has been tackled using SMOTE (Nitesh et al. 2002) [5] for oversampling and Wilsons [26,27] edited Nearest Neighbor rule for under-sampling. For classification, we have used cost sensitive SVM [1,16] and neural network. In our method we have used only one classifier for prediction of both night and day extremes unlike Munir's method.

7.3 Results and Comparison

Extreme rainfall prediction has been carried out with AFM-SVM method as well as our model. The results are as follows.

7.3.1   6 Hours Prior Prediction

Table 1 shows the results obtained with Munir's method and SAE based methods for Mumbai. Here the raw features are taken 48, 36, 24 and 6 h prior to the occurrence of heavy rainfall. The results using Munir's method shown in the table is as available in the Munir's paper, with the false alarms (for night-SVM and day-SVM) added up. The AFM threshold and the SVM classifier used for each of the experiments have been mentioned along with the actual extremes, the model is able to detect. The number of false alarms have also been shown. There were 15 extreme events (both day and night combined) in Mumbai during the period considered. The best result obtained from AFM based method predicts all the 15 extremes and generates 149 false alarm cases (combining the day and night cases together). Though in the AFM-SVM method day and night extremes have been predicted separately, here we present the aggregate results for better comparison. For instance, the AFM-SVM technique with night frequency threshold 29 and SVM with rbf kernel, predicts 12 (out of 12) night extremes correctly.

Table 1. Prediction 6 h prior for Mumbai (Actual number of extremes for Mumbai = 15)

Model	Extreme detected correctly	Extreme detected on previous day	Total true positives	Actual number of false alarms
SAE+SVM(quadratic kernel)	13	1	14	200
SAE+SVM(3rd order polynomial kernel)	13	2	15	128
SAE+Neural Network Classifier	12	2	14	224
AFM(night freq.=28, day freq.=14)+SVM (RBF, quadratic)	14	-	14	190
AFM(night freq.=29, day freq.=14)+SVM (quadratic)	15	-	15	149
AFM(night freq.=30, day freq.=14)+SVM(RBF, quadratic)	13	-	14	292
PCA + SVM(Quadratic)	7	1	8	260
Fisher LDA	3	3	6	205

With day frequency threshold 14 and SVM with quadratic kernel, it is able to detect 4(out of 4) day extremes correctly. The same raw data has been fed to our model. Firstly the data has been reduced with a SAE and then classifier is applied. We find that most of the extremes were correctly predicted for the day on which the extreme event occurred. However in a few cases our model has not been able to predict the extreme on the exact day when it occurred but one day before. This is still a good estimation because if high rainfall is predicted one day before the actual day when it occurs, there will still be enough preparedness to ensure least damage. Warning can be issued much ahead of time, and situation can be skillfully handled even if disastrous condition occurs the next day. The best result with SAE is obtained using SVM of polynomial kernel. It predicts 13 extremes correctly and 2 extremes on the previous day. There are 128 false alarms which is still less than that obtained with the AFM based method. However, we have achieved only a slight improvement with our method over the conventional one, with the data of the past 48-6 h. The results obtained with some other conventional methods like principal component analysis and Fisher linear discriminant analysis are also shown.

7.3.2   24 Hours Prior Prediction

We have then fed only the weather parameters of the previous day, 36 and 24 h prior, to both the SAE based model and AFM based model and tried to predict high rainfall on the following day. Here we have reproduced Munir's method with 36 and 24 h prior features as input. The results have been compared in the Table 2 for rainfall in Mumbai. We find that in this case our model works much better than AFM model. In our model the neural network classifier could predict all the 15 extreme events correctly. It generated 430 false positive cases. However further analysis of the false positives revealed that in some of those cases it rained heavily elsewhere in the surroundings of Mumbai. Therefore these alarms could also be treated as meaningful. Only in 180 instances our model generated false alarms when it did not rain heavily in Mumbai or the surroundings. SVM classifier with quadratic kernel has also been used and the results have been shown. SVM with quadratic kernel detects 10 extreme events correctly and the 5 remaining actual extremes were also predicted but one day before they actually occurred. There were 191 false alarms in total out of which in 37 instances there was heavy rain in the surrounding regions and the rest 154 were actual false alarms. It is found that the our model can predict rainfall in the surrounding regions as well. Thus it can be said that the features obtained from the SAE give a good prediction of rainfall not just for the region concerned but also its

Table 2. Some results for Mumbai. (Actual number of extremes for Mumbai is 15)

Raw Features	Model	Extremes predicted correctly	Extremes predicted on previous day	Total extremes predicted	Total false alarms	False alarms but rain in surroundings	Actual number of false alarms
	24 hours prior prediction						
36, 24 hours prior features	*SAE+SVM (quadratic kernel)*	*10*	*5*	*15*	*191*	*37*	*154*
	SAE+ neural network classifier	**15**	**-**	**15**	**430**	**250**	**180**
	Munir's model (freq threshold=29)	7	3	10	233	40	193
	Munir's model(freq. threshold=28)	6	3	9	235	33	201
	PCA+SVM(quadratic kernel)	7	-	7	225	33	192
	Fisher LDA	3	-	3	121	12	109
	36 hours prior prediction						
48, 36 hours prior features	*SAE+SVM (quadratic kernel)*	*12*	*1*	*13*	*223*	*30*	*193*
	SAE+ neural network classifier	**12**	**1**	**13**	**398**	**180**	**218**
	Munir's method (freq. threshold=29)	6	2	8	256	35	221
	Munir's method (freq. threshold=28)	5	3	8	244	32	212
	PCA+SVM(quadratic kernel)	7	1	8	256	32	224
	Fisher LDA	3	1	4	140	12	128
	48 hours prior prediction						
48 hours prior features	*SAE+SVM (quadratic kernel)*	*11*	*1*	*12*	*152*	*39*	*113*
	SAE+neural network classifier	**12**	**1**	**13**	**413**	**158**	**255**
	Munir's method (freq. threshold=29)	5	3	8	270	38	232
	Munir's method (freq. threshold=28)	7	1	8	254	40	214
	PCA+SVM(quadratic kernel)	6	1	7	268	45	223
	Fisher LDA	2	0	2	122	7	115

surroundings. Whereas the AFM based method in the best case detects 10 total extremes and generates 193 false alarms, leaving out those cases where it rained in the surroundings.

7.3.3   36 and 48 Hours Prior Prediction

Table 2 also shows the prediction of rainfall for Mumbai and it's surroundings with weather features taken 48 and 36 h before. In the best case, SAE followed by SVM classifier of quadratic kernel is able to detect 13 out of 15 extremes correctly, out of which only one is detected on the previous day and generates 193 false alarm cases. Munir's method, reproduced with 48 and 36 h prior features, detects 8 total extremes in the best case and generates 212 false positive cases. The weather features 48 h before rainfall have also been used as raw features that are fed to SAE. SAE followed by SVM classifier (quadratic kernel) predicts 12 total extreme cases and generates 152 total false positives out of which in 39 cases there was rainfall elsewhere. SAE followed by neural network detects 13 extremes, 48 h before and generates 255 false alarms. Munir's method is able to detect only 8 out of 15 extremes correctly and generates 214 false alarms in the best case. PCA and LDA performs poorly in these tasks.

Table 3. Results for Kolkata (Actual number of extremes for Kolkata = 9)

Raw Features	Model	Extremes predicted correctly	Extremes predicted on previous day	Total extremes predicted	Total false alarms	False alarms but rain in surroundings	Actual number of false alarms
		6 hours prior prediction					
48, 36, 24 and 6 hours prior features	SAE+SVM (quadratic kernel)	5	2	7	460	134	326
	SAE+neural network	4	2	6	263	56	207
	Munir's method	5	2	7	600	151	449
	PCA+SVM(quadratic kernel)	3	0	3	231	51	180
	Fisher LDA	1	3	4	210	21	189
		24 hours prior prediction					
36, 24 hours prior features	*SAE+SVM (quadratic kernel)*	*8*	-	*8*	*399*	*231*	*168*
	SAE+ neural network classifier	4	1	5	220	30	190
	Munir's model	5	-	5	497	112	385
	PCA+SVM(quadratic kernel)	2	3	5	228	12	216
	Fisher LDA	2	3	5	228	12	216
		36 hours prior prediction					
48, 36 hours prior features	*SAE+SVM (quadratic kernel)*	*6*	*1*	*7*	*411*	*200*	*211*
	SAE+ neural network classifier	5	1	6	300	25	275
	Munir's method	4	1	5	505	124	381
	PCA+SVM(quadratic kernel)	2	3	5	234	101	133
	Fisher LDA	1	1	2	140	11	129
		48 hours prior prediction					
48 hours prior features	*SAE+SVM (quadratic kernel)*	*5*	*2*	*7*	*420*	*190*	*230*
	SAE+neural network classifier	4	3	7	310	43	267
	Munir's method	4	1	5	512	101	411
	PCA+SVM(quadratic kernel)	5	0	5	227	23	204
	Fisher LDA	1	1	2	149	19	130

7.3.4   Results for Kolkata

We have duplicated Munir's method and have applied it for Kolkata. PCA and LDA techniques have also been used for comparison. We have also used the SAE based method for Kolkata and the results have been compared in Table 3, where the labels have their usual meaning. The raw features are taken in time periods 48, 36, 24 and 6 h before extreme events; 36 and 24 h before extreme event which gives 24 h prior prediction; 48 and 36 h before extreme event, which gives 36 h prior prediction; features 48 h before rainfall which gives 48 h prior prediction. Here also we find that our model is able to predict more extreme events in the test period and generates less false positive cases compared to Munir's method. There are total of 9 extreme events for Kolkata. Using our method, 7 out of 9 could be detected 6 h before, 8 could be detected 24 h before, 7 could be detected 36 h before and again 7 could be detected 48 h before. The raw features and the false alarms generated in each of the cases is mentioned in the table. However the results for Kolkata is not very satisfactory which is because of the fact that the number of training examples are very few for Kolkata. With more training instances the results could improve further. The sensitivity and specificity values for both the models for the performed experiments are shown in Table 4.

Table 4. Sensitivity and Specificity of the models

Prediction Task	Mumbai				Kolkata			
	Munir's method		Our method		Munir's method		Our method	
	Sensitivity	Specificity	Sensitivity	Specificity	Sensitivity	Specificity	Sensitivity	Specificity
6 hrs prior	1	0.845	1	0.867	0.778	0.253	0.778	0.458
24 hrs prior	0.67	0.799	1	0.84	0.556	0.358	0.89	0.72
36 hrs prior	0.53	0.78	0.867	0.799	0.556	0.365	0.778	0.648
48 hrs prior	0.53	0.777	0.8	0.88	0.556	0.314	0.778	0.616

8   Conclusion and Future Work

In this work we have explored a machine learning technique namely deep learning with SAE to learn and represent weather features and use them to predict extreme rainfall events. We found that though our method gives results that are quite similar to the baseline case when prediction is done in 6 h before the event, it gives significantly better performance when prediction is done 24 h and 48 h before the events. The AFM based method does not perform well when it comes to such early prediction. Our experiments show that these intelligently learnt features can improve the performance of the classical approaches. One of the reason why this model works better is that, here we include all the features and try to understand underlying patterns and dependencies unlike other approaches which rely on feature extraction or selective feature reduction. Our experiment shows that Deep Learning can be quite promising in the field of weather forecasting, just like in field of image recognition, speech processing, NLP, etc.

However there is still much scope of improvement. Firstly more and higher resolution data should be used for learning, since more the data available better becomes the learning. Thus higher resolution data with large number of training instances should be used in deep learning. As mentioned earlier weather data has many specific characteristics which depend on time and spatial location. This should be taken into consideration to understand and learn the system properly. This calls for more sophisticated learning methods. We would like to explore the data considering it's temporal and spatial behaviour. Here we have solved only a classification problem where we are only able to predict whether there will be heavy rainfall or not. In future we would also like to predict the amount of rainfall as well with the improved methods.

Acknowledgements. This research was supported and funded by Indian Institute of Technology, Kharagpur, India and MHRD, India under the project named "Feature Extraction and Data Mining from Climate Data (FAD)". We would like to thank IIT Kharagpur, MHRD and also the IMD(India Meteorological Society) for their helpful suggestions and support. Without their help this work would not have been completed.

References

1. Akbani, R., Kwek, S.S., Japkowicz, N.: Applying support vector machines to imbalanced datasets. In: Boulicaut, J.-F., Esposito, F., Giannotti, F., Pedreschi, D. (eds.) ECML 2004. LNCS (LNAI), vol. 3201, pp. 39–50. Springer, Heidelberg (2004)
2. Bengio, Y.: Learning deep architectures for AI. Found. Trends Mach. Learn. **2**(1), 1–127 (2009)
3. Bengio, Y.: Deep learning of representations for unsupervised and transfer learning. Unsupervised Transf. Learn. Chall. Mach. Learn. **7**, 19 (2012)
4. Boser, B.E., Guyon, I.M., Vapnik, V.N.: A training algorithm for optimal margin classifiers. In: Proceedings of the Fifth Annual Workshop on Computational Learning Theory, pp. 144–152. ACM (1992)
5. Chawla, N.V., Bowyer, K.W., Hall, L.O., Kegelmeyer, W.P.: SMOTE: synthetic minority over-sampling technique. J. Artif. Intell. Res. **16**, 321–357 (2002)
6. Erhan, D., Bengio, Y., Courville, A., Manzagol, P.A., Vincent, P., Bengio, S.: Why does unsupervised pre-training help deep learning? J. Mach. Learn. Res. **11**, 625–660 (2010)
7. Goswami, B.N., Venugopal, V., Sengupta, D., Madhusoodanan, M., Xavier, P.K.: Increasing trend of extreme rain events over india in a warming environment. Science **314**(5804), 1442–1445 (2006)
8. Graves, A.: Generating sequences with recurrent neural networks (2013). arXiv preprint arXiv:1308.0850
9. Grover, A., Kapoor, A., Horvitz, E.: A deep hybrid model for weather forecasting. In: Proceedings of the 21th ACM SIGKDD International Conference on Knowledge Discovery and Data Mining, pp. 379–386. ACM (2015)
10. Haykin, S., Network, N.: A comprehensive foundation. Neural Netw. **2**(2004) (2004)
11. Hinton, G.E., Osindero, S., Teh, Y.W.: A fast learning algorithm for deep belief nets. Neural Comput. **18**(7), 1527–1554 (2006)

12. Hinton, G.E., Salakhutdinov, R.R.: Reducing the dimensionality of data with neural networks. Science **313**(5786), 504–507 (2006)
13. Hochreiter, S., Schmidhuber, J.: Long short-term memory. Neural Comput. **9**(8), 1735–1780 (1997)
14. Hong, S.Y., Lee, J.W.: Assessment of the WRF model in reproducing a flash-flood heavy rainfall event over Korea. Atmos. Res. **93**(4), 818–831 (2009)
15. Hong, Y.: Precipitation estimation from remotely sensed information using artificial neural network-cloud classification system (2003)
16. Japkowicz, N., Stephen, S.: The class imbalance problem: a systematic study. Intell. Data Anal. **6**(5), 429–449 (2002)
17. Khaladkar, R., Narkhedkar, S., Mahajan, P.: Performance of NCMRWF Models in Predicting High Rainfall Spells During SW Monsoon Season: A Study for Some Cases in July 2004. Indian Institute of Tropical Meteorology (2007)
18. Liu, J.N., Hu, Y., You, J.J., Chan, P.W.: Deep neural network based feature representation for weather forecasting. In: Proceedings on the International Conference on Artificial Intelligence (ICAI), p. 1. The Steering Committee of The World Congress in Computer Science, Computer Engineering and Applied Computing (WorldComp) (2014)
19. Nayak, M.A., Ghosh, S.: Prediction of extreme rainfall event using weather pattern recognition and support vector machine classifier. Theoret. Appl. Climatol. **114**(3–4), 583–603 (2013)
20. Ranzato, M.A., Huang, F.J., Boureau, Y.L., LeCun, Y.: Unsupervised learning of invariant feature hierarchies with applications to object recognition. In: IEEE Conference on Computer Vision and Pattern Recognition, 2007, CVPR 2007, pp. 1–8. IEEE (2007)
21. Root, B., Knight, P., Young, G., Greybush, S., Grumm, R., Holmes, R., Ross, J.: A fingerprinting technique for major weather events. J. Appl. Meteorol. Climatol. **46**(7), 1053–1066 (2007)
22. Bhowmik, S.R., Durai, V.: Application of multimodel ensemble techniques for real time district level rainfall forecasts in short range time scale over indian region. Meteorol. Atmos. Phys. **106**(1), 19–35 (2010)
23. Sahai, A., Soman, M., Satyan, V.: All india summer monsoon rainfall prediction using an artificial neural network. Clim. Dyn. **16**(4), 291–302 (2000)
24. Schölkopf, B., Smola, A.J.: Learning with Kernels: Support Vector Machines, Regularization, Optimization, and Beyond. MIT Press, Cambridge (2002)
25. Shi, X., Chen, Z., Wang, H., Yeung, D.Y., Wong, W.K., Woo, W.C.: Convolutional LSTM network: A machine learning approach for precipitation nowcasting (2015). arXivpreprint arXiv:1506.04214
26. Wilson, D.R., Martinez, T.R.: Instance pruning techniques. In: ICML, vol. 97, pp. 403–411 (1997)
27. Wilson, D.L.: Asymptotic properties of nearest neighbor rules using edited data. IEEE Trans. Syst. Man Cybern. **3**, 408–421 (1972)

Identifying and Characterizing Truck Stops from GPS Data

Russel Aziz[1]([☒]), Manav Kedia[1], Soham Dan[1], Sayantan Basu[2],
Sudeshna Sarkar[1], Sudeshna Mitra[1], and Pabitra Mitra[1]

[1] Indian Institute of Technology Kharagpur, Kharagpur, West Bengal, India
{russel.aziz,sudeshna}@cse.iitkgp.ernet.in
[2] Institute of Engineering and Management, Kolkata, West Bengal, India

Abstract. Information about truck stops in highways is essential for
trip planning, monitoring and other applications. GPS data of truck
movement can be very useful to extract information that helps us understand our highway network better. In this paper, we present a method
to identify truck stops on highways from GPS data, and subsequently
characterize the truck stops into clusters that reflects their functionality. In the procedure, we extract the truck stoppage locations from the
GPS data and cluster the stoppage points of multiple trips to obtain
truck stops. We construct arrival time distribution and duration distribution to identify the functional nature of the stops. Subsequently, we
cluster the truck stops using the above two distributions as attributes.
The resultant clusters are found to be representative of different types of
truck stops. The characterized truck stoppages can be useful for dynamic
trip planning, behavior modeling of drivers and traffic incident detection.

Keywords: Data mining in logistics · Data analytics · Highway · Clustering · GPS data · Tracking

1 Introduction

Modelling of vehicle stoppages is very important for a comprehensive understanding of trips, the highway network and driver behaviour. Trip planning
systems worldwide generally focus on movement data of the vehicles on-road
to extract characteristic features relevant to trips, the road networks and the
vehicles. But vehicle stoppages are also an important part of trip-scheduling.
The stoppages enroute dictate driver comfort, driver efficiency, vehicle fuel status which in turn influences the driving time, the schedule of stoppages and
stoppage duration.

For the trucking industry planning of truck stoppages is as much critical as
planning of truck movement to optimise fuel costs, driver efficiency and ensure
timely delivery. Transportation Management Systems (TMS) are useful tools for
application of Intelligent Transportation Systems (ITS) in Freight management
systems and logistics industry. They cover route optimization and trip planning among other things. For optimal trip planning information regarding truck

P. Perner (Ed.): ICDM 2016, LNAI 9728, pp. 168–182, 2016.
DOI: 10.1007/978-3-319-41561-1_13

stops are essential [1]. Further to identification of truck stops, characterization of truck stops based on their functionality is very useful. Characterization of stops is essential to determine whether it is an enforced stop, like a checkpoint or toll plaza. It is helpful to know which type of stop is being approached, to make the decision to stop based on the need of the driver and the functionality of the stop. The location, nature and working hours of the truck stoppages can help us in developing an practical trip planning system and provide insights into how drivers behave on roads. In addition this is useful for tracking and monitoring, detecting suspicious behavior, non-conformance, etc. There is a dearth of information about such stopping places on Indian National Highway network. Generally, the truck drivers rely on experience and word-of-mouth to plan their stoppage, and often the plan may not be efficient, timely or cost-effective. Also, in the event of extraordinary circumstances, there is no alternative plan to fall back upon. Thus, there is a dire need of more information regarding location of truck stops on Indian highways along with their functionality. Such information has a huge potential to be used for a dynamic trip planning decision support system to optimize fuel costs, driver efficiency and comfort [2,3]. The challenges are to identify the stoppages and then characterize them reliably into different types. Based on our study, we have characterized truck stops into four fundamental types: meal stops, refuelling stops, rest stops and toll stops/checkpoints. In addition to these, there are mixed use stops, which fulfill the functionality of a combination of the four former fundamental stops.

The lack of information about truck stoppages on the National Highway network can be taken care of by identifying the truck stoppages from remotely sensed GPS data, and then characterizing the stoppages to know of their functionality. Global Positioning System (GPS) device is reliable and useful for obtaining remotely sensed spatial information. Locations where multiple vehicles stoppages are co-located are prime candidates for truck stoppages. The data available with us contains the GPS locations of trucks at a frequency of 10 min. There is an issue of missing out on stops of duration less than 10 min. Also, the durations of stoppages may be off by a few minutes due to this temporal resolution. But by mining stoppage instances from thousands of trips on the same stretch of highway network this can be overcome.

2 Related Work

There is a lack of available literature on identifying and analysing or characterizing truck stops from GPS data. Greaves and Figliozzi [4] describe an approach to establish when an individual trip has ended for commercial vehicles for Melbourne City, Australia. The work proposes that if the geographic distance between two locations of a vehicle at consecutive communications are smaller than a threshold over a prescribed period of time, they will be considered as a stop. McCormack et al. [5] talks about identifying the pick-up and drop-off locations of individual trips, segregating them from the other stops reported in the GPS record of trucks. The work zeroed in on a dwell-time threshold to identify the pickup and drop-off locations. Kuppam et al. [6] presents an approach to identify individual truck stop instances enroute and at trip ends from truck GPS data using speed criteria. Yang et al. [7] talks about delivery stop identification from second-by-second GPS data using support vector machines. The work trains the SVM from ground truth delivery stops and proposes the following features for the proposed SVM model stop duration, distance to city center and distance to closest bottleneck. Arifin [8] talks about determination of trip ends through clustering individual trip ends, while modelling route choices in the city of Jakarta. The work acknowledges that vehicles that have the same destination do not stop at exactly the same location, they stop at different locations nearby. Thus there is a need for clustering those stoppages occur over a specific range to determine trip destination. Sharman and Roorda [9] present a clustering approach to analyse freight GPS data in order to identify trip destinations. The clustering technique they employ is a two-step process, where the first step involves clustering the trip ends, and next the results are refined using property boundaries. The study was done across the Canadian cities of Toronto and Hamilton. Prasannakumar et al. [10] uses similar spatial clustering techniques to determine locations in Thiruvananthapuram city that have problems related to traffic accidents and congestion. Anderson [11] presents an approach using kernel density estimation and subsequent application of K-means clustering to obtain road accident hotspots. K-means clustering is used to classify each hotspot into relatively homogenous types based on their environmental characteristics. Aziz et al. [12] describe an approach where they cluster locations of drastic speed changes, using a modified Density Based Spatial Clustering of Applications with Noise (DBSCAN) clustering technique, across multiple trips to obtain highway segment ends.

The related work has focused on delivery point delays and stops, drop-off locations and trip ends but not enroute stops and their planning. In the work presented here, we have set out to identify the truck stops by clustering stoppage instances mined from GPS transmissions made from the trucks. Subsequently, we wish to characterize the truck stops using as attributes the arrival time distribution and the duration distribution of each truck stop. We validate our identification technique by visualising some identified truck stops on a map and tallying them with existing structures on map likely to be truck stops. We evaluate the clusters obtained from the characterization procedure by their ability to characterize the truck stops into distinct clusters, and how their average arrival time distributions and duration distributions reflect the truck stops of differing functionalities.

3   Objective

The objective of this paper is to develop a method to identify truck stops by mining GPS data collected from trucks making trips across the highway network, and to propose a method for characterizing the truck stops according to their functionality, based on their arrival time distribution and duration distribution.

We wish to apply these methods to GPS data of truck movement in Indian National Highways to identify and characterize the truck stops in this networks.

4 Details of Data Used

The data used in the paper was provided by eTrans Solutions Pvt Ltd. The data is of two types, the trip data, and the GPS data. The trip data corresponds to the record of each trip made between an origin-destination pair. Each record contains the origin and destination of the trip, the start date-time, end date-time and the coded identification of the vehicle making the trip. The GPS data contains the GPS transmissions made by all the vehicles making trips on the National Highway network, made at approximately 10 min intervals. The GPS signal quality provides for an accuracy of better than 2.5 m Circular Error Probable (CEP). The relevant information content of the GPS data consists of the latitude, longitude, timestamp and coded identification of the vehicle and the GPS device transmitting the information. The work was carried out on a total of 60,324 trips made in the year of 2013 by trucks on the National Highway network of India. The number of GPS records that were processed amount to nearly 30 million.

5 System Design

We develop a system that clusters the stoppage instances from multiple trips made on the highway network to get truck stop locations. After obtaining the truck stop locations, we analyze the stoppage instances for each truck stops to construct the arrival time distribution and duration distribution of each stop. We select these distributions as attributes to characterize the truck stops to obtain clusters based on their functionality. The process is described by the flowchart in Fig. 1. The steps in the stop identification process are outlined below:

1. We process our input GPS data to prepare tripwise Location Time Velocity (LTV) tables, as explained in detail in Sect. 5.1.1.
2. The stoppage points from each LTV table are then extracted and consolidated into tripwise Location Time Duration (LTD) tables, the exact procedure for which is described in Sect. 5.1.2.
3. The consolidated stoppage points are subsequently clustered using a modified DBSCAN technique to get the identified truck stops. The method for the same is elaborated in Sect. 5.1.3.

For the truck stop characterization method, the process is as follows:

1. We mine the data for each truck stop obtained and construct the arrival time distribution and duration distribution from the details of the respective stoppage points. This is elaborated in Sect. 5.2.1.

2. To accomplish clustering based on histogram similarity, we normalize each bin of arrival time distribution and duration distribution and select them as a dimension in the feature space for k-means clustering. This procedure is explained in detail in Sect. 5.2.2.

The final output contains clusters of truck stops based on similarity in their arrival times and stoppage durations. We evaluate the obtained clusters in Sect. 6.

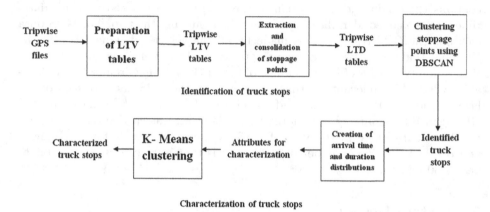

Fig. 1. The flowchart of the system designed for identification of truck stops and characterization of truck stops

5.1   Identification of Truck Stops

The identification of truck stops procedure attempts to extract all the stoppages made by the trucks from the GPS data, and then cluster these stoppages to obtain locations at which sizable number of trucks have stopped. These locations are considered to be truck stops.

5.1.1   Preparation of Location Time Velocity (LTV) Tables

Initially, LTV tables are constructed from the GPS records from each trip. Each row in the LTV table corresponds to two subsequent readings in the tripwise GPS file. This is because each LTV record corresponds to the velocity between two subsequent GPS transmissions, as mentioned earlier. Thus, if the GPS file has n readings, the corresponding LTV table shall have $n - 1$ rows. The location and time attribute of each LTV record points to the latitude and longitude, and the time stamp of when and where it entered the region for which the velocity has been observed. The velocity can be calculated using the haversine formula [12]. A sample LTV table is given in Table 1.

5.1.2 Extracting and Consolidating Stoppage Points

The records in the LTV table contains moving instances as well as stationary instances. Due to a temporal resolution of 10 min, there can be instances when a particular LTV record contains both stationary and moving state of the vehicle. We specify a threshold of 5 Km/hr, such that vehicles travelling less than 830 m in 10 min or between successive GPS readings are considered to be stationary and allowed to remain in the LTV table. For vehicles in stationary state for a long time, subsequent stationary LTV readings, point to the same stationary state. There is a need to consolidate these individual stationary points corresponding to the same stoppage instance. For this, LTD (Location Time Duration) table is constructed by:

1. We take the median location of the stationary LTV records part of the stoppage instance.
2. We take the time attribute of the first record which is a part of the stoppage instance as the arrival time of the stoppage instance.
3. We add the time duration that each stationary record denote to obtain the stoppage duration. As the temporal resolution is 10 min, the lowest stoppage duration is 10 min.

Table 1. A sample LTV table

Latitude	Longitude	Time	Velocity
22.83372	86.24261	16:47:00	23.698
22.82471	86.27991	16:57:00	39.352
22.78723	86.32937	17:07:00	36.232
...	...	...	...

Table 2. A sample LTD table

Latitude	Longitude	Time (arrival time)	Duration (in minutes)
22.83372	86.24261	16:47:00	20.0
21.10931	84.87631	08:40:00	90.0
20.78723	85.09821	21:47:00	260.0
...	...	...	...

A sample LTD table is given in Table 2, where the location attribute gives the location of the stoppage, the time attribute corresponds to the arrival time, and the duration attribute denotes the duration of the stoppage in minutes. This table contains a single record for each stoppage instance ecountered during the trip. A total of 935118 stoppage instances were mined from the GPS records of all the trips.

5.1.3   Clustering Stoppage Points Using DBSCAN

The stoppage instances dispersed across the national highway network of India, represent the locations where the trucks making the trips stopped individually. These stoppages can occur due to location of truck stops, traffic congestion, breakdown or malfunctions, individual needs. The stoppage instances which correspond to truck stops are expected to be densely situated in or around the location of the truck stop, while stoppage instances related to other factors are expected to be situated randomly along the highways not clustered around any particular location. Thus, we need to identify locations around which the stoppage instances are densely located as truck stops while rejecting the sparsely located stoppage instances as noise. This is illustrated in Fig. 2. The DBSCAN algorithm [13] suits our need in this case.

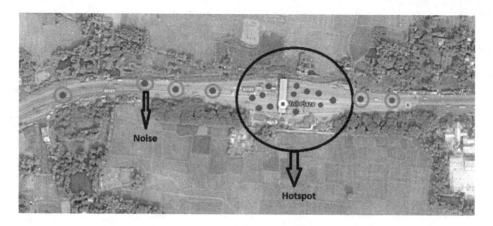

Fig. 2. This figure motivates the use of clustering to distinguish between stationary points that form a hotspot and the noisy observations.

The DBSCAN algorithm leaves the sparsely located points unclustered as noise, while giving us clusters with density-connected data points numbering greater than $minPts$, within a threshold distance e of another data point within the cluster. We apply the DBSCAN algorithm here to cluster the individual stoppage instances mined from the tripwise GPS data. The clusters formed are expected to correspond to truck stops on highways, while stoppages due to individual factors are eliminated as noise. For our experiment, we chose the $minPts$ to be 12, and the e to be 0.2 Km. These values for the constraints were decided after observing the experimental results.

The DBSCAN technique was modified with KD-trees to reduce the computational complexity. The tripwise LTD file records are indexes of consolidated stoppages made during trips. We use these indexes to find out geodetic coordinates of these stoppage points using the previously constructed LTD table. They are converted into Earth Centric Coordinates with x, y, z parameters. These x, y, z values are populated into a KD-tree data structure. The neighbours of each point are queried exactly once, and by using KD-trees, the time complexity for this step reduces to O(log n). Thus, an overall runtime complexity of O(nlog n) is obtained. If we do not use KD-tree for indexing the points, the runtime complexity would be $O(n^2)$. So, this technique scales well for road networks where the value of n can be very high.

After clustering using DBSCAN, we obtain a set of stoppage points with latitude and longitude for each cluster or truck stop. The median latitude and the median longitude of these latitude and longitude records are chosen as representative location of the truck stop. Figure 3 shows the visualisations of the truck stops obtained all over India, and in a particular region. A total of 5,820 truck stops were identified from the data.

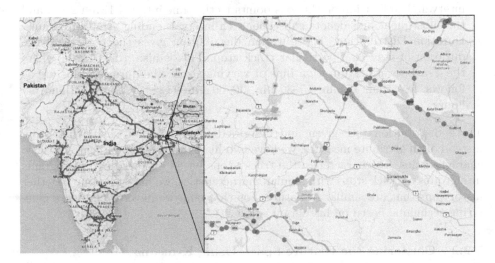

Fig. 3. Map visualisations of the identified hotspots from data. The left part gives the countrywide visualization with red circles representing the identified hotspots and the right part gives the zoomed representation of the identified hotspots in the Durgapur Bankura region of the state of West Bengal (Color figure online).

5.2   Characterization of Truck Stops

We set out to characterize the truck stops obtained from the identification procedure, based on the assumption that the arrival pattern throughout the day and duration distribution of stoppages can estimate the functionality of the truck stop. Thus, the truck stops obtained are clustered into different sets of truck stops with similar arrival patterns and similar duration of stoppages.

5.2.1   Creating Arrival and Duration Distribution Histograms

The truck stops obtained after the clustering step, are of various kinds according to their function. They can be meal stops, refuelling stops, rest stops, stops at toll plazas or check posts, and also they can have mixed use. We try to estimate the functionality of stops by observing the times vehicles arrive at the stop and the amount of time that a vehicle stays in those stops. Each of the clusters obtained contain a number of stoppage instances. From the LTD table described in previous sections, we can get the information regarding the starting time of each stopping instance, and the duration of each stoppage instance. Thus, for each truck stop, we create two distributions which we treat as representative of their functionality.

1. Arrival Time Distribution. This distribution displays the arrival pattern throughout the day at a particular truck stop. This is a histogram with the Y-axis as the Time of Day, and where each bin corresponds to a 2 h time interval. The first bin on the axis points to the time interval 12 am 2 am, and the subsequent bins denote subsequent 2 h intervals ending at 10 pm 12 am.
2. Duration Distribution. The duration distribution gives the pattern of the duration of stay at a particular truck stop. This is also a histogram where the Y-axis contains duration bins of varying sizes. The duration bins are as follows: less than 15 min, 15–30 min, 30–60 min, 1–2 h, 2–4 h, 4–6 h and 6+ hours.

These distributions are given and described in Sect. 6. For each truck stop, instead of an attribute measure, we have two histograms which indicate intrinsic characteristics. To characterize the truck stops based on these two histograms into groups of truck stops with similar characteristics, we have to group together histograms that are similar in nature. For that end, we have to employ a clustering technique based on histogram similarity.

5.2.2   Clustering the Feature Space Using K-means

The unsupervised K-means clustering algorithm is applied to test its applicability to process histogram similarity. As the K-means algorithm is not ideally suited to accomplish clustering based on histogram similarity, we tweak it a little bit to suit our needs. Suppose, each arrival time histogram has A bins and each duration histogram has D bins. Each truck stop is represented as a data point in an $(A + D)$ dimensional space, where, the first A coordinates represent the corresponding arrival bucket values and the next D coordinates represent the corresponding duration bucket values. These values are not absolute frequency values but are normalized against the total vehicle count of each truck stop. Thus, the distance between two histograms is the Euclidean Distance between their represented points in the $(A + D)$ dimensional space. Two data points, or truck stops will be close only if their coordinates are similar, which means that their histograms should be roughly identical.

We then run the K means algorithm on this space using $K = 30$. Each truck stop is thus assigned to one of the 30 resultant clusters. Two average histograms are constructed for each cluster, one for the arrival time and another for the duration. This was created by taking the average value of each bin in the arrival time distribution and duration distribution from each truck stop part of the specific cluster. This average histograms for arrival time and duration are considered as being representative of the characteristics of the cluster, denoting a particular functionality of the truck stops. The K value is taken experimentally keeping in mind the following principles:

1. We ensured that all possible types of functionality were covered, looking at the resultant average histograms of the clusters formed.
2. The instances of different types of truck stops belonging to the same cluster should be minimized.
3. Each cluster of truck stops should have a sizable population to be considered not as spurious instances.

So, we experimented with K = 20, 25, 30 and 35, and K = 30 was chosen as it looked to have covered all possible types of functionalists of stops, and all the clusters formed had almost atleast 100 members. Thus, we obtain the 30 clusters of truck stops based on their arrival time and stoppage duration, a set of truck stops for each cluster, and then calculate the average arrival time distribution and average duration distribution for each cluster. The results of the characterization of hotspots are discussed in the following section.

6   Results and Discussions

The work has been evaluated for its ability to correctly identify truck stops at their proper location, and assign the truck stops into distinct clusters.

6.1   Evaluation of Identification of Truck Stops

A total of 5,820 truck stops were detected from the available data of 60, 324 trips made across the country. Investigations of the detected truck stops show that the detected points correspond to features and structures on maps likely to be truck stops and intersections. Figure 4 shows that the detected truck stops (circled in black) correspond to features on-road that are truck stops.

6.2   Evaluation of Characterization of Truck Stops

We evaluate in detail the top 5 numerous clusters from Table 3, which gives the number of truck stops in each cluster. These clusters are cluster numbers 1, 2, 8, 18 and 28. Figures 5 and 6 give the average arrival time distributions and duration distributions for the clusters mentioned. For each cluster section the plot on the left gives the arrival time distribution and the plot on the right gives the duration distribution. Table 4 gives the type of truck stop each cluster is expected to have after analysis of their average arrival time distribution and average duration distribution.

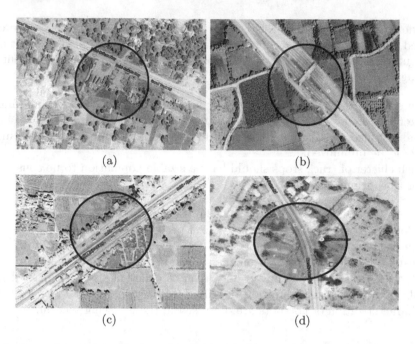

(a)   (b)

(c)   (d)

Fig. 4. Some detected truck stops (circled in black) visualised on map with a radius of 100m. The locations of the detected truck stops are as follows: (a) Rest stop on national highway 2. (b) Toll Plaza on the Ahmedabad-Vadodara Expressway. (c) A collection of snack shops on the national highway 8. (d) Mixed-use food and rest stop on national highway 33

We observe that the distribution plots of each of these clusters resemble the expected behaviour at different types of truck stops. In Fig. 5(a), the arrival time distribution of cluster 1 shows sustained activity throughout the day except at late night, while the stoppages are predominantly of very short durations. This indicates that the truck stops in this cluster are predominantly toll stops, as stated in Table 4. Figure 5(b) shows that arrivals at the stoppages of the cluster 2 happened throughout the day except late nights, and majority of the stoppages were of the durations less than 15 min or of 15–30 min. This indicates that the truck stops in this cluster reflect the behaviour of fuel stops, as stated in Table 4. In Fig. 5(c), while the duration distribution shows that a majority of stoppages are of less than an hour pointing to meal stops, we find the arrivals in cluster 18 have a distinct peak in morning and forenoon suggesting that cluster 18 consists of truck stops which are mostly breakfast places, as in Table 4. If we had peak arrival activity in the afternoon, the cluster would have represented lunch stops, while peaks in the evening indicate dinner meal stops.

Thus far, we have considered clusters which clearly reflect the expected arrival and duration behaviour of specific types of truck stops. But there are clusters which do not solely indicate the behaviour of one specific type of truck stop. In Fig. 6(a) the evening to late evening peak in the arrival time distribution

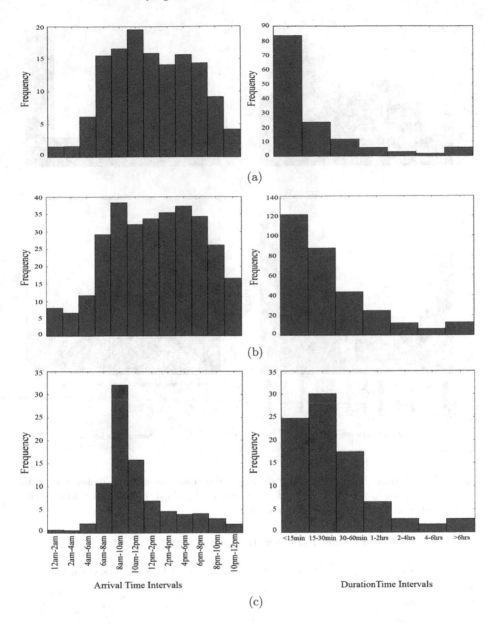

Fig. 5. Arrival time distributions and duration distributions of (a) Cluster 1, (b) Cluster 2, (c) Cluster 18. Each of these clusters reflect the characteristics of specific types of truck stops

and the distinct peak for stoppage durations of 6+ hours indicates that cluster 8 primarily consists of rest stops. But there are stoppages of lesser durations, which indicates that mixed use stops with rest facilities may also be a part of this

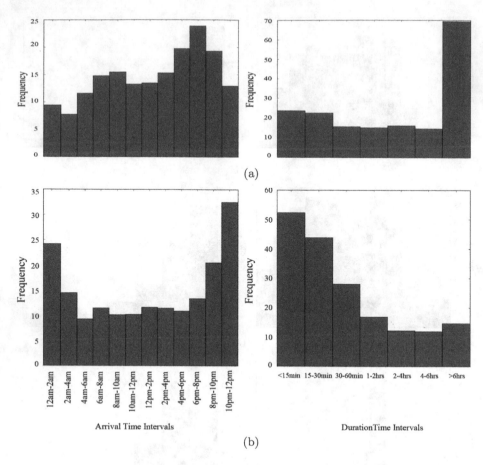

(a)

(b)

Arrival Time Intervals

DurationTime Intervals

Fig. 6. Arrival time distributions and duration distributions of (a) Cluster 8 reflects the characteristics of a specific type of truck stops, (b) Cluster 28 which portrays the characteristics of multiple types of truck stops

Table 3. Number of truck stops assigned to each cluster

Cluster ID	No. of truck stops	Cluster ID	No. of truck stops	Cluster ID	No. of truck stops
1	431	11	133	21	107
2	488	12	185	22	172
3	143	13	130	23	141
4	134	14	110	24	297
5	94	15	124	25	131
6	180	16	165	26	202
7	84	17	149	27	154
8	306	18	418	28	304
9	170	19	184	29	182
10	236	20	95	30	171

Table 4. Types of truck stops assigned to each cluster

Cluster ID	Truck stop type	Cluster ID	Truck stop type	Cluster ID	Truck stop type
1	Toll	11	Fuel	21	Toll
2	Fuel	12	Fuel, meal, mixed	22	Meal
3	Toll	13	Meal, rest, mixed	23	Toll
4	Toll	14	Fuel	24	Meal
5	Toll	15	Fuel	25	Meal
6	Meal, rest, mixed	16	Toll	26	Fuel
7	Toll	17	Meal	27	Fuel
8	Meal, rest, mixed	18	Meal	28	Meal, rest, mixed
9	Toll	19	Toll	29	Fuel, toll
10	Toll	20	Toll	30	Meal

cluster. This is given in Table 4. In Fig. 6(b), the duration distribution of cluster 28 denotes that the majority of stoppage durations are of short and medium durations, with quite a few long stoppages as well. The arrival time distribution of cluster 28 shows sustained activity around the day with distinct peaks in the night hours. This signifies that the cluster contains meal stops as well as mixed use stops for meals and rest, which is given in Table 4.

7   Conclusion and Future Work

In this work, we have presented a novel approach to identify and character-ize truck stops, where we identify truck stops by clustering stoppage points from GPS data using the DBSCAN clustering algorithm with KD-Tree index-ing. Subsequently, we characterize the truck stops according to their arrival time and duration distributions, using the K-means clustering algorithm. The char-acterized stops can be useful for trip planning, travel time prediction and traffic incident detection. For proper testing of the accuracy of this classification, these clusters need to be validated with ground truth data. Clusters that reflect multi-ple truck stop functionalities need to be investigated whether they reflect mixed use stops, or is it a case of mixed clustering where multiple types of stops are being clustered together mistakenly. In that case there will be a need to differenti-ate them by implementing other advanced clustering algorithms, or constructing models for each type of functionality of stops and mixed use stops with training data. Temporal analysis of truck stops can lead to a better understanding traffic patterns and more accurate trip planning decision support systems.

Acknowledgement. The study is supported under the project 'Real Time Traffic Prediction and Traffic Management (RTT)' by the Ministry of Human Resource Devel-opment, Government of India, and IIT Kharagpur. The data was provided by eTrans Solutions Pvt Ltd.

References

1. Tripathi, J.P.: Algorithm for detection of hotspots of traffic through analysis of GPS data. Masters thesis, Thapar University, Patiala, June 2010
2. Deliverable D1.4 - Analysis of ICT solutions employed in transport and logistics. Report. FInest Future Internet enabled optimisation of transport and logistics networks Seventh Framework Program, 30 March 2012
3. Giannopoulos, G.A.: Integrating freight transportation with intelligent transportation systems: some european issues and priorities. Transp. Res. Rec. **1790**, Paper No. 02–3460, 29–35 (2002)
4. Greaves, S.P., Figliozzi, M.A.: Collecting commercial vehicle tour data with passive global positioning system technology issues and potential applications. Transp. Res. Rec. **2049**, 158–166 (2008)
5. McCormack, E., Ma, X., Klocow, C., Currarei, A., Wright, D.: Developing a GPS based-truck freight performance measures platform. Rep. WA-RD 748.1, TNW 2010–02, Washington State Dept. of Transportation, Olympia, WA (2010)
6. Kuppam, A., Lemp, J., Beagan, D., Livshits, V., Vallabhaneni, L., Nippani, S.: Development of a tour-based truck travel demand model using truck GPS data. In: 93rd Annual Meeting of the Transportation Research Board (No. 14–4293) (2014)
7. Yang, X., Sun, Z., Ban, X., Holguin-Veras, J.: Urban freight delivery stop identification with GPS data. Transp. Res. Rec. J. Transp. Res. Board. **2411**(−1), 55–61 (2014)
8. Arifin, Z.N.: Route choice modeling based on GPS tracking data: the case of Jakarta. Ph.D. thesis. ETH Zurich (2012)
9. Sharman, B.W., Roorda, M.J.: Analysis of freight global positioning system data: clustering approach for identifying trip destinations. Transp. Res. Rec. **2246**, 83–91 (2011)
10. Prasannakumar, V., Vijith, H., Charutha, R., Geetha, N.: Spatio-temporal clustering of road accidents: GIS based analysis and assessment. Procedia Soc. Behav. Sci. **21**, 317–325 (2011)
11. Anderson, T.K.: Kernel density estimation and K-means clustering to profile road accident hotspots. Accid. Anal. Prevent. **41**(3), 359–364 (2009)
12. Aziz, R., Kedia, M., Dan, S., Sarkar, S., Mitra, S., Mitra, P.: Segmenting highway network based on speed profiles. In: 2015 IEEE 18th International Conference on Intelligent Transportation Systems (ITSC), pp. 2927–2932. IEEE (2015)
13. Ester, M., Kriegel, H.P., Sander, J., Xu, X.: A density-based algorithm for discovering clusters in large spatial databases with noise. In: Proceedings of the Second International Conference on Knowledge Discovery and Data Mining (KDD 1996), pp. 226–231. AAAI Press (1996)

A Fuzzy Rule-Based Learning Algorithm for Customer Churn Prediction

Bingquan Huang[1($\boxtimes$)], Ying Huang[1], Chongcheng Chen[2], and M.-T. Kechadi[1]

[1] The INSIGHT Centre, University College Dublin, Belfield, Dublin 4, Ireland
bingquan.huang@insight-centre.org
[2] Key Lab of Spatial Data Mining and Information Sharing of Ministry of Education,
Fuzhou University, Fuzhou, People's Republic of China

Abstract. Customer churn has emerged as a critical issue for Customer Relationship Management and customer retention in the telecommunications industry, thus churn prediction is necessary and valuable to retain the customers and reduce the losses. Recently rule-based classification methods designed transparently interpreting the classification results are preferable in customer churn prediction. However most of rule-based learning algorithms designed with the assumption of well-balanced datasets, may provide unacceptable prediction results. This paper introduces a Fuzzy Association Rule-based Classification Learning Algorithm for customer churn prediction. The proposed algorithm adapts CAIM discretization algorithm to obtain fuzzy partitions, then searches a set of rules using an assessment method. The experiments were carried out to validate the proposed approach using the customer services dataset of Telecom. The experimental results show that the proposed approach can achieve acceptable prediction accuracy and efficient for churn prediction.

Keywords: Fuzzy rules · Churn prediction · Rule-based classification

1 Introduction

Service companies in telecommunications suffer from a loss of valuable customers to competitors; this is known as customer churn. In the last a few years, there have been many changes in the telecommunication industry, such as, new services, technologies and the liberalizations of the market increasing competition. For most services, once the number of service subscribes reaches its peak, the company places a special emphasis on the retention of valuable customers because acquiring new customer is difficult and costly. Recently, data mining techniques have been emerged to tackle the challenging problem of customer churn [1–4].

The study of customer churn can be seen as a classification problem. The main goal is to build a robust churn model to predict potential churn customers with the interpretation. Many modelling techniques have been applied to churn prediction problem, including artificial neural networks, CART, Linear Regression, support vector machines, decision trees. For examples, Wei and Chiu [1]

© Springer International Publishing Switzerland 2016
P. Perner (Ed.): ICDM 2016, LNAI 9728, pp. 183–196, 2016.
DOI: 10.1007/978-3-319-41561-1_14

examined an interesting model for churn prediction problem using Decision Trees; Hadden et al. [5] employed Neural Networks, CART, and Linear Regression to build telecommunication churn prediction model; Coussement and Poe [6] investigated the effectiveness of SVM technique over logistic regression and random forests; Wai-Ho et al. [7] proposed the general-to-specific rule learning algorithms for churn prediction. In these techniques, beside obtaining acceptable results, the rule-based modelling techniques only can provide the easiest interpretable concept form of representation.

However, most of rule-based learning techniques were designed and attempt to find an accurate performance over a full range of samples, based on the balanced classes of training dataset [1,6,8–10]. If learning from the dataset with the highly imbalanced distribution of churners and non-churners, these learning techniques tend to be overwhelmed by the non-churn class and ignore the churn class, and consequently might provide poor churn classification results. Therefore, most of rule-based learning techniques may not be more capable towards customer churn classification. Although a number of the techniques were introduced (For example, Huang and Sato [11] introduced the PCA-based and SMOTE [12] sampling methods), the prediction accuracy still need to be improved and the most of them are not based on a learning strategy which is widely used.

In this paper, we introduced a modelling approach – Fuzzy Association Rule-based Classification Learning Algorithm, for customer churn prediction. This proposed learning algorithm finds fuzzy partitions, heuristically searches a set of fuzzy rules, then prunes the unimportant and redundant rules. Finally, the algorithm uses this set of rules with the membership function to classify the observations. In order to handle problem of imbalanced classes in churn prediction, the proposed algorithm adapts CAIM discretization algorithm to obtain fuzzy partitions of the dataset, and uses support weigh factors to calculate the rule support degree during searching rule set. In the experiments, the lift curve and AUC techniques were used to evaluate the performance of the proposed algorithm.

The rest of this paper is organised as follows: Sect. 2 introduces the basic concepts of fuzzy rules. Section 3 describes the proposed algorithm. Section 4 provides the experimental results and discussion. The conclusion and future work are given in Sect. 5.

2 Basic Concepts

Consider a training dataset consists of $|D|$ samples $x_i = (x_{i1}, ..., x_{im})$, $i = 1, 2, ..., D$ and C classes where x_{ip} is the p-th attribute value ($p = 1, 2, ..., m$) of the i-th training sample. A fuzzy rule of classification usually can be presented as the following form for a classifier:

Rule R_j: If $x_i \in A_{1p} \bigwedge ... \bigwedge x_m \in A_{jm}$, then Class $= C_j$;

where R_j is is the label of the j-th rule, $x = x_1, ..., x_m$ is an m-dimension input vector, A_{jp} is an antecedent fuzzy set, C_j is a class label. A fuzzy associative classification rule R_j can be measured directly in terms of support and confidence [13] as follows:

$$FS(R_j) = \frac{\sum_{x_i \in C_j} \mu_{A_j}(x_i)}{|D|} \tag{1}$$

$$FC(R_j) = \frac{\sum_{x_i \in C_j} \mu_{A_j}(x_i)}{\sum_{x_i \in D} \mu_A(x_i)} \tag{2}$$

where FS and FC are respectively the fuzzy support and confident values, $|D|$ is a number of samples in the training dataset, $\mu_{A_j}(x_i)$ is the matching degree of input vector x_p with the antecedent and consequent of the rule R_j. $\mu_A(x_i)$ is the matching degree of x_i with the antecedent part of the rule. The match degree can be calculated by the trap function:

$$\mu_{A_j}(x_i) = \max\{1 - |\frac{x_i}{F} - (k-1)|, 0\} \tag{3}$$

where k is the index of the intervals from 0 to m, m is a number of intervals of the pth attribute, and F is calculated by:

$$F = \frac{1}{m-1} \tag{4}$$

As mentioned above, in a IF-THEN rule, the antecedent part consists of a number of pairs <attribute, interval> and the **consequence** denotes the class label. In this paper, a pair <attribute, interval> is defined as an **item**. If the consequence of a rule is the class c, the items of this rule are called **Class^c items**. For examples, if the consequence of a rule is class "positive", the items of this rule are **positive_items**.

A rule that has only 1 item in the antecedent part is called a **1_Order_Rule**. If a rule that has 2 items in the antecedent part is called a **2_Order_Rule**. Similarly, a rule that has n items is called a **n_Order_Rule**. If a rule that has more than 1 items is called a **higher_Order_Rule**.

If the consequence of a n_Order_Rule is the class label c, this rule is written as a **class^c n_Order_Rule**. Similarly, the **class^k higher_Order_Rules** and **class^k lower_Order_Rules** respectively refer to the higher-Order Rules and lower-Order Rules that have more than 1 items in the antecedent part.

Another basic concept of rule induction is **Cover**. It expresses that a sample satisfies the antecedent part of the rule. That's, when a sample matches the condition of a rule, the rule *Correctly* **Covers** this sample. The task of rule induction is to build a model by generating a set of rules, which either do or do not Cover sample.

General rules and **specific rules** are two important concepts. Usually, a rule with a smaller number of pairs <attribute, interval> in the **antecedent** part is more **general** than the one having more. The reverse sample is more *specific*. The *Most general* rules are the First-Order-Rules, while the *most specific* means the number of antecedent elements is the total number of attributes. Normally, the most specific rules should not be considered since they are too specific and not able to describe a relative big data space.

3 Rule-Based Classification Algorithms

The objective of the new algorithm is to generate a set of rules for churn classification. Generally, this proposed approach (1) uses the adapted fast CAIM Discretization algorithm [14] to find the intervals of each continue attribute; (2) applies the heuristic methods with the fuzzy logic membership function to generate classification rules with the fuzzy support and confidence calculated by using the adapted method; (3) prunes these rules; (4) uses these set of rules to predict the given observations.

3.1 Fuzzy Partition of Continue Attributes

As the first step, a continuous attributes is discretized into a number of intervals. Based on these intervals, the classical triangle membership function is used to calculate the fuzzy degrees. There are two kinds of discretization algorithms: supervised and unsupervised techniques. In the contrast of them, the supervised discretization algorithms are outperform for classification tasks [14]. As one of promising supervised discretization algorithm, the Fast CAIM (F-CAIM) algorithm [14] outperforms most of discretization algorithms (e.g. cluster algorithms) [14]. The goal of fast CAIM is to maximize the class-attribute interdependence and to generate a minimal number of discrete intervals. The detail of this algorithm can be found in [14].

As mentioned above, the distribution of classes (churn and nonchurn) of our application is very imbalanced. However, fast CAIM Algorithm was designed under the hypothesis of the class balanced in the dataset. Therefore, this paper adapts the F-CAIM algorithm to discrete data for our application. Based on the F-CAIM algorithm, we only changes the CAIM discretization criterion into the following:

$$CAIM'(C, D|F) = \frac{\sum_{r=1}^{N} \frac{\max \frac{q_{ir}}{M_{i+}}}{M_{+r}}}{n} \tag{5}$$

where: q_{ir} is the total number of continuous values belonging to the ith class that are within interval $(d_{r-1}, d_r]$. M_{i+} is the total number of objects belonging to the ith class, and M_{+r} is the total number of continuous values of attribute F that are within the interval $(d_{r-1}, d_r]$, for $i = 1, 2..., S$ and, $r = 1, 2, ..., n$. $\frac{q_{ir}}{M_{i+}}$ is the portion of the ith class objects. Therefore, when measuring the dependency between class variable and the discretization variable for an attribute, the modified criterion 5 take the class ratio to account, and consider less of the number of objects for an variable interval. This can lead the adapted discretization algorithm to more successfully discretize the continuous attributes in the imbalanced class data.

3.2   Rule Learning

As mentioned above, there is a very challenge for traditional rule-based learning techniques to learn from the dataset which consists of the highly imbalanced distribution of churners and non-churners, as consequently, the classifiers would easy ignore the minority classes (churners). In order to handling this issue, we adapt the support of rules to strengthen the significant of classes by multiplying weight factors, called support weight factors. The measurement of rule support is replaced by the following formula:

$$FS'(R_j) = w_{C_j} * \frac{\sum_{x_i \in C_j} \mu_{A_j}(x_i)}{|D|} \tag{6}$$

where w_{C_j} is the support weight factor of the j-th class. When the weight factor w_{C_j} increases, the degree of significance of the rule R_j and the rule weight will increase, according to Eqs. (7), (8) and (9). Therefore, this formula can effect the rule evaluation, pruning and classification (see Sects. 3.3 and 3.4). Usually, we expect that the weight factor w_{C_j} for the more interesting class is large in such a way keep related class rules for the classification. The best weight factor w_{C_j} can be selected by using the AUC technique Sect. 4.3: (1) use a number of different value for the weight factor w_{C_j} to build a number of models and validate the training data, (2) calculate AUC values, and (3) select the value which obtain the maximum AUC value as the best weight factor w_{C_j}.

The proposed algorithm applies the strategy of general-to-specific to generate a set of rules from the dataset. According to the class labels of training data, the algorithm firstly generates the first-order rules (the most general rules) of each class. Next, the learning algorithm iteratively constructs the higher-order rules of each class, based on the lower-order rules. Suppose SET^c_{rules} is the set of generated rules of class c, FullSET$_{rules}$ is the full set of rules, k is the number of orders, data-subset$_c$ is the discretised sub-dataset of class c, and Max_num_order is the maximal number of orders. The process of this learning rules can be illustrated by Algorithm 1.

First-Order Fuzzy Rules. As mentioned in Sect. 2, the first-order fuzzy rules are **the most general rules**, which the number of antecedent elements is 1. The first-order rules can be obtained by two steps. First, for each interval of each attribute, any single pair <attribute, interval> is considered as the antecedent of a new rule; then this new rule is evaluated (which is described in Sect. 3.3); if this

---

Algorithm 1. Rule-based Learning Algorithm

```
 1: FullSET_rules ← φ;
 2: c ← 1;
 3: while c ≤ C do
 4:     SET^c_rules ← φ;
 5:     SET^c_rules ← Class^c_1_order_rules(training dataset);
 6:     k ← 2;
 7:     while k ≤ Max_num_order do
 8:         class^c_k_higher_order_rules ← Class^c_high_order_rules;
 9:         SET^c_rules ← SET^c_rules ⋃ class^c_k_higher_order_rules;
10:         k + +;
11:     end while
12:     FullSET_rules ← FullSET_rules ⋃ SET^c_rules;
13:     c + +;
14: end while
15: return FullSET_rules
```

---

new rule is valid, it will be selected to combine a higher-order rules. Algorithm 2 illustrates the details process of generating the first-order fuzzy rules.

In Algorithm 2, $num\_attribute$, $num\_interval$ and num_class respectively denotes the number of attributes, the number of fuzzy intervals of the current attribute and the number of classes. The set of $first\_order\_rules$ is initially empty. $rule_{ijk}$ expresses a rule: "IF f_i is s_j THEN $Class_k$". Evaluate($rule_{ijk}$) is the function which evaluates whether $rule_{ijk}$ is valid or interested. If Evaluate($rule_{ijk}$) returns true, then $rule_{ijk}$ can be accepted as a valid rule (see Sect. 3.3). $Class^k\_1\_order\_rules$ is a set of first-order rules that can cover the samples of class K. The $first\_order\_rules$ (1_order_rules) includes all of $Class^k\_1\_order\_rules$ ($0 \leq k \leq C$).

Algorithm 2. $First\_order\_rules$

```
 1: 1_order_rules ← φ;
 2: for k = 1; k <= num_class; k + + do
 3:     Class^k_1_order_rules ← φ;
 4:     for i = 1; i <= num_attribute; i + + do
 5:         for j = 1; j <= num_interval; j + + do
 6:             if Evaluate(rule_ijk) returns true then
 7:                 Class^k_1_order_rules ← Class^k_1_order_rules ⋃ rule_ijk;
 8:             end if
 9:         end for
10:     end for
11:     1_order_rules ← 1_order_rules ⋃ Class^k_1_order_rules;
12: end for
13: return 1_order_rules
```

High-Order Fuzzy Rules. Basically, the proposed learning algorithm iteratively combines the lower-order rules with items (pairs of ⟨attribute, interval⟩) into higher-order rules by using the heuristical hill-climbing method for each class. The learning algorithm uses the $(n - 1)\_order\_rules$ to construct n_order_rules for each class. The learning algorithm starts to construct 2-order-rules based on the 1-order rules, and stops when completed the rules of maximum order.

Suppose: $Class^c\_item$ is a set of pairs <attribute, interval> that are from the rules of class c; $num\_Classes$ is the number of classes; $num\_Class^c\,(n-1)\_rules$ is the number of rules of class c; all_high_rules are the rules of all of classes; $hill\_climbing()$ is the function of heuristical hill-climbing method; min_FC is the threshold of fuzzy confidence; The procedure of constructing a set of high rules can be illustrated by Algorithm 3. Algorithm 3 first extracts all exclusive *items* from the $Class^c\,1\_order\_rules$, secondly uses $hill\_climbing()$ to combine a lower-order rule $Class^c\,(n-1)\_rule$ with 1 $Class^c\_item$ into a new $high\_rule$ (n_order rule), thirdly evaluates this new rule by the function $Valid()$. If this new rule is valid, the algorithm calculate the confidence by Eq. (2). If the fuzzy confidence FC_i^c of this new rule is not less than the threshold of rule confidence values, then this new rule is added into $Class^c\_n\_rule$. This procedure is iteratively carried out to generate the set of $Class^c\,(n)\_rules$, based on the set $Class^c\,(n-1)\_rules$ and the item set of $Class^c\,1\_order\_rules$. Finally, the algorithm combines all of the high-order rules of different classes $\{Class^c\,(n)\_rules,\ 1 \le c \le C\}$ into one set of all_high_rules.

Algorithm 3. $Higher\_Order\_Rules(first\_order\_rules)$

```
1:  all_high_rules ← φ;
2:  n ← 1;
3:  while n ≤ Maxim_Order do
4:      n_order_rules ← φ;
5:      for c = 1; c ≤ num_Classes; c + + do
6:          Class^c_items ← extract all exclusive items from Class^c 1_order_rules;
7:          Class^c_n_order_rules ← φ;
8:          accuracy_list ← φ;
9:          for i = 1; i ≤ num_Class^c (n − 1)_rules; i + + do
10:             high_rule ← hill_climbing(Class^c (n − 1)_order_rule_i, Class^c_items);
11:             if Valid(high_rule) returns true then
12:                 FC_i^c ← FC(high_rule);
13:                 if FC_i^c ≥ min_FC then
14:                     Class^c_n_order_rules ← Class^c_n_order_rules ∪ high_rule;
15:                 end if
16:             end if
17:         end for
18:         n_order_rules ← n_order_rules ∪ Class^c_n_order_rules;
19:     end for
20:     all_high_rules ← all_high_rules ∪ n_order_rules;
21:     n++;
22: end while
23: return all_high_rules;
```

The heuristical hill-climbing method for generating $high\_order$ rules is illustrated by Algorithm 4. In Algorithm 4, FC_list is a list of different FC values and is initially empty. Basically, one piece of $high\_order$ is based on one piece of low_order rule and a $new\ item$, which has been mentioned earlier. If the $new\ item$ is not included in the antecedent of the low_order rule, then the hill-climbing algorithm combines this $lower\_rule$ and this $new\ item$ into t one new $high\_order$ rule. Meanwhile, the FC value of the newly built this $high\_order$ rule is stored in the $accuracy\_list$. This algorithm returns the $high\_order$ rule with the highest of FC values.

Algorithm 4. $hill\_climbing(one\_lower\_rule, all\_low\_items)$

1: $FC\_list \leftarrow \phi$;
2: **for** $i = 1$; $i <= num\_low\_items$; $i + +$ **do**
3:    **if** low_rule does not include $item_i$ **then**
4:       $high\_rule \leftarrow$ combination of low_rule and $item_i$;
5:       $FC_i \leftarrow$ FC(high$\_$rule);
6:       $FC\_list \leftarrow FC\_list \bigcup FC_i$;
7:    **end if**
8: **end for**
9: $BEST \leftarrow$ one of high order rules with the highest FC value;
10: **return** $BEST$

3.3 Rule Assessment

Due to the redundant information and the over-fitting problem, the proposed rule learning algorithm applies χ^2 statistic test to measure the quality of rules in such way to find the interesting rules. In this paper, χ^2 is used to measure the correlation between the antecedent part and the consequence of a rule. The stronger correlation usually presents more significance. Suppose there is a rule which A is the antecedent, B is the consequent, and $A \Rightarrow B$. The χ^2 statistic value of this rule can be calculated by:

$$\chi^2(A \Rightarrow B) = \frac{(O(AB) \times w_C - E(AB))^2}{E(AB)} \tag{7}$$

where $O(AB)$ is the observed frequency of the rule, w_C is the support weight factor for the consequent B (B is C_j), and $E(AB)$ is the expected frequency which can be calculated by Eq. (8):

$$E(AB) = \frac{(|D| \times FS'_A) \times (|D| \times FS'_B)}{|D|} = FS'_A \times FS'_B \times |D| \tag{8}$$

Where $|D|$ is a number of samples in training data, FS'_B and FS'_A are respectively the fuzzy support values of B and A, which can be obtained by Eq. 6.

Algorithm 5 illustrates the pruning method. α is the critical value for χ^2 significance test, and it is set to be 3.84 in this research. $minS$ and $minC$ are two threshold values, which define the minimal value of support and confidence, respectively. In this research, the minimum value of support and confidence were set to be 0.01 and 0.5, respectively. A rule can be retained if it satisfies all the threshold values.

Algorithm 5. Evaluate($rule$)

1: $Flag \leftarrow$ false;
2: **if** chi-square statistics $> \alpha$ AND $Support\_rule >$ minS AND $Confidence\_rule >$ minC **then**
3:    $Flag \leftarrow true$;
4: **end if**
5: **return** Flag;

3.4 Classification

Once the proposed algorithm obtains a set of different class rules from the given training dataset, the algorithm can use this set of rules to classify the observations. Suppose the observation is x_i, the classification procedure can be described as: (1) the algorithm calculates weight values of x_i in each rule by Eq. (9), then (2) assigns the class label with the highest weight value to the observation x_i.

$$weight(R_j, x_i) = \mu(x_i) \times FC(R_j) \times FS'(R_j) \tag{9}$$

where $\mu(x_i)$ is the aggregate membership function of the multiple conditions of R_j, and is calculated by $\mu_{A_j}(x_{ip}) \cdot ... \cdot \mu_{A_j}(x_{ip}) \cdot ... \cdot \mu_{A_j}(x_{im})$;

4 Experiments and Discussion

4.1 Dataset

Experiments were conducted by the telecommunication dataset of the Ireland Telecoms [15]. The telecommunication dataset also was divided into one training dataset and one testing datasets. There are 6000 churners, 94000 nonchurners and total 100000 customers in the training dataset. In the testing dataset, there are 39000 customers which includes 2000 churners and 37000 nonchurners. the 122 features were extracted using constant non-time series data (e.g. demographic profiles, payment method, etc.) and time series data (e.g. call-details that consists of five-month call-details M 4, ..., M 1, M. The time series training data lags one month behind testing data. The detail of these features can be found in [3,16].

4.2 Experiment Set-up

In order to evaluate the proposed learning algorithm, two sets of experiments were carried out in this paper. The first set of experiments is to compare the performance of the proposed learning algorithm and traditional learning algorithms. The second set of experiments is to evaluate the performance of the different rule-based learning algorithms which employ different discretization methods or non-fuzzy rules.

Experiment Set-up I. In this set of experiments, six modelling techniques (DMEL [7], Logistic regression, CN2, MLP, C4.5 and the new proposed algorithm) were used to make a prediction for the telecom dataset. The related software tools used in the experiments are Weka and CN2 software packages of R. Boswell [17]. All the predictors were trained by 10 folds of cross-validations in each experiment. The Wrappers for Performance Enhancement and Oblivious Decision Graphs algorithm (WPEODG) [18] was used to select the best learning parameters for C4.5, MLP and Logistic regression.

The three-layer MLPs, each of which only contains one hidden layer, were used. The number of input neurons in the MPL is the same as the number of dimensions in a feature vector. The number of output neurons of the network is the number of classes. The sigmoid function is selected as the activation function for all MLPs. Each MLP was trained by BP learning algorithm with learning rate 0.1, maximum cycle 1800 and tolerant error 0.05. The number of training cycles to yield the highest accuracy is about 800.

Each DMEL model was trained by the following parameters: population size of 30, number of generations of 300, Z-value of 1.96 for finding interested features, probabilities of mutation and crossover which are 0.1 % and 60 % respectively. The best parameters for C4.5 are when the pruning confidence threshold is 0.1 and the minimum number of samples per leaf is 2 for the Telecom dataset.

For the new proposed rule learning algorithm, the threshold value of fuzzy support (min_fs) is set to 0.01, which is a commonly used value; while the critical value of fuzzy confidence (min_fc) is set in the range [0.5: 0.9] with the step size of 0.02, the optimal min_fc would be the one that leads to the largest AUC value for both datasets. The support weight factors for churner and nonchurner classes (w_{C_0} and w_{C_1}) are 0.8 and 0.2.

Experiment Set-up II. In the second set of experiments, the 4 rule-based learning algorithms (which are the new proposed learning algorithm, CAIM-Fuzzy algorithm, CAIM-Rules algorithm and kmeans-Fuzzy algorithm) were carried out. Comparing to the new proposed algorithm, the CAIM-Fuzzy and kmeans-Fuzzy algorithms use different discretization techniques. The CAIM-Fuzzy algorithm uses the supervised CAIM algorithm, and the kmeans-Fuzzy algorithm uses the unsupervised clustering algorithm – Kmeans algorithm, to discrete the data. Contrasting to new proposed learning algorithm, the CAIM-Rule learning algorithm uses the supervised CAIM to discrete data and the traditional rules for the classification. The value K of the Kmeans algorithm was 20. The parameters used in the learning algorithm are as the same as the ones used in the first set of experiments.

4.3   Evaluation

Generally, the service operators aim to relatively identify more potential churners by spending less cost (i.e., contacting a relatively small amount of customers). Accordingly, it is important for the telecom operators to estimate how much percent of potential churners can be identified by contacting a certain small proportion of customers (e.g. 5 % or 10 %). Therefore, Lift Curve [19] is applied in this work to evaluate the classification models.

The *Lift Curve* method first scores each data sample, then sorts the samples according to the scores, finally, it calculates the portion of samples and the True Positive rate (true churn in this paper), respectively.

However, it is difficult to use the lift-curve for comparing two or more results, since it is difficult to say whether one result is better than the other results.

To overcome this problem, the AUC is also used to evaluate the models. The area under a Lift Curve can be calculated by the following (if it was calculated based on trapezoid):

$$AUC(t) = \sum_{i=1}^{N} 0.5 \times (Y_i(t) + Y_{i+1}(t)) \times (X_{i+1}(t) - X_i(t)) \tag{10}$$

where N is the number of the trapezoid, $X_i(t)$ are the portion of selected samples and $Y_i(t)$ is the true positive rate at time t. The area under lift curve is the sum area of a series of trapezoids, the details of AUC can be found in [19].

4.4   Results and Discussion

In order to retain the customers with less costs, the service operators select a small portion of customers to offer special consideration or attractive services. The lift curve technique was used to find a relative small amount of customers that are the relative more potential churners. In order more accurately evaluate those modelling techniques, the AUC technique was used to analyses the experimental results and the learning modelling algorithms. Figures 1 and 2 show the experimental results which refer the lift charts with AUC from the first and second set of experiments, respectively.

In each lift chart figure, the horizontal axis represents the portion of the considered customers, the vertical axis represents the true positive rate, and the different colours represent the different modelling techniques used in the experiments. In each AUC figure, the x-axis represent different modelling techniques and the y-axis represents the values of AUC, and the different colors presents different portion of customers (10 %, 20 %, 30 %, 40 %, 50 %, 60 % and 100 %) were selected.

Result I. In a lift chart, the lift curve which is closer to the left-top corner (the area under it is the larger) is better than the ones that are farther to the left-top corner. In Fig. 1(a), the theory with the red curve is the best one, while the yellow curve that was obtained by randomly selecting samples is the worse. From these two figures, when a small portion of the selected samples are about 10 % less, the best curve is from DT and the better one is from the new proposed technique, but both of them are very similarly covered together. It also shows, generally, the block curve from the new proposed learning algorithm is closer than others, excepting these theory curves (the red one).

The related AUC is plotted plotted into Fig. 1(b). It shows the AUC values obtained by different modelling algorithms when the different portions of samples were selected. From the AUC, we can obtain that: (1) when the sample portions from 30 % to 100 % were selected for the Telecom dataset, the new proposed technique obtained the largest AUC values. (2) when the sample portion of 10 % or 20 % was selected for the Telecom dataset, the largest AUC values are from the new proposed technique while DT and AUC values were obtained by the

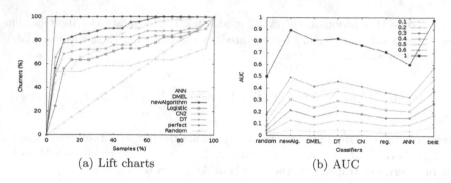

(a) Lift charts   (b) AUC

Fig. 1. Lift curves and the AUC of traditional and new modelling techniques (Color figure online)

new proposed technique is approximately equal to the ones obtained by DT. Therefore, the new proposed modelling technique which obtained larger AUC values are more suitable for churn prediction problem than others that obtained lower AUC values.

Result II. The comparative results from the 4 rule-based learning algorithms are showed in Fig. 2. These two figures show: (1) which rule-based learning algorithm obtained larger AUC values than others for a specified portion of the considering samples. For examples, among these 7 learning algorithms (including the new proposed learning algorithm, CAIM-Fuzzy algorithm, CAIM-Rule algorithm and kmeans-Fuzzy algorithm, DMEL, CN and DT), the AUC values of DT and the new proposed algorithm generate are larger when the sample portion of 10 % of considering samples is 20 % less. (2) when the sample portion of 10 % was selected from the Telecom dataset, the AUC values of the 4 rule learning algorithms are approximately equal. (3) when the sample portion increasing, among these 4 learning algorithms, the AUC values generated by the new proposed algorithm increase largest.

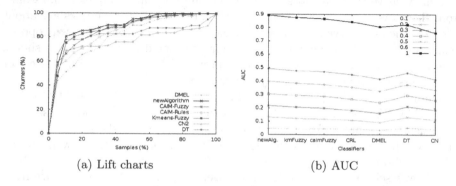

(a) Lift charts   (b) AUC

Fig. 2. The Lift curves and the AUC of rule-based learning techniques

5   Conclusion and Future Works

In this paper, we used the fuzzy rule-based classification technique to predict the customer churn for Telecom services. Lift Curve and AUC techniques were applied to evaluate the predicton techniques in the experiments. The results highlight that the rule-based classification technique is more efficient than the traditional rule-based methods which were designed and attempt to find An accurate model over a full range of samples using on the balanced classes of training dataset.

When using fuzzy-based rules to deal with large number of features, the computational overhead may increases for generating high order rules. In the future, we need to find a method to control the number of high order rules rather than setting the size arbitrarily. In addtion, theoretically, the cost-sensitive learning algorithms and sampling methods (e.g. PCA, SMOTE) can be used with fuzzy-based classification methods to improve results for the churn prediction. This will be studied in the future.

Acknowledgement. The authors gratefully acknowledge the research support from the EU Framework Programme 7, Marie Curie Actions under grant No. PIRSES-GA-2009-247608.

References

1. Wei, C., Chiu, I.: Turning telecommunications call details to churn prediction: a data mining approach. Expert Syst. Appl. **23**, 103–112 (2002)
2. Hung, S.-Y., Yen, D.C., Wang, H.-Y.: Applying data mining to telecom churn management. Expert Syst. Appl. **31**, 515–524 (2006)
3. Huang, B.Q., Kechadi, M.-T., Buckley, B.: Customer churn prediction for broadband internet services. In: Pedersen, T.B., Mohania, M.K., Tjoa, A.M. (eds.) DaWaK 2009. LNCS, vol. 5691, pp. 229–243. Springer, Heidelberg (2009)
4. Huang, B.Q., Kechadi, M.-T., Buckley, B.: A new feature set with new window techniques for customer churn prediction in land-line telecommunications. Expert Syst. Appl. **37**, 3657–3665 (2010)
5. Hadden, J., Tiwari, A., Roy, R., Ruta, D.: Churn prediction: does technology matter? Int. J. Electr. Comput. Eng. **1**, 6 (2006)
6. Coussement, K., Van den Poel, D.: Churn prediction in subscription services: an application of support vector machines while comparing two parameter-selection techniques. Expert Syst. Appl. **34**(1), 313–327 (2008)
7. Au, W., Chan, C.C., Yao, X.: A novel evolutionary data mining algorithm with applications to churn prediction. IEEE Trans. Evol. Comput. **7**, 532–545 (2003)
8. Liu, B., Hsu, W., Ma, Y.: Integrating classification and association rule mining. In: KDD, pp. 80–86 (1998)
9. Show-Jane, Y., Yue-Shi, L.: Cluster-based under-sampling approaches for imbalanced data distributions. Expert Syst. Appl. **36**(3), 5718–5727 (2009)
10. Burez, J., Van den Poel, D.: Handling class imbalance in customer churn prediction. Expert Syst. Appl. **36**(3), 4626–4636 (2009)

11. Sato, T., Huang, B.Q., Huang, Y., Kechadi, M.-T., Buckley, B.: Using PCA to predict customer churn in telecommunication dataset. In: Cao, L., Zhong, J., Feng, Y. (eds.) ADMA 2010, Part II. LNCS, vol. 6441, pp. 326–335. Springer, Heidelberg (2010)

12. Han, H., Wang, W.-Y., Mao, B.-H.: Borderline-SMOTE: a new over-sampling method in imbalanced data sets learning. In: Huang, D.-S., Zhang, X.-P., Huang, G.-B. (eds.) ICIC 2005. LNCS, vol. 3644, pp. 878–887. Springer, Heidelberg (2005)

13. Yi-Chung, H., Chen, R.-S., Tzeng, G.-H.: Finding fuzzy classification rules using data mining techniques. Pattern Recogn. Lett. **24**(13), 509–519 (2003)

14. Kurgan, L., Cios, K.: Fast class-attribute interdependence maximization (caim) discretization algorithm (2003)

15. http://www.eircom.ie/cgi-bin/bvsm/mainPage.jsp

16. Huang, B., Kechadi, M.T., Buckley, B.: Customer churn prediction in telecommunications. Expert Syst. Appl. **39**(1), 1414–1425 (2012)

17. http://www.cs.utexas.edu/users/pclark/software/#cn2

18. Kohavi, R.: Wrappers for performance enhancement and oblivious decision graphs. Department of Computer Science, Stanford University (1995)

19. Vuk, M.: Roc curve, lift chart and calibration plot (2006)

DDoS Attacks Detection in Cloud Computing Using Data Mining Techniques

Konstantin Borisenko[1], Andrey Smirnov[1], Evgenia Novikova[2],
and Andrey Shorov[1(✉)]

[1] Saint Petersburg State Electrotechnical University LETI,
Saint Petersburg, Russia
{borisenkoforleti, ashxz}@mail.ru,
sdsavll@gmail.com
[2] Saint Petersburg Institute for Informatics and Automation of the Russian
Academy of Sciences, Saint Petersburg, Russia
novikova.evgenia123@gmail.com

Abstract. Cloud computing platforms are developing fast nowadays. Due to their increasing complexity, hackers have more and more opportunities to attack them successfully. In this paper, we present an approach for detection internal and external DDoS attacks in cloud computing using data mining techniques. The main features of the cloud security component that implements suggested approach is an ability to detect both types of DDoS attacks and usage of data mining techniques. The component prototype is implemented in OpenStack cloud computing platform. The paper presents the results of the experiments with different types of DDoS attacks.

Keywords: Cloud security · Cloud computing · Cloud security architecture · Data mining · Ddos attack

1 Introduction

Cloud computing systems are developing fast nowadays. There were no such systems 30 years ago, however, according to the report [16] in 2013 47 billion dollars were spent on cloud services all over the world. And the sum is expected to be doubled by 2017 as companies invest in cloud services for creating new competitive offerings. Today life without cloud for many people would be unthinkable as there were no Facebook, Twitter, and Google. The cloud has recently adopted for business needs too. Millions of organizations around the world strongly rely on cloud services starting with document management to using cloud resources.

When starting usage of the cloud technologies in business processes, one should be aware that availability of cloud services depends on the work of cloud platform (infrastructure). It is extremely important for every company to have a cloud platform working up to 24/7. But sometimes that does not happen because hackers are trying to get unauthorized access or just trying to damage the services for different benefits. That is why it is important to have defensive methods implemented against malicious activities.

© Springer International Publishing Switzerland 2016
P. Perner (Ed.): ICDM 2016, LNAI 9728, pp. 197–211, 2016.
DOI: 10.1007/978-3-319-41561-1_15

The international scientific group in the cloud computing security area published threats report in 2013 [4]. According to it the cloud infrastructure attacks were placed to the 5th position in the list of the actual threats to clouds. Moreover, such infrastructure attacks as "distributed denial of service" (DDoS attacks) represent a huge threat for every element of the cloud computing service standard model (Infrastructure as a Service - IaaS, Platform as a Service - PaaS, Software as a Service - SaaS). Interestingly that in the report for 2010 year these attacks were not mentioned in the list of the biggest threats to clouds [4].

Distributed Denial of Service attacks are especially harmful to the companies providing services for the customers. Massive DDoS attacks often affect websites of the government bodies in various countries, websites of leading IT-corporations, including Amazon, Yahoo, Microsoft, etc. It is essential for the companies providing cloud services to be protected against DDoS attacks, because successful attacks can lead to a big loss of money [8].

In this paper, the authors present a novel approach for protection cloud computing against DDoS attacks. We suggest to distinguish external and internal DDoS attacks depending on the location of the attack source relative cloud infrastructure. Such attack classification allows selecting correct counteraction measures. For example, blocking all network traffic from internal virtual node of cloud system can affect the efficiency of all business processes the suspicious node participates in. In this case when the detection module detects that the attack originates from the cloud, the counteraction module will try to block network traffic coming from the specific ports of the suspicious node only. We also propose an architecture of security component that is able to detect both kinds of the attack. The architecture does not assume installation of any sensors on client-side and thus all processes in the cloud are kept confidential. At the same time it analyzes not only incoming external traffic like it is done in some commercial tools [13].

Our contribution is also in developing detection techniques based on data mining and machine learning techniques, including self-learning models. We use supervised models in order to classify network traffic. Experiments showed that the developed algorithms are working fast and malicious traffic is caught within 5 s after the attack starts. We collect attributes for models using Netflow protocol. In addition usage of self-learning algorithms make it easier to maintain cloud security, because models are learning on new types and scenarios of DDoS attacks.

All modules of the security component are flexible and can be deployed on nodes where cloud platform is installed, or on separate ones. The component prototype was implemented in OpenStack cloud computing platform.

The structure the paper is described as follows. Section 2 presents related work. The common approach for detection DDoS attacks in cloud computing is shown in Sect. 3. Section 4 considers the architecture of OpenStack cloud computing platform with particular focus on the components needed to implement suggested approach. Section 5 explains how the experimental environment was set up and describes how training and test data sets were collected, and how the classification models were learned and tested. Section 6 shows and discusses results of the experiments. Section 7 analyzes the paper results and provides an insight into the future research.

2   Related Work

Nowadays, researchers are developing and implementing different defensive techniques for detecting malicious traffic and protecting cloud computing platforms against DDoS attacks.

2.1   Security Solutions for Clouds

Elastic Cloud Security System (ECS^2) [17] provides complex security methods against malicious traffic. ECS^2 has antibot IP reputation tables and antivirus engines. Firewalls are rule-based and working in real time. The disadvantage of using this approach is the time of updating filter tables and antivirus signatures. Late tables updating can lead to receiving malicious traffic from the new unlabeled IP addresses. That is why our approach is not based on signatures, malware IP tables and other features that have to be updated rapidly.

In [5], the authors propose the defense method that places detection processes into virtual machines (VMs). The authors tested security methods on 108 services launched on different VMs. Those methods were based on data mining techniques and launched applications were analyzed on VMs. The drawback is the placement of the detection system. Some customers do not want to have background security processes inside their VMs due to specific security company rules.

Confidence-based filtering method is presented in [7]. The method focuses the probe on transport and network layers, creating correlation characteristics of co-appearance between attributes in the IP header and TCP header. Few attack types were tested. The authors conclude that the model has high effectiveness and low storage space when working with high-loaded networks.

In [20] authors outline neural network model for malicious traffic detection in cloud networks. Their model searches anomalies in the traffic flows and creates alerts for administrators to prevent damages. The authors notice that increasing the sample period for learning phase improves results, so higher accuracy can be reached not in real time.

The paper [15] presents a novel flow-based anomaly detection scheme based on the K-mean clustering algorithm. Training data containing unlabeled flow records are separated into clusters of normal and anomalous traffic. The corresponding cluster centroids are used as patterns for computationally efficient distance-based detection of anomalies in new monitoring data. The authors state that applying the clustering algorithm separately for different services (identified by their transport protocol and port number) improves the detection quality.

The authors of [21] propose an algorithm for detection of Denial of Service attacks that utilize SSL/TLS protocol. The algorithm based on filtering noise data and clustering detects malicious traffic. They trained models on the data obtained from realistic cyber environment. The authors conclude that the proposed model allows detecting all intrusive flows with very low number of alarms.

2.2   Cloud Computing Systems

Nowadays there are many cloud computing systems. One of the most popular is AWS Amazon [1], a proprietary vertically integrated solution for IaaS, PaaS, SaaS.

Another example is OpenNebula [14], a cloud management system using the functions of hypervisors for providing IaaS. ESXi (vSphere Hypervisor), KVM (Kernel-based Virtual Machine), XEN (OpenSource virtualization platform), Hyper-V (Windows Server Virtualization) are hypervisors supported by the system.

OpenStack [18] is one of the most popular solution for managing cloud services. It is open-source project and is sponsored by IBM, HP, Intel, Ubuntu and others. OpenStack developers implemented universal architecture of cloud computing platform. It allows providing different services on all cloud service models. However, the system has not enough components for detection and mitigation DDoS attacks. The main security method in OpenStack is a security group service. Obliviously, that this is not enough for successful counteraction to DDoS attacks in OpenStack. That is why, OpenStack [18] was chosen as environment for testing developed components.

3   Common Approach for DDoS Attack Detection in Cloud Computing Platforms

3.1   Taxonomy of DDoS Attacks on Cloud Computing

We suggest differentiating DDoS attacks depending on the attack source location relative the attack target located in the cloud. Therefore, there are two possible attack types: *external* attacks when the attack source is located outside cloud infrastructure; and *internal* one, when attack source is located inside the cloud.

When considering web server installed on ordinary computer there exists only one type of DDoS attacks — outside the computer network. Having a web server as a cloud instance gives possibilities to attack it both from outside the cloud and from other instances situated in the cloud.

In the IaaS service model customers receive not only service (like in SaaS service model) but also a network consisting of virtual machines and routers. The external attacks are similar to the attacks against ordinary web service. They can be deflected by filtering traffic incoming from attacking sources. The internal attacks are the attacks originating from the virtual machines running in this cloud. Network traffic coming outside can be easily rejected. Rejecting customer's internal traffic together with malicious one may cause serious problems as a cloud provider does not able to provide full time access to one instance from another one. That is why DDoS attacks on cloud-based services are different comparatively to attacks on simple server-based services.

3.2   Cloud Security Component Architecture

The analysis of the common security problems in the cloud computing infrastructure showed that security system should meet the following requirements:

- be capable to serve high-loaded networks;
- mitigate attacks with high accuracy as soon as possible, almost in real-time;
- meet the requirements of the customer's security policies;
- should not consume many resources of the cloud platform.

According to these requirements, the following security architecture of cloud infrastructure is proposed. The key modules of the security component are *gate sensor*, *security controller* that consists of *collector, analyzer* and *counteraction module*. They are shown on the Fig. 1. All traffic coming from outside the Cloud Network and inside it goes through the gate. This means that every instance communicates with another inside the cloud network using gate.

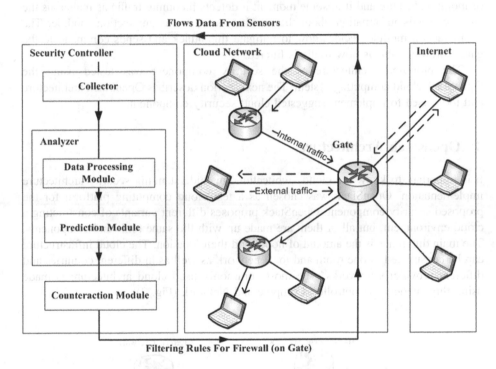

Fig. 1. Security component architecture for cloud networks

The *gate* has sensor that monitors traffic flows passing through the gate and sends data about them to security controller. The *security controller* is a module that processes incoming flow data by defining type of network traffic – benign or malicious - and blocks potentially malicious traffic by sending commands to firewall. The *collector* stores the received data. The *analyzer* prepares input data for the prediction module. The *prediction module* uses data mining classification models. The output of the module is the type of traffic flow. This result goes to the *counteraction module* that in its turn sends commands to firewall according to the received information.

The proposed architecture allows creating an analyzer component without analyzing payload of the packets in order to meet requirements of the company's security policy.

The gate sensor and the controller have to operate fast to produce monitoring in real time. The response time of the controller on monitored traffic should be less than 10 s to mitigate attacks timely.

The analyzer of the security controller is the part of data mining process. Firstly, the analyzer prepares data vectors from collector's data. Currently, the period of capturing traffic parameters is set to four seconds. Experiments showed that such time slot is enough to obtain high prediction accuracy and produce responses on the malicious traffic flows quickly. The input vector is generated from collector's data every four seconds and is sent to the prediction module. The model produces a prediction on type of monitored traffic and if at some moment it detects incoming traffic as malicious the analyzer sends information about the attack type to the counteraction module. The counteraction module decides how to mitigate this attack and sends command to the gate's firewall. This is how traffic is filtered.

The proposed architecture of the security component was tested using the OpenStack cloud computing system. The next section describes OpenStack architecture and tools used to implement suggested cloud security component.

4   Openstack Architecture

It is important to know the cloud architecture for understanding security architecture implementation. OpenStack was chosen as a test cloud computing platform for the proposed security component. OpenStack proposes different variants of constructing a cloud environment, but all of them are made up with the same tools and components. The main difference is the amount of nodes and their location. The cloud infrastructure can be constructed in one room and local network as well as in different countries and different provider's networks. The most commonly used cloud architecture is made using three nodes — Controller, Compute and Network (Fig. 2).

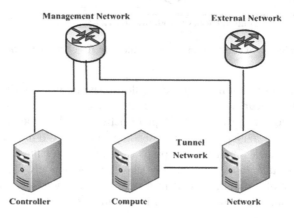

Fig. 2. OpenStack topology layout

The Controller node runs the Identity service, Image Service, management utilities for the Compute and Networking nodes, Networking plug-in, and the dashboard. It also includes supporting services such as a SQL database, message queue, and Network Time Protocol (NTP). The proposed Security Controller (Fig. 1) is implemented on the Controller node.

The Compute node runs the Compute hypervisor that operates tenant virtual machines or instances. By default, Compute uses KVM as the hypervisor. The Compute node also runs the Networking plug-in and an agent that connects tenant networks to instances and provides firewalling (security groups) services. It is possible to run more than one compute node.

The Network node runs the Networking plug-in and several agents that provide tenant networks, switching, routing, NAT, and DHCP services. This node also handles external (Internet) connectivity for tenant virtual machine instances.

If the architecture of the cloud infrastructure does not include Network node all network processes are handled by the Compute node using the same tools as in the Network node.

In our experiments to construct OpenStack infrastructure, we use three identical servers with following characteristics: processor: Intel Xeon i5520, 2.26 GHz * 4 cores; operation system: Ubuntu 14.04; RAM: controller node has 24 GB, network node — 16 GB and compute node — 72 GB.

In OpenStack the route of packet from instance to external nodes or to other instances always goes through the special software router called *qrouter* [6]. The qrouter is the part of Open Virtual Switch (OVS) [12]. This component is a good place for the gate sensor and firewall. In our experiments the gate sensor was Cisco's Net-Flow sensor [3]. It can be placed at the entry point of the cloud's internal network. In the case of OpenStack the gate to the cloud network is OVS which has the sensor already implemented for launching in the network infrastructure. Therefore, we needed only to reconfigure OVS to send the data to the collector.

The collector of the security component can be placed on a distinct node for saving clouds resources or on the controller's node of cloud system as it was done in our case. We used nfdump [10] as a collector. OpenFlow [19] can be used as an automated interface for cloud network management.

5 Data Mining Process and Experiment Setup

Data mining algorithms are implemented using scikit-learn [9]. It is open-source machine learning library for the Python programming language.

We use different classification models to make decisions on the incoming network traffic. Every four seconds data processing module makes a new data vector. We use supervised learning in order to learn and test data mining models. Currently in the analysis process we do not determine the location of the attack source. In the future, we will focus on the second detection layer determining the attack source. Depending on attack source location different data will be collected for making decisions and different counteraction measures will be applied. If it is internal source, the algorithm will try not to block whole IP address, but some ports, in order to keep the legitimate user alive.

However, we set up experimental environment suitable for detection both internal and external attacks. Next subsections describe experimental environment used to train and test classification models.

5.1   Environment for Experimenting with External Attacks

To make predictions, the prediction module needs training datasets. According to that, we created Real Service in Virtual Network Framework (RSVNet) [2]. The framework allows connecting real nodes to virtual network. This feature improves the accuracy of experiments in comparison with simulation in case when modeling the infrastructure attacks on cloud computing platforms. Also, the framework allows implementing any known protection mechanisms or create new ones, including hybrid protection mechanisms. The RSVNet was successfully verified. The framework helps to construct quickly multi-level topology consisting of virtual routers, clients, and real services. The users are able to define new attack scenarios, defensive methods and other features. During experiments, traffic can be logged into pcap files for further analysis. The experimental network consisted of 504 clients and 20 routers. Figure 3 shows the network topology used in the experiments.

The virtual workstation is represented by the OpenStack instance with following characteristics:

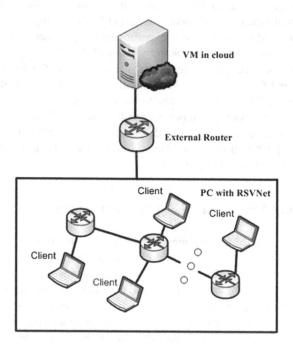

Fig. 3. Topology made using RSVNet

- CPU: 1 VCPU, RAM: 2048 MB;
- operating system: Ubuntu 14.04;
- web-server: Apache/2.4.7, PHP version: PHP 5.5.9-1.

The router is a D-LINK DIR-615 router. To make packets go through virtual network to real and backwards, the routing tables are set inside the router. The range of IP addresses in virtual network is set to 1.0.0.0/8. So all packets with the destination IP address in this subnet are forwarded to the computer with the RSVNet installed. The framework captures needed packets and forwards them into virtual network.

5.2 Environment for Experimenting with Internal Attacks

To simulate internal attacks on the cloud, we setup cloud network that consisted of 6 routers, 10 subnets and, 104 instances. All instances were launched on a Compute node with following characteristics: processor: 24 * Intel(R) Xeon(R) CPU X5680 @ 3.33 GHz, RAM: 62 GB.

One instance was set to manage attack deployment, and one instance was chosen as victim. Attacking instance was launched with following characteristics: 2 GB RAM available and 1 VCPU. All other instances had 1 VCPU and 512 MB RAM available.

To manage all nodes Ansible Opensource Software was installed [11]. Ansible is an IT automation tool, and is used to configure systems, deploy software, and orchestrate more advanced IT tasks such as continuous deployments or zero downtime rolling updates. It is uses ssh to connect to selected nodes and deploy commands.

To generate HTTP traffic, Siege-3.1.0 software was installed on every instance. Siege is an http load testing and benchmarking utility. To generate TCP Flooding attacks, UDP Flooding attacks, ICMP Flooding attacks, hping software was installed on every instance.

Each specific attack is independent to others. They can start and end simultaneously or one after another; and use the same computers to attack.

Figure 4 shows an example of settings file for experiment.

```
[attack0]
name=http0
desc=http_Page1_210victim
command=siege -c 1000 -t60S -b http://192.168.100.210/Page1.html
start_hunter_ip=111
end_hunter_ip=180
start_time=20

[attack1]
name=http1
desc=http_Page1_211victim
command=siege -c 1000 -t60S -b http://192.168.100.211/Page1.html
start_hunter_ip=151
end_hunter_ip=209
start_time=40
```

Fig. 4. Example experiment's scenario

Here you can see two attacks on different victim servers. One attack starts at 20th second and lasts 60 s (-t60S) and other starts at 40th second and last 60 s too. Parameters start_hunter_ip and end_hunter_ip are the ranges of IP addresses listed in additional settings file. It contains every client's IP address, which is able to attack. Therefore, is it can be seen, some of clients will participate in both attacks.

5.3   Learning Network Flow Classification Models

To build up training datasets, the following experiments were made: HTTP Flooding, SYN Flooding and NTP Flooding.

The empirical study showed that the most important attributes for detecting DDoS attacks are amount of bytes, amount of packets, amount of unique pairs of source IP address with port and destination IP address with port. This information is enough to make assessment of the network traffic quickly. For example, traffic of the SYN Flooding attack is characterized by a huge amount of packets having small length. HTTP Flooding causes big amount of packets (however, less than in case of SYN Flooding). The amount of bytes depends on the size of pages being requested by attackers. For NTP Flooding the amount of bytes and packets are extremely big, and it is not difficult to identify such attack. The number of unique pairs can also show whether the network is under the attack. If web server receives lots packets from many IP addresses using different ports then this can be a sign of the attack. The benign traffic is very similar to HTTP Flooding, but the power is significantly lower. In our experiments its power was set to 15 % power of HTTP Flooding traffic settings and the experiment lasted 80 min. The sensor was sending data to the collector immediately after the experiment was launched. The data processing module generated data vectors for classification model every 4 s. This time unit showed to be enough to make clear difference between benign and malicious traffic. In the training mode generated data vectors were stored in the file. By the end of simulation, they were automatically sent to train data mining models. We used cross-validation method to find optimal results.

6   The Experiment Results

6.1   Testing the Classification Models on External Attacks

We have tested several classification models on datasets collected using RSVNet: k-NN, Decision Tree, SVM, and Naïve Bayes.

We made the experiments with different attack power depending on the ratio of the clients involved in the attack: 50 %, 80 % and 100 % of all clients. Total number of the clients was 504.

The results of false positives and false negative rates of the experiments are shown in Tables 1, 2, 3.

Table 1. Results of experiments with 50 % attack power

	False positive rate			False negative rate		
	SYN	HTTP	NTP	SYN	HTTP	NTP
k-NN	1.76	0.8	0.47	0.6	0.34	0.23
Decision tree	0.43	0.37	0.28	0.09	0.1	0.17
SVM	2.47	1.85	1.89	2.98	1.73	0.37
Naïve Bayes	1.46	1.54	1.3	0.8	1.02	0.22

The first series of experiments were made with the 50 % attack power (Table 1).

Almost every model showed a low false positive rate, but Decision Tree had the best result.

The second series of experiments were made with 80 % of all clients involved in the attack (Table 2).

Table 2. Results of experiments with 80 % attack power

	False positive rate			False negative rate		
	SYN	HTTP	NTP	SYN	HTTP	NTP
k-NN	1.23	0.33	0.24	0.42	0.29	0.18
Decision tree	0.35	0.27	0.16	0.07	0.09	0.11
SVM	1.2	0.78	0.83	2.4	1.42	0.34
Naïve Bayes	1.43	1.7	1.7	1.31	1.07	0.25

Again the Decision Tree model was the best in detecting attacks. Less than 1 benign flow per 100 flows was detected as malicious. This means that almost all benign traffic reached its destination. In case of SYN Flooding attack other algorithms showed almost equal amount of false positives. The k-NN algorithm showed the second result in detecting HTTP Flooding attack.

The third series of experiments were made with all clients involved in the attack (Table 3).

Table 3. Results of experiments with 100 % attack power

	False positive rate			False negative rate		
	SYN	HTTP	NTP	SYN	HTTP	NTP
k-NN	1.05	0.27	0.18	0.37	0.24	0.13
Decision tree	0.12	0.16	0.08	0.05	0.04	0.03
SVM	1.34	0.8	0.85	1.99	0.87	0.18
Naïve Bayes	1.27	1.54	1.3	1.12	0.65	0.17

As it can be seen the accuracy of traffic classification increases with the power of the attack. This happens because the higher attack power the greater differences in attribute values of vectors for different types of traffic. Decision Tree showed better results in false positive rate as in the previous experiment.

The speed of data processing and making predictions with all data was less than one second. That means that every five second security controller has resulted about traffic flow type in cloud network. So if the attack starts immediately after last prediction it is identified not more than in 5 s.

The next experiments using Decision Tree as a prediction algorithm in the cloud security component. The same attack types were used: SYN Flooding, HTTP Flooding, and NTP Flooding. The first series of experiments were made with the same benign model, but with different amount of clients involved in the attack. There were from 25 to 400 attack clients with the increment of 25 clients per experiment. Each experiment lasted 10 min. To analyze the results, we calculated F-measure for each type of traffic.

F-measure can be a better single metric when compared to precision (TP/(TP + FP)) and recall (TP/(TP + FN)); as precision and recall give different information that can complement each other when combined. If one of them excels more than the other, F-measure reflects it. Weights of recall and precision can be changed due to their importance in results. In this paper traditional F-measure was chosen as:

$$F = 2 \cdot \frac{precision \cdot recall}{precision + recall} \tag{1}$$

Results of F-measure values are shown in the Fig. 5.

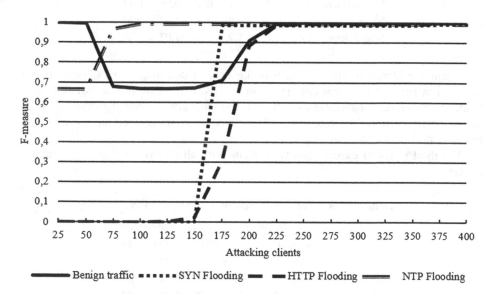

Fig. 5. F-measure of experiments with different attacking clients amount

As it follows from the plot Decision Tree has very low F-measure until 200 clients involved in SYN Flooding and HTTP Flooding attacks. Often SYN Flooding attack was classified as HTTP Flooding and HTTP Flooding was classified as NTP flooding traffic. That is not a problem for cloud services due to low attack power: it was just 16 000 flows per 10 min for HTTP Flooding, and 200 000 SYN requests per 10 min for SYN Flooding. Such attack power does not affect cloud services. The reduction of F-measure was detected when 75 to 200 clients were involved in the benign traffic modeling. That happened because HTTP Flooding flows were interpreted as benign traffic since the model of legitimate traffic generation is similar to 100 clients involved in HTTP Flooding. NTP Flooding had the highest F-measure due to the features of the attack.

It should be mentioned that Decision Tree was trained on network consisted of 504 attacking clients. That means that the model predicts traffic type adequately with high-loaded networks.

We also made several experiments with different power of legitimate traffic. Results show that when there is almost no traffic in the cloud network the false positive rate grows up. The solution of that problem is better benign traffic modeling.

6.2 Testing the Classification Models on Internal Attacks

We implemented series of experiments using different attack scenarios to test classification models against internal attacks. We have made experiments with different attack types: HTTP Flooding, SYN Flooding, UDP Flooding and with benign traffic scenario. We learned the same models as in the case of external attacks. We have reached the power of 1 Gb/s in HTTP Flooding attack in the virtual network of OpenStack.

We built up scenario consisting of each attack with increasing power. The whole experiment lasted 5 h: 2 for training dataset and 3 h for testing dataset.

Each attack type and legitimate scenario lasted almost the same time. Differences between training and testing scenarios were the following: amount of attackers during particular period of time, amount of attacking threads coming from one client.

To analyze effectiveness of each classification algorithm we calculated false positive and false negative rates (FP and FN) and F-measure. HTTP Flooding and benign traffic scenario results are provided in Table 4.

Table 4. Results of experiments with inside HTTP Flooding attack

	FP rate	FN rate	Recall	Precision	F-measure
kNN	0.33	0.25	98.5	99.9	99.1
Decision tree	0.05	0.246	99.8	99.8	99.8
SVM	4.1	2.7	96.2	97.9	97.0
Naïve Bayes	15.74	75.1	0.7	25	1.4

Decision Tree model was the best classification model again. False positive rate is 0.05 % of all. So 99.95 % of legitimate traffic is not rejected. 99.75 % of malicious traffic is rejected. Practically that means that only 250 Mb/s goes through attack detection system when 100 Gb/s attack power will be on the cloud network. Naïve Bayes showed the worst results with very high false negative rate.

Comparing results of the experiments with outside and inside attacks are both very good. Using such detecting techniques can improve any cloud defense system.

7 Conclusion and Future Work

In this paper, we presented the approach for detection internal and external DDoS attack in cloud computing using data mining techniques. We also proposed the architecture of the cloud security component for detecting DDoS attacks. The architecture consists of sensor, controller storing information about traffic flow, analyzer processing data for data mining model and counteraction module sending commands to firewall to prevent attack's damage. Proposed architecture does not affect customer's data to meet the requirements of company's security policies, and can be used in constructing protection of services in the cloud computing platforms from DDoS attacks on them. The authors implemented the component prototype in the OpenStack cloud computing system and carried out a set of experiments with DDoS attacks. The experiments with external DDoS attacks were done using RSVNet framework, to analyze internal DDoS attacks we created experimental environment using python, siege and hping was made. The results of the tested models show high accuracy and small false positive and false negative rate with high F-measure. The Decision Tree is the best according to series of experiments with different attack scenarios. Decision tree shows better results than other models because it constructs well-defined dependencies between different features for detecting the most popular DDoS attacks. However, we expect that testing decision tree on more complicated data could show different results. And our future work will be analysis of model's efficiency on more sophisticated data sets.

The future research will be devoted to the enhancement of the prediction module and development of the self-training models. This can help to react correctly on changes in traffic power or in the amount of public services. Next papers will be focused on the DDoS attacks mitigation based on information gained by the improved prediction module, described in this paper. In addition, we plan to improve legitimate traffic model and implement experiments involving many instances facing DDoS attacks.

Acknowledgments. The paper has been prepared within the scope of the state project "Organization of scientific research" of the main part of the state plan of the Board of Education of Russia, the project part of the state plan of the Board of Education of Russia (task 2.136.2014/K) as well as supported by grant of RFBR # 16-07-00625, supported by Russian President's fellowship, as well as with the financial support of the Foundation for Assistance to Small Innovative Enterprises in the scientific and technical spheres.

References

1. Amazon Web Services, Inc.: Amazon Web Services (AWS) - Cloud Computing Services. https://aws.amazon.com
2. Bekeneva, Y., Borisenko, K., Shorov, A., Kotenko, I.: Investigation of DDoS attacks by hybrid simulation. In: Khalil, I., Neuhold, E., Tjoa, A.M., Xu, L.D., You, I. (eds.) ICT-EurAsia 2015 and CONFENIS 2015. LNCS, vol. 9357, pp. 179–189. Springer, Heidelberg (2015). doi:10.1007/978-3-319-24315-3_18

3. Choudhary, S., Srinivasan, B.: Usage of netflow in security and monitoring of computer networks. Int. J. Comput. Appl. **68**, 17–24 (2013)
4. Cloudsecurityalliance.org.: Top Threats: Cloud Security Alliance. https://cloudsecurityalliance.org/research/top-threats
5. Delimitrou, C., Kozyrakis, C.: Security implications of data mining in cloud scheduling. IEEE Comput. Archit. Lett. 1-1 (2015)
6. Docs.openstack.org.: OpenStack Docs: Scenario: Legacy with Open vSwitch. http://docs.openstack.org/networking-guide/scenario_legacy_ovs.html
7. Dou, W., Chen, Q., Chen, J.: A confidence-based filtering method for DDoS attack defense in cloud environment. Future Gener. Comput. Syst. **29**, 1838–1850 (2013)
8. Editor, L.: Cyberattacks and Distributed Denial of Service (DDoS) threats on financial firms produce big pay offs - Beyond Bandwidth. http://blog.level3.com/finance/cyberattacks-and-distributed-denial-of-service-ddos-threats-on-financial-firms-produce-big-pay-offs/
9. Garreta, R., Moncecchi, G.: Learning scikit-learn
10. Haag, P.: Watch your flows with NfSen and NFDUMP. 50th RIPE Meeting (2005)
11. Hall, D.: Ansible configuration management
12. Justin, P.: Virtual switching in an era of advanced edges. In: 2nd Workshop on Data Center–Converged and Virtual Ethernet Switching (2010)
13. Kaspersky DDoS Protection: Kaspersky Lab. http://media.kaspersky.com/kaspersky-ddos-protection-data-sheet.pdf
14. Milojičić, D., Llorente, I., Montero, R.: OpenNebula: a cloud management tool. IEEE Internet Comput. **15**, 11–14 (2011)
15. Munz, G., Li, S., Carle, G.: Traffic anomaly detection using K-means clustering. In: GI/ITG Workshop MMBnet (2007)
16. Salesforce.com: What is Cloud Computing? - Salesforce UK. http://www.salesforce.com/uk/cloudcomputing/#where
17. Secucloud Web-Site: Secucloud. https://secucloud.com/en/company/about-us
18. Sefraoui, O., Aissaoui, M., Eleuldj, M.: OpenStack: toward an open-source solution for cloud computing. Int. J. Comput. Appl. **55**, 38–42 (2012)
19. Vaughan-Nichols, S.: OpenFlow: the next generation of the network? Computer **44**, 13–15 (2011)
20. Vieira, K., Schulter, A., Westphall, C., Westphall, C.: Intrusion detection for grid and cloud computing. IT Prof. **12**, 38–43 (2010)
21. Zolotukhin, M., Hamalainen, T., Kokkonen, T., et al.: Data mining approach for detection of DDoS attacks utilizing SSL/TLS protocol. In: 15th International Conference, NEW2AN 2015, pp. 274–285. St. Petersburg, Russia (2015)

Coordinate Refinement on All Atoms of the Protein Backbone with Support Vector Regression

Ding-Yao Huang[1], Chiou-Yi Hor[2], and Chang-Biau Yang[1]([⊠])

[1] Department of Computer Science and Engineering,
National Sun Yat-sen University, Kaohsiung 80424, Taiwan
cbyang@cse.nsysu.edu.tw
[2] Iron and Steel Research Development Department, China Steel Corporation,
Kaohsiung 81233, Taiwan

Abstract. For the past decades, many efforts have been made in the fields of protein structure prediction. Among these, the protein backbone reconstruction problem (PBRP) has attracted much attention. The goal of PBRP is to reconstruct the 3D coordinates of all atoms along the protein backbone for given a target protein sequence and its C_α coordinates. In order to improve the prediction accuracy, we attempt to refine the 3D coordinates of all backbone atoms by incorporating the state-of-the-art prediction softwares and support vector regression (SVR). We use the predicted coordinates of two excellent methods, PD2 and BBQ, as our feature candidates. Accordingly, we define more than 100 possible features. By means of the correlation analysis, we can identify several significant features deeply related to the prediction target. Then, a 5-fold cross validation is carried out to perform the experiments, in which the involved datasets range from CASP7 to CASP11. As the experimental results show, our method yields about 8 % improvement in RMSD over PD2, which is the most accurate predictor for the problem.

Keywords: Protein backbone · Bioinformatics · Three-dimensional coordinates · Support vector regression · Prediction

1 Introduction

Proteins are required for growth and development in the human body. A protein, called a polypeptide, is composed of a chain of amino acids. A series of amino acids are linked together by peptide chains to form a protein backbone. An *amino acid* is the fundamental unit of a protein. There are twenty kinds of standard amino acids and each kind of amino acid can be differentiated by its R group. Each protein has its own specific functions and unique structure which can cooperate with other proteins to achieve some required functionalities. Because protein structure and protein function are closely related, in order to find out these protein functions, most biologists adopt the approaches that predict protein tertiary structures by means of amino acid sequences or other related information.

© Springer International Publishing Switzerland 2016
P. Perner (Ed.): ICDM 2016, LNAI 9728, pp. 212–222, 2016.
DOI: 10.1007/978-3-319-41561-1_16

There are two main approach types of the protein structure prediction. The first one is the experimental method, including X-ray diffraction and nuclear magnetic resonance (NMR) [10,14]. The other one is the computational method, including *homology modeling* [5], *folding recognition* [6], and *ab initio* [9]. As for these two categories, the former one requires a lot of time and cost while the latter one doesn't. Therefore, this motivates us to adopt the computational method to predict the 3D protein structure.

The *all-atom protein backbone reconstruction problem* (PBRP) is to utilize a protein sequence and its 3D coordinates of α-carbon (C_α) for predicting the 3D coordinates of all atoms (N, C and O atoms) on the protein backbone. There are several related studies which fall into this category, such as SABBAC [11], Wang's method [17], Chang's method [2], BBQ [4], Chen's method [3], Wu's method [18], PD2 [12] and so on. For a complete survey on the methods (or software) of the protein backbone prediction, one can refer to the thesis written by Yuan [19].

In order to improve the accuracy of protein structure prediction, we propose a new method to refine the 3D coordinates of all backbone atoms by means of *support vector regression* (SVR) [15,16]. Our prediction target is the differences of N and O atoms' coordinates between the predicted results of PD2 and the real 3D coordinates of PDB, and the differences of C atoms' coordinates between the predicted results of BBQ and the real 3D coordinates of PDB. Our training features are generated from the predicted results of PD2 and BBQ.

The experimental datasets range from CASP7 to CASP11, where CASP stands for the *Critical Assessment of Protein Structure Prediction* [13]. We perform a 5-fold cross validation experiment for performance evaluation. For each fold of validation, a CASP dataset is extracted for testing and the remaining CASP datasets are involved for training. Coordinates in the training datasets are first predicted by PD2 and BBQ. Then these predicted coordinates are compared with their real coordinates to produce the differences. The differences are the learning objective for the SVR. In the feature selection stage, we analyze the correlations between the objective value and available features, and then select the most representative features. To ease the training process, we partition the amino acids into twenty groups (twenty datasets) and then predict these differences by each individual SVR. Finally, we combine our predicted differences, their corresponding predicted N and O coordinates with PD2, and predicted C coordinates with BBQ to export our predicted coordinates of N, C and O. The performance is evaluated by the RMSD values. The experimental results show that our prediction results yield about 8 % improvement over the results predicted by PD2, which is the most accurate predictor for the problem [19].

The rest of this paper is organized as follows. In Sect. 2, we will introduce experimental datasets, root-mean-square deviation (RMSD) and features used in this paper. In Sect. 3, we will describe our proposed method in detail. In Sect. 4, we will present our experimental results. Finally, in Sect. 5, the conclusion will be given.

2   Preliminaries

2.1   Datasets and Performance Evaluation

Critical Assessment of Protein Structure Prediction (CASP) [13] is an international competition held every two years since 1994. The main goal of CASP is to evaluate the capabilities of the methods for identifying three-dimensional structure of the protein from its amino acid sequence. In order to assess the performance of a method, CASP examines the predicted 3D structures in many different ways, such as the accuracy of a model, accuracy of a quaternary structure, and so on. Because our research also focuses on the 3D structure prediction, we use CASP datasets to perform our experiments.

Root-mean-square deviation (RMSD) [7,8] is an evaluation method of molecular modeling which computes the average distance between the predicted values and the ground truths.

$$RMSD = \sqrt{\frac{1}{l} \sum_{i=1}^{l} (X_i^A - X_i^B)^2}, \tag{1}$$

where X_i^A and X_i^B denote the coordinates of the ith atom on the backbone in the proteins A and B, respectively, and l denotes the length of the proteins. Generally, a lower RMSD indicates the higher similarity, which means that the predicted coordinates is close to the real ones. RMSD has been widely used in structural biology. In this paper, we also use RMSD to evaluate the quality of the prediction models.

2.2   Feature Generation and Feature Selection

This subsection describes the features we use to build the SVR models. All features are extracted within a fragment, as illustrated in Fig. 1. That is, once the prediction target, like the C atom, is determined, we define a window around this atom. We say this atom and its surrounding atoms constitute a fragment. The features required by SVR are calculated as follows.

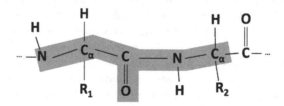

Fig. 1. The fragment of $L_C = 6$ consecutive atoms on the protein backbone, where the C atom is assigned to be the fragment center.

Coordinate: The predicted x, y and z coordinates of N, C, O and N_{next} atoms are obtained from BBQ and PD2, denoted as $N_x(BBQ)$, $N_y(BBQ)$, $C_z(PD2)$, etc. The real coordinates of two C_αs are also involved as the features. Thus, there are totally 30 features.

Coordinate difference: The coordinate differences are calculated from the predicted x, y and z coordinates of N, C, O and N_{next} atoms. Only the difference of each corresponding pair of coordinates is computed, such as $N_x(PD2) - N_x(BBQ)$, $N_y(PD2) - N_y(BBQ)$, etc. Thus, 12 features are obtained.

Euclidean distance: The Euclidean distance measures the amount of space between the two predicted positions, obtained from BBQ and PD2, of the same atom on the Euclidean space. The formula for calculating Euclidean distance is given as follows.

$$d(x_{p_i}, x_{b_i}, y_{p_i}, y_{b_i}, z_{p_i}, z_{b_i}) = \sqrt{\left(x_{p_i} - x_{b_i}\right)^2 + \left(y_{p_i} - y_{b_i}\right)^2 + \left(z_{p_i} - z_{b_i}\right)^2} \quad (2)$$

In Eq. 2, x_{p_i}, y_{p_i}, z_{p_i} are the 3D coordinates of the ith atom on the protein backbone predicted by PD2 and x_{b_i}, y_{b_i}, z_{b_i} are the 3D coordinates of the ith atom on the protein backbone predicted by BBQ.

Bond length: Two adjacent atoms form a bond on the protein backbone. There are five bond lengths in a fragment, including N-C_α, C_α-C, C=O, C-N_{next}, and N_{next}-$C_{\alpha_{next}}$. The bond length is calculated in terms of the Euclidean distance. Because bond lengths associated with BBQ and PD2 are calculated separately, it follows that 10 features are obtained.

Bond length difference: Each bond length difference is derived by the two corresponding bond lengths, predicted from the two methods.

Bond angle: Since three atoms form an angle, the bond angle can thus be obtained by *law of cosines*. Consequently, for a fragment with $L_C = 6$, we can get five different angles, including N-C_α-C, C_α-C=O, C_α-C-N_{next}, O=C-N_{next}, and C-N_{next}-$C_{\alpha_{next}}$.

Bond angle difference: Each bond angle difference is calculated from the two corresponding predicted bond angles of two methods.

Torsion angle: The torsion angle is computed by four consecutive atoms on the main chain. In addition to $\phi(C_{prev}$-N-C_α-C$)$, $\psi(\text{N-}C_\alpha\text{-C-N})$ and $\omega(C_\alpha$-C-N_{next}-$C_{\alpha_{next}})$, we also choose the other features in our fragment, including N-C_α-C=O, O=C-N_{next}-$C_{\alpha_{next}}$, C_α-C=O-N_{next}, C_α-C-N_{next}=O (planes of C_α-C-N_{next} and C -N_{next}=O), N_{next}-C-C_α=O and C-N_{next}-$C_{\alpha_{next}}$-C_{next}. Since PD2 and BBQ are used for the computation of torsion angles, 18 features are obtained.

Torsion angle difference: This feature is obtained from the torsion angles. We compute the differences from the torsion angles obtained by PD2 and BBQ.

So far, we have defined nine kinds of features. Table 1 shows the feature names and their sizes.

The above feature extraction method is performed around the C atom in a fragment-by-fragment manner, as shown in Fig. 1. We assign the C atom as the

Table 1. The names and sizes of all feature subsets.

Feature index	Feature name	Size
F1	Coordinate	30
F2	Coordinate difference	12
F3	Euclidean distance	4
F4	Bond length	10
F5	Bond length difference	5
F6	Bond angle	10
F7	Bond angle difference	5
F8	Torsion angle	18
F9	Torsion angle difference	9
Total		103

fragment center because this arrangement is most suitable for predicting both C and O atoms. Nevertheless, one may wonder whether this is also suitable when N atom is served as the prediction target. Therefore, we assign the N atom as the fragment center and perform another experiment. The experimental results (not shown in this paper) exhibit that the prediction accuracies of N atoms with the C-center window and those with the N-center window only show little difference. Thus, we still use the C-center window for predicting N atoms here.

Since the performance of models depends heavily on the selected features, we have to consider which one is relevant to the coordinate prediction. In order to identify important features, we calculate the Pearson's correlation coefficient between the objective value and each feature value. For a given feature, its correlation coefficient with the objective variable is given in Eq. 3.

$$r = \frac{\sum_{i=1}^{n}(x_i - \bar{x})(y_i - \bar{y})}{\sqrt{\sum_{i=1}^{n}(x_i - \bar{x})^2}\sqrt{\sum_{i=1}^{n}(y_i - \bar{y})^2)}},$$ (3)

where n denotes the number of data elements, x_i denotes the ith element of data instances $(x_1, x_2, \ldots, x_n)$, y_i denotes the ith instance of the objective values $(y_1, y_2, \ldots, y_n)$, $\bar{x}$ and $\bar{y}$ represent the means of x and y, respectively.

3   The Coordinate Difference Prediction Method

In order to improve the predicted results, we adopt SVR to predict the x-difference, y-difference and z-difference of each of N, C and O atoms on the backbone of a target protein. Then, these differences are combined with the predicted results of BBQ and PD2 to yield our predicted coordinates. Our coordinate difference prediction procedure is described as follows.

Algorithm: The Coordinate Difference Prediction Method.

Input: 1. One training set T, containing the predicted coordinates obtained from PD2 and BBQ and the real coordinates in PDB.

2. One target protein, containing the predicted coordinates of PD2 and BBQ, along with real C_α coordinates.

Output: The predicted coordinates of N, C and O atoms along the target protein backbone.

Step 1 (Extract features): Partition the residues of all proteins in T into 20 groups, corresponding to 20 types of standard amino acids. Calculate the 103 feature values associated with each kind of residue, defined in the previous section.

Step 2 (Perform correlation analysis): For each of the nine objective values (O_x-difference, O_y-difference, O_z-difference, N_x-difference, etc.) in T, calculate the Pearson's correlation coefficient between each feature value and the objective value. Since the p-value represents the confidence level associated with its correlation coefficient, we thus can adopt a thresholding method to identify significant features.

Step 3 (Predict the difference by SVR): For each kind of objective values and amino acid groups, we use the selected features to train an SVR model. Thus, 180 models (20 kinds of residues, 3 kinds of atoms, 3D coordinates) are obtained. Then, these models are invoked to perform prediction of the target protein based on the residue and atom types.

Step 4 (Combine the predicted difference with PD2/BBQ): Combine the predicted differences with their corresponding predicted positions obtained by BBQ and PD2 to generate the final coordinates.

Step 5 (Merge all residues together): Bring the predicted coordinates of all residues together to reconstruct the 3D positions of all atoms (N, C and O atoms) on the target protein backbone.

The flow chart is shown in Fig. 2.

4   Experimental Results

For evaluating the performance of our method, we adopt CASP7, CASP8, CASP9, CASP10 and CASP11 as the experimental datasets, which contain 65, 52, 63, 39 and 55 proteins, respectively. We use only the information of chain A of proteins to carry out the experiments. If there is no chain A, the next chain is used. All features are scaled into the range of $[a, b] = [-1, 1]$ by Eq. 4.

$$\frac{x_i - x_{min}}{x_{max} - x_{min}}(b - a) + a, \tag{4}$$

where x_i denotes the value of a certain feature of the ith training data element, x_{max} is the maximum value in the feature, x_{min} is the minimum value, a and b are the lower and upper bounds of the range, respectively.

In the 5-fold cross validation, each CASP is selected as the testing dataset for one time. Once a CASP is determined for testing, the remaining CASPs serve

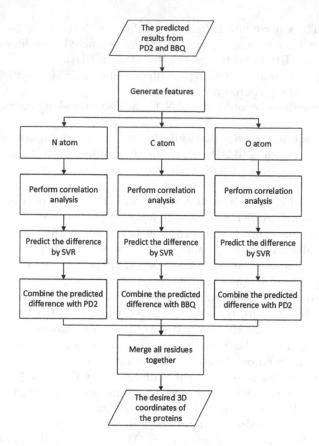

Fig. 2. The flow chart for predicting the 3D structure of a protein.

as the training datasets. For example, if we select CASP7 as the testing dataset, then the rest ones, from CASP8 to CASP11, are used as the training dataset. The testing procedure is performed for each CASP dataset.

As mentioned in the previous section, in our algorithm, Step 1 generates the feature candidates and Step 2 calculates correlation coefficients of these features and the objective values. From the correlation analysis, we find that some of the features are indeed significant.

Figure 3 shows the correlations of the 103 features and the O_x-differences of all 20 amino acids obtained from CASP8 to CASP11. We count the number of significances, which means that p-value (the complement of confidence level) is less than 0.05. The value of 0.05 is a widely adopted standard cut-off. It denotes that the test shows strong evidence against the null hypothesis that no correlation between the objective value and the feature value. Note that maximum count of significances is 20, because there are 20 types of amino acids. Here, if one of the following criteria is satisfied, a feature is considered as a significant one, and it is selected for training.

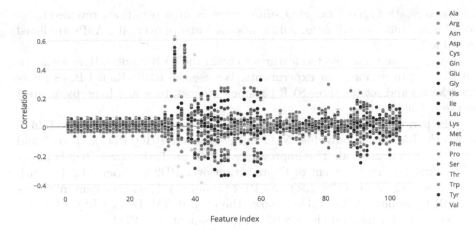

Fig. 3. The correlations of 103 features and O_x-differences of 20 amino acids in the training set consisting of CASP8 to CASP11. (Color figure online)

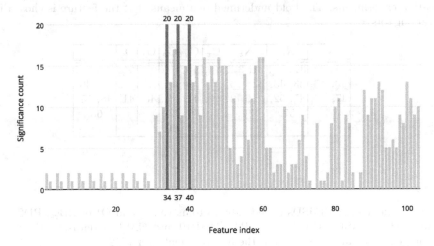

Fig. 4. The counts of significant occurrences in various features with respect to O_x-differences in the training set consisting of CASP8 to CASP11.

1. The feature has correlation value greater than or equal to 0.15 and the number of its significance counts is greater than or equal to 15.
2. The feature has correlation value greater than or equal to 0.4 and the number of its significance counts is greater than or equal to 10.

Furthermore, Fig. 4 illustrates the count of significances of the 103 features with respect to O_x-differences obtained from CASP8 to CASP11. According to our criteria, we can extract three significant features (indices 34, 37 and 40), which are the predicted differences $C_x(PD2) - C_x(BBQ)$, $O_x(PD2) - O_x(BBQ)$ and $N_{next,x}(PD2) - N_{next,x}(BBQ)$, respectively. That is to say, these features highly correlate with our target O_x-differences. Similar results are revealed for y-coordinates and z-coordinates. All significant features over all CASPs are listed in Table 2.

Next, we use these selected features to train the SVR models. Here we adopt LibSVM [1] to perform our experiments. We use the RBF (Radial Basis Function) kernel and set the three SVR parameters (cost, tube and hyperparameter) to the default values.

The performance of our experimental results is evaluated by RMSD. Table 3 shows the RMSDs of individual N, C and O atoms of BBQ, PD2 and our 5-fold cross validation method. The improvement of O prediction over PD2 is about 7.7 %. And the improvement of C prediction over PD2 is about 13.2 %, which combines the gain from the BBQ over PD2 (about 6.4 %) and the gain from ours over BBQ (about 7.8 %). Table 4 shows the overall RMSDs of BBQ, PD2 and our method. Our method yields 8.03 % improvement over PD2.

Table 2. The significant features selected in the nine objective values of the 5-fold cross validation experiments. The bold underlined one means that the feature is chosen in all training sets.

N_x	N_y	N_z	C_x	C_y	C_z	O_x	O_y	O_z
31, 32, **33**, **43**, 47, **52**, **57**.	31, **32**, 33, 43, 47, 52, 57.	31, **32**, **33**, **43**, **47**, **52**, **57**.	**34**, **37**, **40**.	**35**, **38**. 41, 60.	**36**, **39**, **42**. 55, 60.	**34**, **37**. 40.	**35**, **38**. 41.	**36**, **39**, **42**, 46, 48, 55, 60.

Table 3. The average RMSDs of individual atoms (N, C and O) of BBQ, PD2 and our method in CASP7, CASP8, CASP9, CASP10 and CASP11 datasets. Here, the percentage inside parentheses means the improvement over PD2.

Dataset	Method								
	BBQ			PD2			Our method		
	N	C	O	N	C	O	N	C	O
CASP7	0.2401	0.2395	0.6347	0.1897	0.2416	0.5879	0.1853	0.2209(8.57 %)	0.5443(7.42 %)
CASP8	0.3775	0.2668	0.7051	0.2375	0.3057	0.6968	0.2336	0.2486(18.68 %)	0.6415(7.94 %)
CASP9	0.4229	0.2569	0.6632	0.2462	0.2864	0.6115	0.2434	0.2367(17.35 %)	0.5672(7.24 %)
CASP10	0.2429	0.2444	0.6513	0.2042	0.2298	0.5991	0.2015	0.2227(3.09 %)	0.5515(7.95 %)
CASP11	0.2992	0.2310	0.6252	0.2197	0.2525	0.5994	0.2144	0.2127(15.76 %)	0.5497(8.29 %)
Arithmetic mean	0.3165	0.2477	0.6449	0.2195	0.2632	0.6189	0.2157	0.2283(12.69 %)	0.5708(7.77 %)
Weighted arithmetic mean	0.3205	0.2477	0.6482	0.2198	0.2646	0.6179	0.2161	0.2284(13.17 %)	0.5701(7.73 %)

Table 4. The RMSDs of all atoms of BBQ, PD2 and our method for CASP7, CASP8, CASP9, CASP10 and CASP11 datasets.

Dataset	Method			Improvement over PD2 %
	BBQ	PD2	Ours	
CASP7	0.3632	0.3335	0.3102	6.99 %
CASP8	0.4412	0.4030	0.3672	8.88 %
CASP9	0.4344	0.3635	0.3341	8.09 %
CASP10	0.3726	0.3386	0.3159	6.70 %
CASP11	0.3775	0.3491	0.3166	9.31 %
Arithmetic mean	0.3978	0.3575	0.3288	7.99 %
Weighted arithmetic mean	0.3986	0.3574	0.3286	8.03 %

5   Conclusion

In the past decades, lots of efforts have been devoted to the study of the protein backbone reconstruction problem. Until now, the methods, such as BBQ, PD2 and so on, have already been successfully developed and applied to the problem. Since these methods have their individual strengths and weakness, the prediction accuracy can thus be improved by taking advantage of their strengths.

In this paper, we propose an algorithm to refine the 3D coordinates of all atoms on a protein backbone with SVR. The objective values of our prediction is the differences between the predicted coordinates and the real ones. We first define a set of feature candidates extracted from the predicted coordinates of BBQ and PD2. It is well-known that the key factor to affect the prediction performance is the feature relevance. Thus, we perform the correlation analysis to identify significant features. The experimental datasets range from CASP7 to CASP11. As the experimental results show, the three most significant features for predicting the O_x-differences and C_x-differences are the differences of the predicted x-coordinates of PD2 and BBQ in C, O and N_{next} atoms. Similar results are exhibited for y-coordinates and z-coordinates. In summary, our method yields about 8 % improvement in RMSD over PD2, which is the best previous predictor in this problem up to now.

Acknowledgments. This research work was partially supported by the Ministry of Science and Technology of Taiwan under contract MOST 104-2221-E-110-018-MY3.

References

1. Chang, C.C., Lin, C.J.: LIBSVM: a library for support vector machines. ACM Trans. Intell. Syst. Technol. **2**(3), 1–27 (2011)

2. Chang, H.Y., Yang, C.B., Ann, H.Y.: Refinement on O atom positions for protein backbone prediction. In: Proceedings of the 2nd WSEAS International Conference on Biomedical Electronics and Biomedical Informatics (BEBI 2009), Moscow, Russia, pp. 99–104 (2009)

3. Chen, K.Y., Yang, C.B., Huang, K.S.: Prediction of protein backbone structure by preference classification with SVM. In: Proceedings of the 9th International Conference on Information Systems and Technology Management, Sao Paulo, Brazil, pp. 1193–1206 (2012)

4. Gront, D., Kmiecik, S., Kolinski, A.: Backbone building from quadrilaterals: a fast and accurate algorithm for protein backbone reconstruction from alpha carbon coordinates. J. Comput. Chem. **28**, 1593–1597 (2007)

5. Holm, L., Sander, C.: Database algorithm for generating protein backbone and side-chain coordinates from a C alpha trace application to model building and detection of coordinate errors. J. Mol. Biol. **21**(1), 183–194 (1991)

6. Eisenberg, D., Bowie, J.U., Luthy, R.: A method to identify protein sequences that fold into a known three-dimensional structure. Science **253**, 164–170 (1991)

7. Kabsch, W.: A solution for the best rotation to relate two sets of vectors. Acta Crystallogr. Sect. A **32**, 922–923 (1976)

8. Kabsch, W.: A discussion of the solution for the best rotation to relate two sets of vectors. Acta Crystallogr. Sect. A **34**, 827–828 (1978)

9. Kazmierkiewicz, R., Liwo, A., Scheraga, H.A.: Energy-based reconstruction of a protein backbone from its α-carbon trace by a Monte-Carlo method. J. Comput. Chem. **23**, 715–723 (2002)

10. Krasnogor, N., Hart, W.E., Smith, J., Pelta, D.A.: Protein structure prediction with evolutionary algorithms. In: Proceedings of the Genetic and Evolutionary Compution Conference, Orlando, USA, pp. 1596–1601 (1999)

11. Maupetit, J., Gautier, R., Tuffery, P.: SABBAC: online structural alphabet-based protein backbone reconstruction from alpha-carbon trace. Nucleic Acids Res. **34**, W147–W151 (2006)

12. Moore, B.L., Kelley, L.A., Barber, J., Murray, J., MacDonald, J.T.: High-quality protein backbone reconstruction from alpha carbons using Gaussian mixture models. J. Comput. Chem. **34**, 1881–1889 (2013)

13. Moult, J., Fidelis, K., Kryshtafovych, A., Rost, B., Tramontano, A.: Critical assessment of methods of protein structure prediction (CASP) x Round IX. Proteins **79**, 1–5 (2011)

14. Ruczinski, I., Kooperberg, C., Bonneau, R., Baker, D.: Distribution of beta sheets in proteins with application to structure prediction. Proteins: Struct. Funct. Genet. **48**, 85–97 (2008)

15. Smola, A.J., Scholkopf, B.: A tutorial on support vector regression. Stat. Comput. **14**, 199–222 (2004)

16. Vapnik, V., Golowich, S.E., Smola, A.: Support vector method for function approximation, regression estimation, and signal processing. In: Advances in Neural Information Processing Systems 9, pp. 281–287. MIT Press (1996)

17. Wang, J.H., Yang, C.B., Tseng, C.T.: Reconstruction of protein backbone with the α-carbon coordinates. J. Inf. Sci. Eng. **26**(3), 1107–1119 (2010)

18. Wu, H.F., Yang, C.B., Hor, C.Y., Peng, Y.H., Tseng, K.T.: Protein backbone reconstruction with tool preference classification for standard and nonstandard proteins. In: Proceedings of the 12th Conference on Information Technology and Applications in Outlying Islands, Kinmen, Taiwan, pp. 175–182 (2013)

19. Yuan, H.C.: A survey of computational methods for protein structure prediction. Master's thesis, National Sun Yat-sen University, Kaohsiung, Taiwan, July 2015

The Computational Wine Wheel 2.0 and the TriMax Triclustering in Wineinformatics

Bernard Chen$^{(\boxtimes)}$, Christopher Rhodes, Alexander Yu, and Valentin Velchev

Department of Computer Science,
University of Central Arkansas, 201 Donaghey Ave, Conway, AR 72034, USA
bchen@uca.edu

Abstract. Even with the current state of technology, data growth is increasing so fast that without proper storage and analytical techniques, it is challenging to process and analyze large datasets. This applies to knowledge bases from all fields and all kinds of data. In Wineinformatics, various kind of data related to wine, including physicochemical laboratory data and wine reviews, are analyzed by data science related researches. In the previous work, we proposed the Computational Wine Wheel, derived from 2011's top 100 wine, to automatically process and extract key attributes from human-language-format wine expert reviews. In this work, past 10 year's top 100 wines are collected and formed a 1000 excellent wines dataset to further improve the Computational Wine Wheel. The extraction process led to the creation of what we call a Computational Wine Wheel 2.0, which is a wine attribute dictionary consisting of 985 categorized and normalized wine attributes. After the Computational Wine Wheel 2.0 is formed, we experiment it on a region- and grape type- specific dataset to seek new types of information in Wineinformatics. A novel TriMax Triclustering algorithm specifically used for the dataset processed by the Computational Wine Wheel is proposed and applied to discover three dimensional clusters (Wine × Attributes × Vintage) in wine. We found that the TriMax Triclustering algorithm produced promising and cohesive results that can be used in various aspects of the wine industry, such as defined palate grouping and wine searching.

Keywords: Wineinformatics · The computational wine wheel · Biclustering · Trimax Triclustering

1 Introduction

There is an intrinsic notion that the computational power of today is essentially limitless, especially when we realize that today's cell phones have more computational power than all of NASA had when it landed two astronauts to the moon in 1969 [1]. We can only imagine what future computational power will be like given said power is supposed to double every eighteen months according to Moore's Law. Even with contemporary capabilities though, it would seem that we could process anything imaginable. However, with more computational power comes the ability to actually

© Springer International Publishing Switzerland 2016
P. Perner (Ed.): ICDM 2016, LNAI 9728, pp. 223–238, 2016.
DOI: 10.1007/978-3-319-41561-1_17

generate new and vastly-growing data every single day. So much data in fact that it is estimated we will have generated 40 zettabytes of data by the year 2020 [2]. With ever-growing sizes in raw data, we have problems not only parsing the data itself, but pulling out meaningful information from it as well. At its core, Data Science is the study that incorporates varying techniques and theories from distinct fields, such as Data Mining, Scientific Methods, Math and Statistics, Visualization, natural language processing, and the Domain Knowledge, to discover useful information from domain-related data. Among all fields in the study of data science, the domain knowledge is the starting point as well as the ending point since all data science researchers need to start with the domain problem, and end with useful information within the domain.

Wine was considered as a luxury in old days; however, it is more and more popular and enjoyed by a wide variety of people today. U.S. consumers bought 29.1 million hectoliters of wine in 2013, a rise of 0.5 % on 2012, while French consumption fell nearly 7 % to 28.1 million hectolitres [3]. Because of the popularity, the demand for luxury and high quality wines produced in great years is high despite their high price; for example, Chateau Petrus 2009 costs $45,600 per case before tax and sold out before the release date. Fortunately, for consumers' point of view, tens of thousands of wines are produced per year and the quality of the wine is not reflecting merely based on the price.

Kosta Browne Pinot Noir Sonoma Coast 2009

Chemical analysis

Sensory analysis

PRIMARY FERMENTATION DETAILS
HARVEST DATES
Gap's Crown: September 19, 21, 23
Terra de Promissio: September 19, 23
Walala: September 23
COLD-SOAK TIME 5 days average
FERMENTATION TIME 14 days average
FERMENTATION TEMP 86° F peak
BARREL PROGRAM
PERCENTAGE OF NEW FRENCH OAK 45%
BARREL AGING 16 months
FINISHED WINE DETAILS ALCOHOL 14.5%
PH 3.63
TITRATABLE ACIDITY 5.3 g/L
BOTTLING DATES
January 26-28, 2011

Ripe and deeply flavored, concentrated and well-structured, this full-bodied red offers a complex mix of black cherry, wild berry and raspberry fruit that's pure and persistent, ending with a pebbly note and firm tannins. Drink now through 2018. 5,818 cases made. (Spectator.)

Fig. 1. The review of the Kosta Browne Pinot Noir Sonoma Coast 2009 (scores 95 pts) on both chemical and sensory analysis

The quality of the wine is usually assured by the wine certification, which is generally assessed by physicochemical and sensory tests [4]. Figure 1 provides an example for a wine review by both perspectives. Physicochemical laboratory tests [5, 6] routinely used to characterize wine include determination of density, alcohol or pH values, while sensory tests rely mainly on human experts [6]. Most of the existing data mining/data science researches related to wine [6–8] focus on the physicochemical laboratory tests data, which is stored in the UCI Machine Learning Repository. However, in wine economics point of view, sensory analysis is much more interesting to consumers and industrial perspective than chemical analysis since they describe aesthetics, pleasure, complexity, color, appearance, odor, aroma, bouquet, tartness, and the interactions with the senses of these characteristics [9] of the wine.

The source of the dataset is always an important factor to the success of a research. Chemical analysis data comes from the lab and costs about $1000 per wine; Sensory analysis produces by prestigious experts who generate consistent wine sensory reviews. In United States, several popular wine magazines provide widely accepted sensory reviews toward wines produced every year, such as Wine Spectator [14], Wine Advocate [15] and Decanter [16] etc. All of those wine magazines review thousands of wines through the 100-point scale and testing notes, which is in the human language format as showed in Fig. 1. Currently, the Wine Spectator database holds more than 300,000 wine reviews. Unquestionably, from these large amount of data, it is interesting to discover meaningful information from those sensory testing notes for answering the questions such as "What makes wine achieve a 90 + rating and considered as a outstanding wine?", "What are the common characteristics shared by 90 + Napa Cabernet sauvignon?", "What characteristics differ between wines from Bordeaux, France and Napa, United States?"

In our previous work [10] published in December 2014, the term "Wineinformatics" was proposed to apply data science techniques and natural language processing on professional wine reviews. A Computational Wine Wheel based on 2011's top 100 wines is proposed to automatically extract wine attributes from professional reviews. The work has been cited in different wine related researches, including mobile app development [11], financial prediction [12] and accessing wine quality [13] research area. In this paper, we would like to redefine the Wineinformatics as a study that incorporate data science in any wine related dataset, including physicochemical laboratory data and wine reviews.

This paper will present the Computational Wine Wheel 2.0 for extracting key attributes from wine reviews like the Fig. 1 example. We will detail the formation of a Computational Wine Wheel 2.0, which will serve as a basis for future, automated extraction of attributes from wine reviews. Given the dictionary and a couple datasets of wine reviews, we explore varying clustering techniques in an attempt to show that it is possible to group similar wines together using only the sensory attributes. A novel tri-cluster in wine (Wine × Attributes × Vintage) is also proposed in this paper. We believe our examination and subsequent evaluation of wine sensory information can advance Wineinformatics researches.

2 The Computational Wine Wheel 2.0

2.1 Wine Sensory Reviews

The wine testing process can be very delicate as a wine is examined not only for its tasting quality, but for physical appearance and physiochemical properties as well. A taster will usually evaluate the appearance of the wine, how it smells in the glass before tasting, the different sensations once tasted, and finally how the wine finishes with its aftertaste. The taster will be looking for how complex the wine is, how much potential it has for aging for drinkability, and if there are any faults present. The experience required can be expansive as any given wine needs to be carefully assessed within comparable wine standards according to its price, region, varietal, and style. Also, if known, the actual wine production techniques will allow the taster to examine further characteristics. To show an example of what might result from a professional tasting, below is an example wine tasting review for Wine Spectator's number one wine of 2014.

Dow's Vintage Port 2011
Powerful, refined and luscious, with a surplus of dark plum, kirsch and cassis flavors that are unctuous and long. Shows plenty of grip, presenting a long, full finish, filled with Asian spice and raspberry tart accents. Rich and chocolaty. One for the ages. Best from 2030 through 2060.

Similar reviews are provided by various prestigious wine magazines [14–16]. Among those, we chose Wine Spectator as our primary data source to start aggregating our wine reviews because of their strong on-line wine review search database and consistent wine reviews. These reviews are mostly comprised of specific tasting notes and observations while avoiding superfluous anecdotes and non-related information. They review more than 15,000 wines per year and all tastings are conducted in private, under controlled conditions. The magazine has been in production since 1976. The company has their reviews available for subscribers directly to their website. Wines are always tasted blind, which means bottles are bagged and coded. Reviewers are told only the general type of wine and vintage. Price is also not taken into account. Their reviews are straight and to the point. For each reviewed wine, a rating within a 100-points scale is given to reflect how highly their reviewers regard each wine relative to other wines in its category and potential quality. The score summarizes a wine's overall quality, while the testing note describes the wine's style and character. The overall rating reflects the following information recommended by Wine Spectator about the wine [14]: $95 \sim 100$ Classic; $90 \sim 94$ Outstanding; $85 \sim 89$ Very good; $80 \sim 84$ Good; $75 \sim 79$ Mediocre; $50 \sim 74$ Not recommended.

In the review example listed above, the attributes are neatly stated without much confusion to what constitutes a proper wine tasting note. For example, to manually process the review, all the terms that are bold will be extracted and considered characteristics of the wine:

Dow's Vintage Port 2011
Powerful, **refined** and **luscious**, *with a surplus of* **dark plum**, **kirsch** *and* **cassis** *flavors that are* **unctuous** *and* **long**. *Shows plenty of* **grip**, *presenting a* **long, full finish**, *filled with* **Asian spice** *and* **raspberry tart** *accents.* **Rich** *and* **chocolaty**. *One for the ages. Best from 2030 through 2060.*

We have bolded key attributes, and these attributes range from actual savory properties, such as "chocolate" and "Asian spice", to subjective properties, such as "powerful" and "refined." One of our major research goals in this paper is to extract as many key attributes as possible from these professional wine reviews automatically.

2.2   The Computational Wine Wheel 2.0

The tasting notes given in a review are very important as they describe the heart and soul of a wine. Even without knowing the producer or varietal, a well-described review can adequately sway a potential consumer into a purchase. Our idea is to build a Savory Wine Dictionary where common, yet important attributes can be stored and referenced as needed. Luckily, this idea was introduced by a sensory chemist and professor named Nobel [17]. She created the Wine Aroma Wheel which is composed of twelve categories of overall wine aromas someone might experience when tasting a wine. Without being overly specific there are times when certain distinct flavor attributes are not unique enough to encapsulate all flavors. An example of this would be the FRU-ITY → (TREE) FRUIT → APPLE attribute. As we will show later with our expansion attributes, things like APPLE and GREEN APPLE are unique enough to warrant a distinction in the (TREE) FRUIT subcategory. Besides, Nobel's wine aroma wheel describes only actual savory attributes; the adjectives and wine body attributes are not included. If we map Novel's wine aroma wheel to the previous example, the processed wine savory review will be:

Dow's Vintage Port 2011
*Powerful, refined and luscious, with a surplus of **dark plum**, **kirsch** and **cassis** flavors that are unctuous and long. Shows plenty of grip, presenting a long, full finish, filled with **Asian spice** and **raspberry tart** accents. Rich and **chocolaty**. One for the ages. Best from 2030 through 2060.*

By expanding the wine aroma wheel, we developed the first version of the Computational Wine Wheel based on Wine Spectator 2011's Top 100 wine and presented it in 2014 International Workshop on Domain Driven Data Mining [10]. The work has been referenced in several related area including mobile app development, financial prediction and accessing wine quality [11–13]. In order to automatically capture as many important characteristics as possible of wine reviews, we advanced the Computational Wine Wheel into the next level in this paper. To achieve the goal, we build our new Computational Wine Wheel based on TEN times more wine reviews from Wine Spectator's Top 100 Wines of 2003 to 2013 for a much comprehensive dictionary. The extraction process for these reviews was purely manual as we handpicked key attributes as well as noted secondary information about the wine. The idea of the Computational Wine Wheel is to memorize the results human labor works and compose a domain specific dictionary for human language processing. In total we gathered the following information: name, vintage, review, varietal, regional information, and price. However, it is worth noting that for our processing purposes the review is the single most important piece of information for a wine. For the review and attributes themselves, there were a few types of attributes we are concerned with. Besides actual biological flavor attributes, we also tried to include anything corresponding to a wine's physical structure, including things like acidity, body, structure, weight, tannins, and finish. These are properties of wine that a taster will physically taste or feel, such as how acidic the wine tastes or how well the wine coats the tongue. Lastly, we also decided to keep generic, subjective terminology that may or may not be the same between two different tasters. For example, one taster may find a wine "vivid" and "beautiful" while another taster may make no mention.

Showing the previous example review again, we want to highlight how we would extract the review's key attributes into the three mentioned categories: savory, body, and descriptive.

Dow's Vintage Port 2011

Powerful!, *refined!* and *luscious!*, *with a surplus of dark plum**, *kirsch** and *cassis* flavors that are unctuous!* and *long!*. *Shows plenty of grip+*, *presenting a long+*, *full finish+*, *filled with Asian spice** and *raspberry tart* accents. Rich!* and *chocolaty**. *One for the ages. Best from 2030 through 2060.*

For this review, **red(*)** words indicate specific flavors and aromas that could possibly be found on Nobel's wine aroma wheel. Orange(+) words indicate traits corresponding to the physical wine itself like its body and finish. That is, how the wine feels physically to a taster. Lastly, **blue(!)** words indicate subjective adjectives used by the taste to describe the overall wine. Should a word or phrase not exist in the original wine aroma wheel, we would add it. Also, if a word or phrase does not fit into any previous categories or subcategories, we would create one for it.

After we process all 1000 wine reviews, we found out there was some contextual overlap between different reviews. That is, there would be two different reviews using slightly different words to express the same tasting notes. A simple example would be one review using the word "distinctive" and another review saying a wine was "very distinct." The human thought process would naturally assume these two differences are the same thing, but computationally, we might miss the connection. For this reason, we added a FOURTH level to the wine aroma wheel that we like to call a **normalized attribute** name. This portion of the wheel would represent a base, or normalized, word to encompass a variety of word usages. This is extremely important not only for differences in word tense or suffixes, but especially the verbiage used when describing biological elements like fruits and their descriptions. A good example of this would be "blueberry", "blueberry fig", and "blueberry jam." Even though all three are components of the same fruit, the taste and consistency of each item convey different connotations and perceptions. All of these normalized processes require domain expert to make judgments. Luckily, our team has a domain expert to assist us.

The Computational Wine Wheel proposed in this paper ended up with 14 distinct categories and a total of 34 distinct subcategories, which is the same with the first version of the Computational Wine Wheel. From all wines mined, we found a total of 1881 specific wine attributes, and of those attributes we were able to finalize 985 distinct normalized attributes. Table 1 provides a detail comparison between the original Computational Wine Wheel and the new one. We also identify the plurals problem in this paper, "BLUEBERRY" and "BLUBERRIES" should be treated as the same specific term as well as normalized attributes. This required program to identify the plurals for specific terms. The full Computational Wine Wheel is available under: https://dl.dropboxusercontent.com/u/13607467/CWW2.0_nonplural.txt.

Table 1. Comparison of the old and new computational wine wheel

	The computational wine wheel 2.0	The computational wine wheel
Data source	Past 10 years top 100 wines	2011 top 100 wines
Categories	14	14
Subcategories	34	34
Specific terms	1881	635
Normalized attributes	985	444
Plurals	Yes	No

2.3 How to Use the Computational Wine Wheel

In order to clarify the usage of the computational wine wheel, we provide an example in this subsection. Table 2 gives a simplified computational wine wheel, which contains only 6 specific attributes.

Table 2. Simplified computational wine wheel

Category	Subcategory	Specific Attribute	Normalize Attribute
FRUITY	BERRY	RASPBERRY	RASPBERRY
FRUITY	BERRY	RASPBERRY TART	RASPBERRY TART
FRUITY	TROPICAL FRUIT	DARK PLUM	PLUM
FRUITY	TROPICAL FRUIT	PLUM	PLUM
OVERALL	FLAVOR/DESCRIPTORS	RICH	RICH
OVERALL	FLAVOR/DESCRIPTORS	RICH AROMAS	RICH

Here is the process of how we apply the simplified computational wine wheel on the *Dow's Vintage Port 2011* wine review: The very first step is to use the words in the Specific Attribute column, which is the 3rd column in Table 2, to scan the review starting with the longest number of combination word. Since the longest number of combination word in the example is 2, we start with Raspberry Tart, followed by Dark Plum, and Rich Aromas. For every word scan, if we had a hit, the wine will have a positive attribute in the corresponding Normalized Attribute and remove the word from the review. Therefore, after the scan of "Raspberry Tart", we got a hit from our review; the wine will have a positive value of "Raspberry Tart" attribute. After the scan of "Dark Plum", we got a hit from our review; the wine will have a positive value of "Plum" attribute. After the scan of "Rich Aromas", we got a miss from our review; the wine will have a negative value of "Rich" attribute.

Once the highest number of combination word is processed, we scan the next number of combination word; in this example, we scan the single word Specific attribute with the same logic. Table 3 represents the Dow's Vintage Port 2011's wine attributes in binary format after the process mentioned above. Please note that the RASPERRY attribute is still negative since we delete the word "RASBPERRY TART" from the review during the first scan. The readers may notice that many important attributes in the example are NOT included, such as ASIAN SPICE, CHOCHLATY... etc. It is because the computational wine wheel is the simplified version. The more SPECIFIC and NORMALIZED attributes included in the computational wine wheel, the more attributes can be picked up from the wine reviews to produce more accurate results. This is also the main reason that we proposed the Computational Wine Wheel to provide higher quality of natural language processing on wine reviews.

Table 3. Attributess of the processed wine example

RASPBERRY	RASPBERRY TART	PLUM	RICH
0	1	1	1

2.4 New Napa Cabernet Sauvignon Dataset Automatically Processed by the Computational Wine Wheel

The quality of the wine is based on various influences; however, two of the most well know and probably most important factors are soil and weather. Soil (or terror) reflects the characteristics of the region and depend on the composition of the soil. Weather controls the quality of the grape production and it changes every year. To recognize and study both factors in Wineinformatics, we collect region specific wine savory reviews over years and compose a dataset processed by the proposed Computational Wine Wheel automatically.

The new dataset encompasses 50 Cabernet Sauvignon wines from the Napa Valley region in California, which is one of the most famous wine regions in United States. For every wine in this set, we retrieved its review for every year from 2006 to 2010. In other words, 250 (50 wines × 5 years) wine reviews are processed by our Computational Wine Wheel. In this way, we control the soil factor and discuss the wines with different weather condition over different years (vintages). Although the dataset may look small, it is caused by our strict criteria: grape type (Cabernet Sauvignon), wine production region (Napa Valley) and years (a wine must have complete wine reviews throughout all years in research). Some wines share the same producer, but each wine has a distinct designation and is technically considered as a different wine production. For this dataset it is best to imagine it as a three dimensional cube of reviews, where the height, width, and depth are the wine name, attributes, and vintage, respectively. This dataset is special as there was nothing manual about attribute extraction. We used the computational wine wheel and scripted the output of only matched attributes. The result of this was 50 wines with 259 attributes across 5 years. The main purpose of this dataset is to discover similar wines over years under specific conditional control criteria. More detail is discussed in the following section.

3   Triclustering

Clustering is generally considered an unsupervised learning and analysis tool. The Wineinformatics information retrieved by clustering can be beneficial to various roles involved in wine. Based on wine consumers' favorite wine, wines in the same cluster can be recommended. Wine retailers can decide to purchase similar wines as the strength of the store or choose representative wines in different cluster to increase the diversity. Wine makers will be able to find wines with similar characteristic within the winery or the wine region to review their wine making process.

The classical clustering algorithms, such as hierarchical clustering and k-means clustering, are usually very good places to start when attempting to explore data. Hierarchical clustering has been successfully applied and visualized in Wineinformatics researches [10]. However, they are flawed in a sense as both algorithms are attempting to detect patterns in observations across all given attributes of a dataset. Sometimes it might be more important to find patterns that consist of a subset of attributes.

In Wineinformatics, the dataset processed through the Computational Wine Wheel is clearly a sparse binary dataset. As the result, dimension selection plays an important role in this research. A bicluster is equivalent to a biclique in a corresponding bipartite graph. This essentially means that all of a bicluster's rows, or observations, are all connected to every column, or attribute, presented in the bicluster. The idea of a bicluster should be explored as it presents the opportunity to find subspaces in our data where subsections of columns define a cluster instead of all attributes contained from that cluster's wines.

The BiMax BiClustering algorithm was a reference method developed by Prelic et al. for baseline comparison of biclustering algorithms in general [19]. The process is fairly simple in that it searches for biclusters that consist entirely of 1 s in a binary matrix. This is perfect for datasets generated with the computational wine wheel in mind because a wine fits the binary requisite; a wine either has an attribute or it does not. With this in mind, our goal is to use the BiMax algorithm to find all inclusion-maximal biclusters of wines and attributes. This means a bicluster cannot be fully contained within another bicluster.

Just as with biclustering, triclustering is becoming a popular method to explore gene expression microarray data with additional "time" dimension. Instead of working with two dimensional matrices, triclustering focuses on finding behavioral patterns between row and columns along a time series. Existed works in the gene expression field for triclustering include Zhao and Zaki [20] and Bhar et al. [21]. However, none of their dataset is considered as sparse binary dataset; hence, the developed algorithm cannot be applied in Wineinformatics directly.

This paper proposes a novel TriMax TriClustering reference algorithm specifically for sparse binary dataset. Trimax Triclustering should be considered a reference algorithm in that it attempts to cluster on the most basic level and makes no assumptions of differing values in the data. That means it expects all values to either be zero or non-zero, so completely binary in nature. For our specific dataset, all data values are either 1 or 0. We consider a tricluster (W, A, T) to correspond to a subset of wines $W \subseteq \{1 \dots n\}$ that jointly share a subset of wine attributes $A \subseteq \{1 \dots m\}$ across a subset of time $T \subseteq \{1 \dots o\}$. The tuple (W, A, T) $(W, A, T) \in 2^{(1 \dots n)} \times 2^{(1 \dots m)} \times 2^{(1 \dots o)}$ is considered inclusion maximal if and only if it meets the following two criteria:

(1) $\forall i \in W, j \in A, k \in T : e_{ijk} = 1$
(2) $\exists (W', A', T')$ with (a) meets criteria (1) and (b)

$W \subseteq W' \wedge A \subseteq A' \wedge T \subseteq T' \wedge (W', A', T') \neq (W, A, T)$ Criteria (1) states that given a possible tricluster, every possible value must be a 1 across all rows, columns, and time slices. Criteria (2) is the inclusion-maximal stipulation that says a tricluster A is considered inclusion-maximal as long as there does not exist another tricluster B in which the grouping of wines, attributes, and time slices of A are a subset of B. If a tricluster A is found, there also cannot be a tricluster B, such that A = B. Now that we have defined a tricluster, we can discussed the algorithm to find them. However, there should be two points noted before we discuss the algorithm. (1) Our proposed algorithm uses the BiMax algorithm, so a good understanding of the algorithm, as we discussed in the previous section, is necessary to proceed. (2) We believe our program is able to find all triclusters, but unlike BiMax which knows at runtime whichs biclusters to ignore thanks to its column callstack, TriMax has to filter out duplicate or subset triclusters after finding all possible triclusters. We will examine an example dataset that shows how duplicates arise, but first we will run through the algorithm itself.

The pseudocode for our proposed TriMax TriClustering algorithm is shown in Fig. 2. As a base concept, we want to take biclusters found in each time slice and see if they can extend across any and all other time slices. To accomplish this, we start with our dataset D and process each of D's t time slices iteratively. For a given bicluster b that is found in a given time slice t, we form a new dataset D', which consists of the rows and columns of b, along every time slices of the input data. That new dataset is then recursively processed using the same methodology until the resulting dataset D' consists only of values of 1. Naively, we can consider a completely 1-valued dataset as a tricluster if it passes the minimum row, column, and time slice amounts set in $mWAT$. Since our process does not have any callstacks like the BiMax algorithm, TriMax will natively introduce duplicate triclusters or triclusters that are subsets, or non-maximal. To combat part of this problem, we introduce a visited array vT, which is populated with the index of a time slice once that time slice's recursive processing has finished. This allows any tricluster found to be ignored if it includes a time slice within vT at any recursive level. If this occurs, ideally it means that the tricluster has already been found previously. However, this only attempts to filter out triclusters between given time slices. It does not work on duplicate or non-maximal subsets formed from partially overlapping biclusters originating from the same time slice. Figure 3 shows an example of duplication issues caused by overlapping biclusters in a given time slice T1.

In Fig. 3, we can say we would process time slice T1 first by expanding the three biclusters found within it: {W1, W2, W3, W4} × {A1}, {W2, W3} × {A1, A2, A3}, and {W2, W3, W4} × {A1, A2}. As shown by the blue squares, all three biclusters share the following subset of rows and columns: {W2, W3} × {A1}. By expanding all

TriMax TriClustering Reference Algorithm
$ID = D = \{\{w_i\}_{i=1}^a\}_{i=1}^t$
$TriList = \{\emptyset\}$
$TriMaxTriClust(D, mWAT, vT)$:
1   **for** any $t \in D_T$ where $t \in vT$ and all $D_{W,A,t} = 1$
2 | **return**
3   **for** $t \in D$ **do:**
4 | **if**($t \in vT$ or (all values in $t = 1$ or 0)):
5   **if**(all values in $t = 0$):
6          $append(vT, t)$
7   **continue to next time slice**
8      $B = BiMaxBiClust(t, mWAT)$
9   **for** $b \in B$ **do:**
10 | $newD' = \{b\}_{t=1}^{ID_T}$
11 | $TriMaxTriClust(D', mWAT, vT)$
12      $append(vT, t)$
13 **if**($len(D_W, D_A, D_T) \geq \{mWAT\}$):
14    $append(TriList, D_{W,A,(D_T - vT)})$: unless $D_T - vT = \{\emptyset\}$
15 **return**
16
17 $\forall t \in TriList$: remove duplicates/subsets

Fig. 2. Proposed TriMax TriClustering reference algorithm pseudocode

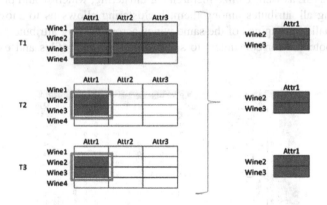

Fig. 3. Tricluster found from multiple intra-timeslice biclusters

three, the same tricluster, as presented on the right of Fig. 4, will be found three times, and thusly will have to be filtered down to one instance afterwards. Even with the slight timing inefficiency here in the post processing, we believe this method will still find all maximal triclusters between all-time slices given in a three dimensional data set. The next section will detail the results when applying triclustering to our multi-vintage 50-wine dataset.

4 Results

4.1 Biclustering 50 Wines

Due to the fact that the biclustering algorithm works on two dimensional data, we arbitrary choose 2010 vintage as the input dataset. For the 50 wines in the 2010 vintage, we implemented and applied the BiMax biclustering. The overall bicluster summarization is described in Fig. 4. This figure represents the total number of maximal biclusters found for the dataset. The table values represent the total number of biclusters that share a specific number of wines (vertical axis) versus a specific number of savory attributes (horizontal axis). For example, in this vintage there are 16 clusters that have exactly four wines and four attributes. In the table, there are darkened, rectangular borders that are meant to be a visual reference to show all biclusters where the minimum number of rows equals the minimum number of columns. For a dataset with 50 wines and 259 possible attributes, this may seem like a low combination, but it makes sense as the number of total possible attributes in any given wine is fairly small.

We then attempt to explore those biclusters that fall into the category of at least 4 wines and 4 attributes. In total, there were 17 (16 + 1) total biclusters that fell into this group, which represents some of the most robust biclusters from this vintage, region, and varietal. Table 4 shows an example of a bicluster that has four wines and share four common attributes. In this example, distinctive flavors that a taster might be accustomed to when sampling a Cabernet Sauvignon. This bicluster was also described as RICH and DENSE as well. Unlike hierarchical clustering, which would present groups of wines using all attributes among them, biclustering allows us to show many different, but smaller, groupings of the same wines across varying attribute patterns. This would give potential for consumers to select small flavor profiles and expect higher

	1	2	3	4	5	6	7	8	9	10	11	12	13	14	15	16	17	18	19	21	TOTAL
1	0	0	0	0	0	1	1	1	7	3	8	6	3	6	4	3	2	2	1	1	49
2	1	31	64	108	50	18	10	0	1	0	0	0	0	0	0	0	0	0	0	0	283
3	7	50	74	38	9	1	0	0	0	0	0	0	0	0	0	0	0	0	0	0	179
4	9	39	33	16	0	0	0	0	0	0	0	0	0	0	0	0	0	0	0	0	97
5	4	26	23	1	0	0	0	0	0	0	0	0	0	0	0	0	0	0	0	0	54
6	2	16	5	0	0	0	0	0	0	0	0	0	0	0	0	0	0	0	0	0	23
7	7	6	7	0	0	0	0	0	0	0	0	0	0	0	0	0	0	0	0	0	20
8	3	5	1	0	0	0	0	0	0	0	0	0	0	0	0	0	0	0	0	0	9
9	1	5	0	0	0	0	0	0	0	0	0	0	0	0	0	0	0	0	0	0	6
10	1	8	0	0	0	0	0	0	0	0	0	0	0	0	0	0	0	0	0	0	9
11	2	2	0	0	0	0	0	0	0	0	0	0	0	0	0	0	0	0	0	0	4
12	2	2	0	0	0	0	0	0	0	0	0	0	0	0	0	0	0	0	0	0	4
14	0	2	0	0	0	0	0	0	0	0	0	0	0	0	0	0	0	0	0	0	2
15	1	1	0	0	0	0	0	0	0	0	0	0	0	0	0	0	0	0	0	0	2
16	1	0	0	0	0	0	0	0	0	0	0	0	0	0	0	0	0	0	0	0	1
17	1	0	0	0	0	0	0	0	0	0	0	0	0	0	0	0	0	0	0	0	1
18	1	0	0	0	0	0	0	0	0	0	0	0	0	0	0	0	0	0	0	0	1
19	1	0	0	0	0	0	0	0	0	0	0	0	0	0	0	0	0	0	0	0	1
21	1	0	0	0	0	0	0	0	0	0	0	0	0	0	0	0	0	0	0	0	1
22	1	0	0	0	0	0	0	0	0	0	0	0	0	0	0	0	0	0	0	0	1
27	1	0	0	0	0	0	0	0	0	0	0	0	0	0	0	0	0	0	0	0	1
28	1	0	0	0	0	0	0	0	0	0	0	0	0	0	0	0	0	0	0	0	1
TOTAL	48	193	207	163	59	20	11	1	8	3	8	6	3	6	4	3	2	2	1	1	749

Fig. 4. Summarization of biclusters of 50 Wines in 2010 vintage

quality results since the biclusters might have filtered out unneeded attributes. The business value of this specific cluster might be "the customer who enjoy one of the four wines, may also enjoy other three wines" and "the customer who like the combination of the four wine attributes can be recommended those four wines".

Table 4. An example of bicluster that has four wines and share four common attributes.

Wine	Shared attributes
Chappellet signature	Black licorice
Beringer private reserve	Rich
Araujo eisele vineyard	Dark berry
Cavus stags leap district	Dense

We can also look for interesting wine and attribute combinations in those clusters that have either low number of wines and high number of attributes, or those with high number of wines and low number of attributes. The former suggests a smaller subset of wines stayed consistent across a majority of attributes across vintages, while the later suggests a larger subset of wines that might share a small pool of distinctive attributes.

4.2   Triclustering 250 Wines

To test out the TriMax Triclustering algorithm, we used the full dataset described in Sect. 2.4 and we found 23,225 possible triclusters. Since we knew a large percentage of these would actually be duplicates or non-maximal, we performed the pairwise subset comparison and pulled out a total of 7,296 superset triclusters. Of all the triclusters found, 6,357 of them only exist in a single time slice, which means these clusters are discoverable simply through Biclustering algorithm. We found 735 triclusters that spanned 2 time slices (Fig. 5A). We found 166 triclusters that spanned 3 time slices (Fig. 5B), and 31 triclusters that spanned 4 time slices (Fig. 5C). Lastly, we found 7 triclusters that spanned all 5 time slices (Fig. 5D). In Fig. 5, the darkened, rectangular borders that are meant to be a visual reference to show all clusters where the minimum number of rows equals the minimum number of columns in each time slice.

To understand the meaning of Fig. 5, let us look into an example. The "1" circled in Fig. 5D indicates there is one tricluster where 8 wines share a single attribute across all five vintages. Through the example in Table 5, the tricluster lets us know that all eight of these wines are considered GREAT for five years in a row. We can also see that four of the wines share the same producer, so it is probable that any other wine produced by BOND would also probably be considered great. One may argue that the attribute GREAT maybe not a significant word. However, if a wine can be reviewed with GREAT in their review sentences, the wine usually scores pretty well. Furthermore, this tricluster example also shows a good reason of why we categorize our attributes in

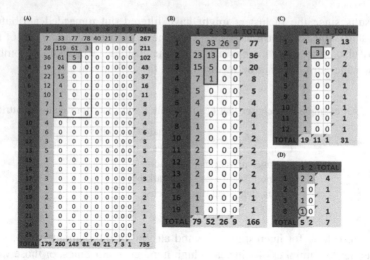

Fig. 5. Summarization of Triclusters of 250 wines in (A) two years (B) three years (C) four years (D) five years time slices

the computational wine wheel into flavor and aroma, body and finish, overall adjectives (red, orange, blue in Sect. 2.2 respectively). If the user is NOT interested in certain category of attributes, they may turn off the attribute detection during the attribute retrieve process.

Table 5. Tricluster example where eight wines share one common attribute across five different vintages

Wine	Attributes	Vintage
BARNETT spring mountain district rattlesnake hill Beringer private reserve, bond melbury, bond quella, bond st. eden, bond vecina, diamond creek gravelly meadow, diamond creek volcanic hill	Great	2010 2009 2008 2007 2006

Table 6. Tricluster example where eight wines share one common attribute across five different vintages

Wine	Attributes	Vintage
Casa piena	Blackberry	2006, 2007, 2008, 2009
Dancing hares	Great	

Table 6 shows another tricluster that has two wines containing two attributes across four consecutive vintages. It is one of three triclusters appeared in the middle rectangular in Fig. 5C. While both wines and attributes are still fairly small in number, this just further provides opportunity for specialized searching and classification. Since the dataset used for this paper contained only a specific varietal from NAPA, we were able to get highly defined cluster results. We believe that the triclusters discovered from a variety of types and sources should produce interesting results and it will be worth exploring those datasets in the future.

5   Conclusion

Data Science is a successful study that incorporates varying techniques and theories from distinct fields. This paper propose the Computational Wine Wheel 2.0 composed by 1000 outstanding wine reviews to support the new data science application field named Wineinformatics, which is a novel data science application area proposed by this paper. We also make the Computational Wine Wheel 2.0 publically available. A new Napa Valley Cabernet Sauvignon across five vintages dataset is automatically generated via the Computational Wine Wheel 2.0. BiMax Biclustering algorithm is applied to the dataset to find similar wines with precise amount of wine attributes. We also develop a new TriMax Triclustering algorithm specifically designed for Wineinformatics. It is the first time that tricluster is applied to Wineinformatics field to form Wine × Attribute × Vintage clusters. The novel idea is able to discover similar wines under different weather condition in different years. We believe this paper helps define the role of Wineinformatics. Many other similar fields such as coffee, whisky and chocolate with professional reviews can also follow the concept and methods in this paper to construct new data science application fields.

References

1. Kaku, M.: Physics of the Future: How Science Will Shape Human Density and Our Daily Lives by the Year 2100. Doubleday, New York (2011)
2. Gantz, J., Reinsel, D.: The digital universe in 2020: big data, bigger digital shadows, and biggest growth in the far east. IDC December 2012
3. USA becomes world biggest wine market as French drinkers cut down, 13 May 2014. http://www.reuters.com/article/2014/05/13/us-wine-usa-france-idUSKBN0DT0YO20140513. Accessed March 2015
4. Ebeler, S.: Linking flavor chemistry to sensory analysis of wine. In: Flavor Chemistry - Thirty Years of Progress pp. 409–422. Kluwer Academic Publishers (1999)
5. Chemical analysis of grapes and wine: techniques and concepts. Patrick Iland Wine Promotions, Campbelltown, Australia (2004)
6. Cortez, P., Cerdeira, A., Almeida, F., Matos, T., Reis, J.: Modeling wine preferences by data mining from physicochemical properties. Decis. Support Syst. 47(4), 547–553 (2009)
7. Ishibuchi, H., Nakashima, T., Nii, M.: Classification and Modeling with Linguistic Information Granules: Advanced Approaches to Linguistic Data Mining. Springer, Heidelberg (2005)
8. Ishibuchi, H., Yamamoto, T.: Rule weight specification in fuzzy rule-based classification systems. IEEE Trans. Fuzzy Syst. 13(4), 428–435 (2005)
9. Olkin, I., Lou, Y., Stokes, L., Cao, J.: Analyses of wine-tasting data: a tutorial. J. Wine Econ. 10(01), 4–30 (2015)

10. Chen, B., Rhodes, C., Crawford, A., Hambuchen, L.: Wineinformatics: applying data mining on wine sensory reviews processed by the computational wine wheel. In: 2014 IEEE International Conference on Data Mining Workshop (ICDMW), pp. 142–149. IEEE, December 2014

11. Kiselev, A., Kuznetsov, A.: Developing a mobile application for wine amateurs (2015)

12. Zhou, Y.: Research on the applications of data mining in financial prediction (2015)

13. Lee, S., Park, J., Kang, K.: Assessing wine quality using a decision tree. In: 2015 IEEE International Symposium on Systems Engineering (ISSE), pp. 176–178. IEEE, September 2015

14. Wine Spectator Magazine. http://www.winespectator.com/. Accessed March 2015

15. rRobertParker. http://www.erobertparker.com/info/wineadvocate.asp. Accessed March 2015

16. Decanter.com. http://www.decanter.com/wine. Accessed March 2015

17. Nobel, A.C.: N.d., Wine Aroma Wheel. http://winearomawheel.com/. Accessed 29 March 2015

18. Wine Spectator: N.p., n.d., Top 100 List. http://top100.winespectator.com/lists/. Accessed 29 March 2015

19. Prelic, A., Bleuler, S., Zimmermann, P., Wille, A., Bühlmann, P., Gruissem, W., Hennig, L., Thiele, L., Zitzler, E.: A systematic comparison and evaluation of biclustering methods for gene expression data. Bioinformatics 22(9), 1122–1129 (2006)

20. Zhao, L., Zaki, M.J.: TRICLUSTER: an effective algorithm for mining coherent clusters in 3D microarray data. In: SIGMOND 2005 (2005)

21. Bhar, A., Haubrock, M., Mukhopadhyay, A., Wingender, E.: Application of a novel Triclustering method (delta-TRIMAX) to mine 3D gene expression data of breast cancer cells. In: GCB 2013 (2013)

Understanding the Wine Judges and Evaluating the Consistency Through White-Box Classification Algorithms

Bernard Chen[1(✉)], Hai Le[1], Christopher Rhodes[1],
and Dongsheng Che[2]

[1] Department of Computer Science,
University of Central Arkansas, Conway, AR 72034, USA
bchen@uca.edu
[2] Department of Computer Science, East Stroudsburg University,
East Stroudsburg, PA 18301, USA

Abstract. Wine is a broad field of study and is more and more popular today. However, limited amounts of data science and data mining research are applied on this topic to benefit wine producers, distributors, and consumers. According to the American Association of Wine Economics, "Who is a reliable wine judge?" and "Are wine judges consistent?" are typical questions that beg for formal statistical answers.

This paper proposes to use the white box classification algorithms to understand the wine judges and evaluate the consistency while they score a wine as 90+ or 90−. Three white box classification algorithms, Naïve Bayes, Decision Tree, and K-nearest neighbors are applied to wine sensory data derived from professional wine reviews. Each algorithm is able to tell how the judges make their decision. The extracted information is also useful to wine producers, distributors, and consumers. The data set includes 1000 wines with 500 scored as 90+ points (positive class) and 500 scored as 90− points (negative class). 5-fold cross validation is used to validate the performance of classification algorithms. The higher prediction accuracy indicates the higher consistency of the wine judge. The best white box classification algorithm prediction accuracy we produced is as high as 85.7 % from a modified version of Naïve Bayes algorithm.

Keywords: Wineinformatics · Wine judges evaluation · Decision tree · Naïve Bayes · K-nearest neighbors · SVM

1 Introduction

Data mining is the search for new, valuable, and nontrivial information in large volumes of data. It is most useful in an exploratory analysis scenario in which there are no predetermined notions about what will constitute an "interesting" outcome. It is a cooperative effort of humans and computers. Best results are achieved by balancing the knowledge of human experts in describing problems and goals with the search capabilities of computers [1].

© Springer International Publishing Switzerland 2016
P. Perner (Ed.): ICDM 2016, LNAI 9728, pp. 239–252, 2016.
DOI: 10.1007/978-3-319-41561-1_18

With the development of society, and as quality of life rises, the qualities and varieties of wines are increasing year by year. According to OIV (International Organization of Wine and Vine) [2] estimates, 2011 global production (not taking into account must and grape-juice) is around 2,558 million hectoliters, 700,000 more than in 2010 [3]. OIV also estimates that 2011 global wine consumption is at about 2, 419 million hl (1 hl = 100,000 ml), which is an increase on the previous year, of 1.7 million hl [3]. In accordance with this information, wine is one of the most widely consumed beverages in the world and has very obvious commercial value as well as social importance. Therefore, the evaluation of the quality of wine plays a very important role for both manufacture and sale [4]. An established approach to investigate which aspects have significant effects on willingness to pay for food products is to focus on objective characteristics (such as price, brand, and appearance), consumer demographics (such as age, income and education level), and frequency of consumption. Sensory properties such as taste, aroma, texture, and flavor are typically not included. However, sensory qualities are often the major factors that affect consumers' perception of a product. Therefore, it is necessary to include them in accessing consumer's preference [5].

To better analyze wines, which vary considerably even within the largest categories, reputable wine reviewers from professional wine magazines use human language to describe them in great detail. For example, Wine Spectator contains more than 300,000 wine reviews available for paid members. These reviews are based on the sensory attributes conveyed by a wine, and they cover a broad range of detail; acidity, flavor, color, and smell are just a few examples of the attributes that wine reviewers take into consideration to describe a wine. Some of these attributes are applicable to all bottles of a certain type of wine and can be used to determine its quality compared to other wines, or even its origins. Hundreds of thousands different wine reviews are stored in each magazine's database. However, very limited amount of data mining researches are applied in this interesting field.

According to American Association of Wine Economics, *"Who is a reliable wine judge? How can we aggregate the will of a tasting panel? Do wine judges agree with each other? Are wine judges consistent? What is the best wine in the flight?"* are typical questions that beg for formal statistical answers [6]. Some researchers work on this problem by looking into ranking, rating, and judging of the wine through traditional statistical methods [6–9]. In this paper, our purpose is to use white-box classification algorithms to understand and evaluate the consistency of wine judges. It is a continuing work on a new data science research area named Wineinformatics, which uses the understanding of wine to serve as the domain knowledge [23]. In this work, we convert the wine savory reviews through the computational wine wheel, and then we apply different data mining white-box classification algorithms to the same dataset [23]. Our goal is to find the best white box classification algorithm to understand and evaluate the consistency of wine judges.

Three different white-box classification algorithms can provide distinct useful information. Classification consists of predicting a certain outcome based on a given input. In order to predict the outcome, the algorithm processes a training set containing a set of attributes and the respective outcome, usually called goal or prediction attribute [12]. Decision tree uses a predictive model to determine consequences. The application of boosting procedures to decision tree algorithms has been shown to produce very accurate classifiers [11]. Second, Naïve Bayes is a statistical classifier to predict class membership probabilities, such as the probability that a given tuple belongs to a particular class [10, 20]. Some of the variations of Naïve Bayes models are used for text retrieval and classification, focusing on the distributional assumptions made about word occurrences in documents [12]. Finally, K-nearest neighbor (k-NN) focuses on how each wine is similar to each other, divides all similar wines into clusters, and predicts the accuracy of the data [18]. These three algorithms will combine with the real data to classify the wine into different categories. Although these classification algorithms are considered textbook algorithms, it is the first time that they are applied in Wineinformatics to the best of our knowledge. We also provide our insight of how to use white-box classification models to benefit wine makers, distributors, and consumers. Finally, we compare the results generated from SVM with all white box classification algorithms.

The framework of this paper is organized as follows: Sect. 2 introduces the data of wine in detail; Sect. 3 describes methods—Decision Tree, Naïve Bayes, and k-NN, and how they work on these data; Sect. 4 demonstrates the result and the accuracy between different classification algorithms; finally, we include the conclusion and future works.

2 Wine Sensory Data

2.1 Wine Data

Wine is traditionally evaluated via two methods. The first is of an analytical instrumental sequence using spectroscopic and chromatographic methods where the wine is analyzed for its chemical compounds [15]. The second is of sensory qualifications which is a professional wine reviewer perceives via organoleptic properties – these being the aspects as experienced by the senses of taste, sight, and smell [13].

Wine sensory analysis involves tasting a wine and being able to accurately describe every component that makes it up. Not only does this include flavors and aromas, but characteristics such as acidity, tannin, weight, finish, and structure. Within each of those categories, there are multitudes of possible attributes or forms that each can take. What makes the wine tasting process so special is the ability for two people to simultaneously view the same wine while being able to share and detect all the same attributes.

2.2 Wine Spectator

Wine Reviews are made of the most sensitive and critical sensory evaluation techniques, which have little room for error, and quality control is critical [16]. Among all different wine expert reviews, the data in this paper is derived from the Wine Spectator magazine's wine sensory data. We used the Wine Spectator data source primarily for its impact on the wine culture due to its extensive wine reviews, ratings and general consistency not to logomachy in wine review. Their database consists of wine reviews that derive from their publication 15 issues a year, in which there are between 400 and 1000 wine reviews per issue [15]. The reviews are direct and specific to the sensory perception of the wine. The wine tests are blind tests in controlled environments, and reviewers are only aware of the type of wine and vintage. Reviews consist of the 50 – 100 point scale in which wine professionals grade each wine against other wines in its same category for overall quality. Reviews also consist of the sensory attributes of each individual wine. These sensory attributes are where we pull our data set from. Wine Spectator tasters review wines on the following 100-point scale:

- 95–100 Classic: a great wine
- 90–94 Outstanding: a wine of superior character and style
- 85–89 Very good: a wine with special qualities
- 80–84 Good: a solid, well-made wine
- 75–79 Mediocre: a drinkable wine that may have minor flaws
- 50–74 Not recommended

Following is an example of sensory attributes contributed by the wine reviewer in a wine review sample:

Kosta Browne Pinot Noir Sonoma Coast 2009 95pts *Ripe and deeply flavored, concentrated and well-structured, this full-bodied red offers a complex mix of black cherry, wild berry and raspberry fruit that's pure and persistent, ending with a pebbly note and firm tannins. Drink now through 2018. 5,818 cases made* [16].

Wines receiving higher than 90 points are usually considered an honor in evaluation systems. Consistency of evaluation is the key to maintain the reputation of wine judges. The goal of this paper is to use classification algorithms to build models based on Wine Spectator's reviews. The performance of the models can be considered as the criteria to evaluate Wine Spectator as the wine judge; the more consistent the wine judge, the higher performance classification models can perform. Since there are many different classification algorithms available, this paper also try to identify the best algorithm for understanding the wine judges and evaluating their consistency.

2.3 Dataset for Experiments

Each year Wine Spectator publishes thousands of wine reviews. In this research, the data set includes a multi-year span that consists of a 1000 wine sensory reviews, including 500 wines scored 90+ and another 500 wines scored 90−. The reviews are scanned word by word through the computational wine wheel [14, 23]. If there is a match word in the review with the "specific name" in the Computational Wine Wheel, the "categorized name" attribute are assigned positive to the wine. For example, if a wine review has "FRESH-CUT APPLE" or "RIPE APPLE" or "APPLE," these wine attributes are categorized into a single category "APPLE." However, according the Computational Wine Wheel, GREEN APPLE is considered as GREEN APPLE, which is not in the APPLE category since the flavor is different. According to [23], the Computational Wine Wheel contains 304 normalized attributes. The dataset can be visualized as Table 1. If a wine review for an individual wine contained an attribute, a 1 was listed in the column for that attribute for that wine to indicate 'true'; otherwise a 0 was listed for 'false'. Also, the wines were given a classification on the 100-point scale. If a wine scores equal or higher than 90 points, we consider it as a positive class; on the other hand, if a wine scores below 90 points, we consider it as a negative class. By using the Kosta Browne Pinot Noir Sonoma Coast 2009 wine mentioned earlier as an example, the attributes extracted are: RIPE, CONCENTRATED, FULL-BODIED, BLACK CHERRY, WILD BERRY, RASPBERRY FRUIT, PURE, PERSISTENT, PEBBLY, and FIRM TANNINS. Plus, this wine is considered as a positive class.

Table 1. Simplified example for our wine dataset

Name	Attribute 1	Attribute 2	...	Attribute 304	Grade
Wine1	1	0	...	0	90+
Wine2	0	0	...	0	90+
Wine3	1	0	...	0	90+
Wine4	0	1	...	0	90-
Wine5	1	0	...	1	90-
Wine6	1	1	...	0	90-

3 Methods and Results

3.1 Decision Tree

Decision Tree induction is the learning of decision trees from class-labeled training tuples [17, 22]. The tree consists of nodes that form a rooted tree, meaning it is a directed tree with a node called "root" that has no incoming edges. All other nodes, called internal nodes, have exactly one incoming edge that denotes a test on an attribute, but it splits or branches to represent an outcome into two edges according to the input variable. Each leaf node holds a class label or an attribute. The Decision Tree algorithm is a tree that is constructed in a top-down recursive divide and conquer manner. In the beginning, all attributes are listed at the root. To determine which attribute is to become the root, we used a statistical measure called information gain. The attribute with the highest information gain is the root of the tree.

Dataset on Table 2 has 6 wines and 4 attributes. Among 6 wines, the first 3 are graded 90+, and the last 3 are graded 90−. Numbers of the first row represent wine attributes. A: CHERRY. B: APPLE. C: PURE. D: BERRY.

After we apply the decision tree algorithm to the dataset showed in Table 1, the decision tree is shown on Fig. 1. Since the dataset has 2 classes, the decision tree becomes a binary tree. Due to the fact that attribute D (BERRY) gets the highest gain information, it becomes the root of the tree. Next is attribute A (CHERRY), and then attribute B (APPLE).

The decision tree can be used to predict the grade of a testing wine. For example, a testing wine has the following attribute: A(0), B(1), C(0), D(1). Since the root of the tree is attribute D, we check the value of attribute D of the Testing wine. Since it is 1, we follow path 1 and reach attribute A. We check value of attribute A of the Testing

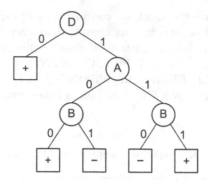

Fig. 1. The decision tree from Table 1

wine; since it 0, we follow path 0 and reach attribute B. Again, we check the value of attribute B of the testing wine; since it is 1, we follow path 1 and reach the bottom of the tree. At this point, it stops and predicts that the testing wine has a 90− grade.

The benefit of the decision tree is that the mined information has high readability. Important attributes are displayed on top of the tree. The prediction results are based on the combination of the attributes. Wine makers can use this information to decide their fermentation method and tools (such as French barrel or American barrel) to avoid bad combinations and improve the quality of the wine.

After we apply the 5-fold cross validation to our 1000 wine dataset with the Decision Tree algorithm, the average accuracy just barely passes 50 % (50.6 %). Since the problem we are facing is a typical bi-class classification problem, the accuracy is just better than guessing "heads" or "tails" when we flip a coin. We noticed a significantly lower percentage of 90− wines that were predicted compared to the 90+ wines. This could be due to the fact that 90− wines do not have as many of the attributes listed as the 90+ wines, which would cause problems with classifying by an attribute. As mentioned above, depending on the datasets, some classification algorithms will generate high accuracy predictions, and some will not; thus, decision tree is not suitable for the wine dataset tested in this paper. As a result, it gives us motivation to try and test more classification models, and k-nearest neighbors is our next choice.

3.2   k-Nearest Neighbors (k-NN)

k-Nearest Neighbors (k-NN) is "a non-parametric method used for classification and regression. In both cases, the input consists of the k closet training examples in the feature space [19]. The output of the algorithm is a class membership, and an object is classified by a majority vote of its neighbors, with the object being assigned to the class most common among its k nearest neighbors (k > 0) [18]. In other words, k-NN does not build any model. k values are chosen, and the algorithm calculates distances between instances and then predicts labels directly.

For our wine dataset, the prediction of a test wine is based on the majority label vote of its k "nearest" wines. In other words, the algorithm chooses k wines that are the most similar to the test wine, counts how many of them are "90+" and "90−", and then predicts the test wine label based on the majority vote. As mentioned in previous section, our wine dataset is in binary format. For that reason, Jaccard's distance formula is used.

Jaccard's distance formula:

$$J = \frac{Q+R}{P+Q+R}$$

Q is the number of positive attributes in wine 1 but not in wine 2; R is the number of positive attributes in wine 2 but not it wine 1; and P is the number of positive attributes in both wine 1 and wine 2. The smaller the value is, the more similar the two wines are. For example, the Jaccard's distance between wine 5 and 6 in Table 1 is 3/4 (Q:1, R:2, P:1).

We tested our k parameter from 1 to 21 with an interval of 1. Because the wine dataset has 2 class labels (90+ and 90−), we choose k being an odd number to prevent equal voting. Figure 2 shows the 5-fold cross validation results of k-NN for each fold and its average when k = [1, 21].

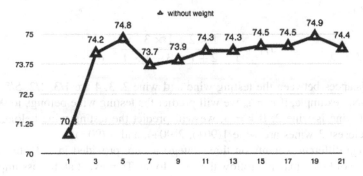

Fig. 2. The averages accuracy of k from 1 to 21 (k-NN)

The highest accuracy is 74.9 % (k = 19), which is much better than Decision Tree result (50.6 %). Overall, the accuracy results of k-NN are similar to each other except when k = 1. Because k-NN predicted the result base on majority vote, with k = 1, the algorithm will be completely based on the label of only one wine. As a result, it leads to bias when the algorithm does not consider more instances to vote. As a result, the result of k = 1 is an outlier.

In this paper, we further study the weight contribution of different attributes in k-NN algorithm. The Computational Wine Wheel [23] has one column about the attributes' category, where attributes are weighted differently (3, 2, and 1) based on their category. "1" is non-flavor descriptions (PURE, BEAUTY, WONDERFUL, etc.). "2" is the non-flavor wine characteristics (TANNINS, ACIDITY, BODY, etc.). "3" is the food and wine flavor characteristics (specific fruit, woods, flavors, etc.). Table 2 gives an example of it.

Table 2. Simplified example for our wine dataset with weight

Name	LONG FINISH	APPLE	PURE	Grade
Weight	2	3	1	
Wine1	0	1	1	90+
Wine2	1	0	0	90+
Wine3	1	1	1	90−
Wine4	0	1	0	90−
Test Wine	1	0	1	?

In Table 2, since we added the weight concept, the Jaccard's distance needs to be adjusted to include it:

$$J = \frac{weight_q \times Q + weight_r \times R}{weight_q \times Q + weight_r \times R + weight_p \times p}$$

For example, the weighted Jaccard's distance between the testing wine and wine 1 in Table 2 is

$$J_{T1} = \frac{1 \times 2 + 1 \times 3}{1 \times 2 + 1 \times 3 + 1 \times 1} = \frac{5}{6}$$

The distances between the testing wine and wine 2, 3, 4 are 1/3, 1/2, 5/5. For this specific k-NN example, if k = 1, we will predict the testing wine belongs to 90+ since the closest wine is wine 2. If k = 3, we will predict the testing wine belongs to 90+ since the closest 3 wines are wine 1(90+), 2(90+), and 3 (90−).

Although different weights on three categories are provided in [23], the accuracy performance of k-NN may not follow the same logic. To prevent all the assumption and bias, we switch the values of weights between attributes, and we create all possible combinations between them. Table 3 shows how it is done and Fig. 3 depicts the 5-fold cross validation prediction accuracy results among all weigh combinations.

Among 6 combinations, combination 3 gives the lowest accuracy (67.1 %) while combination 4 gives the highest result (76.5 %). Compared to the highest accuracy without weight (74.9 %), the original weight (combination 1) generates lower accuracy with the highest result being 70.6 %. Combinations 4 and 6 are the two that perform better than the dataset without weight.

Based on the result of the experience, combination 4, which is generated the best accuracy among all, suggests that attributes with weight 3 are kept the same; but attributes that are weighted 1 are actually more important that those attributes that are weighted 2, so we need to switch them. Combination 6 follows the same routine, but it says the attributes weighted 1 are the most important. In both cases, even though there is a conflict between the original weight 1 and 3, all combinations agree that the attributes that are weighted 2 should be the least important attribute.

Table 3. All combinations of weight for different category attributes

	Category 1 Attributes	Category 2 Attributes	Category 3 Attributes
original weight	1	2	3
combination 1	1	2	3
combination 2	3	2	1
combination 3	2	3	1
combination 4	2	1	3
combination 5	1	3	2
combination 6	3	1	2

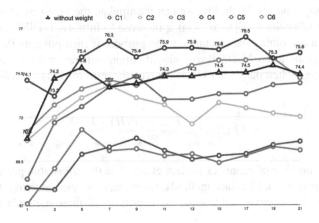

Fig. 3. The accuracy comparison chart of all weight combinations and without weight

3.3 Naïve Bayes

A Naïve Bayes classifier is a simple probabilistic classifier based on applying Bayes' theorem [20] with strong (naive) independence assumptions. A more descriptive term for the underlying probability model would be "independent feature model". In other words, a Naïve Bayes classifier assumes that the presence (or absence) of an instance of a class is unrelated to the presence (or absence) of any other instance [20]. For example, a wine may be considered to be a "90+" wine if it has BLUE BERRY, APPLE, and LONG FINISH. Even if these attributes depend on each other or on other attributes, when a Naïve Bayes classifier generates the probability of the wine, it considers all of these attributes independently. As a result, depending on the precise nature of the probability model, Naïve Bayes classifiers can be trained very efficiently in a super-vised learning setting.

In Naïve Bayes algorithm, zero frequency happens when none of the training instances have the same value as testing instances; therefore, the result will equal zero and ignore all the effects of other instances. There are several solutions to minimize the effect of zero frequency problems. We applied two methods: (1) add penalty and (2) Laplace methods to see how the wine dataset result is modified.

With the Add Penalty method, when computing the probabilities, it substitutes each 0 with: $(1/k)^n$. As a result, Bayes theorem formula (1) will be modified to:

$$P(H|X) = \left(\frac{1}{k}\right)^n \times \frac{P(X|H)P(H)}{P(X)}$$

Where P(H|X) is the probability that the hypothesis holds given the observed data sample X; P(H) is (prior probability), the initial probability; P(X) is the probability that sample data is observed; P(X|H) (posteriori probability) is the probability of observe the sample X, given that the hypothesis holds. n is numbers of 0 value; k is a adjustable parameter (k not equal 0). With this, we can manipulate the result by changing k. The larger k, the smaller the P (H|X) is, so if k increase to infinity, P(H|X) will become 0.

Laplace is a smoothing data technique; the purpose is manipulating the value of the data at the beginning, so Naïve Bayes classification will never have zero frequency problems. (Except when parameter k = 0). Bayes theorem formula (1) will be modified to:

$$P(H|X) = \frac{P(X|H)P(H) + k}{P(X) + b}$$

Where b is number of instances in dataset and k is the adjustable parameter. When applying Add penalty and Laplace methods, we manipulate the value of k from 1 to 20. After we apply 5 fold cross validations, the results of all three methods are shown in Fig. 4.

Overall, Naïve Bayes generates very good results. The accuracy is better than 80 %, which is quite high for a real dataset. The satisfactory accuracy indicates that Wine Spectator does have a consistent review. For Naïve Bayes classification without penalty, since there is no k parameter in the formula, there is only one accuracy result = 79.6 %. Add penalty achieves the highest accuracy of 70.4 % when k = 20. Add penalty achieves the highest accuracy of 85.7 % when k = 2. Figure 4 is provided to display the comparisons.

With Add Penalty method, at first the accuracy increased rapidly until k = 15; after that, the result is still increasing, but much slower and bound around 70 %. Laplace on the other hand, shows that accuracy decreases when k increases. In conclusion, the zero frequency problem usually needs to be prevented when applying Naïve Bayes classification, but for the Wineinformatics dataset, zero value makes the accuracy increase. We apply two different methods to resolve the zero frequency problem, and the results of both methods show that the nearer P(H|X) is to zero, the greater the accuracy will be.

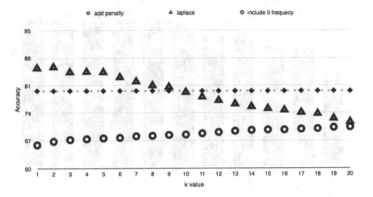

Fig. 4. The comparison between three methods: include 0 frequency, Add penalty, Laplace.

4   Prediction Accuracy Comparison with Published Results and SVM

All of the classification models that we have covered so far (Decision Tree, k-NN, and Naïve Bayes) are white box testing. In other words, we can analyze the prediction accuracy and draw out useful information from how the models react to the database. Opposite to white box testing is black box testing, which tests the algorithm functionality. It usually works faster or generates better results than white box testing, but we will not be able to explain how it obtains the conclusions. To complete our results comparison, Support Vector Machine (SVM), which is one of the most popular black box testing methods, is applied to our wine dataset for benchmark purpose.

While we used lib-SVM as our SVM implementation, some minor improvements to the prediction accuracy of SVM is suggested by [21]: scale dataset and choose the best parameter. Scaling dataset is to "to avoid attributes in greater numeric ranges dominating those in smaller numeric ranges" and "avoid numerical difficulties during the calculation" [21]. As the paper suggests, we apply linearly scaling each wine attribute to the range [−1, +1]. For the best parameter method, SVM used grid search to scans through the whole dataset and tried to pick the best C and Υ (C is penalty parameter, and Υ is kernel parameters). Table 4 shows the results produced by SVM and Fig. 5 gives the final comparisons results with all methods mentioned in this paper and results in [23].

Table 4. Prediction accuracy of 3 support vector machine methods

Support vector machine methods	Accuracy
SVM	81.9 %
SVM scale	86.1 %
SVM parameter	88.0 %

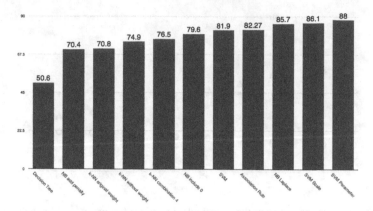

Fig. 5. The comparison chart between association rule [23], decision tree, Naïve Bayes (3 methods), k-NN (3 methods), and support vector machine (3 methods) classifications.

Decision Tree achieves the lowest accuracy with the prediction of only 50.6 %, and that is just better than guessing heads or tails when flipping a coin. For all other methods, the accuracy results are above 70 %, which is acceptable. Among our implementation algorithms, Naïve Bayes Laplace archives the highest accuracy of 85.7 %. Compare the Support Vector Machine; the accuracy of Naïve Bayes Laplace beat the original SVM method (85.5 % compares to 81.9 %). However, the other two SVM methods generated even better results, especially SVM Parameter with an accuracy of 88 %. Compared to our previous work, the results of Association Rules use a 1 % support with 90 % confidence that includes 61 % coverage (61 % of the testing data is used). In this work, all the results include 100 % coverage. That means that more of the wines were predictable by the Naïve Bayes algorithm than by the Association Rules, and Naïve Bayes Laplace also gets the better results. With an accuracy of 85.7 %, it is a successful achievement. In summary, the results suggest that the Naïve Bayes Laplace might be most suitable white box classification algorithm for understanding wine judges and evaluate the consistency.

5   Conclusion and Future Works

In this paper, we propose the idea of using white box classification algorithms to understand wine judges and evaluate their consistency. We introduced classifying wines by grade with the sensory data provided from well respected, established wine review sources. We explained how we processed the normalized data set with three white box classification algorithms, the Naïve Bayes classifier, the Decision Tree classifier, and K-nearest neighbors. As mentioned in introduction section, our purpose is using other data mining classifications to further analyze the dataset. The comparison in Fig. 4 shows that we are able to utilize classification techniques that generate the better accuracy than association rules, which is the original method that is used in our Wineinformatics paper [23]. We identify Naïve Bayes Laplace as possibly the most suitable white box classification algorithm for understanding wine judges and evaluating their consistency.

In this work, we evaluate Wine Spectator as a composite professional wine review source. Since Wine Spectator has ten reviewers, each reviewer puts his or her initials in the end of each review. The most important future work of this paper is to evaluate every review and possibly rank them according to the consistency. Dimension selection is probably the next most important future work. The dataset in Wineinformatics is considered as high dimensional sparse binary dataset. It is obvious that not all attributes contribute equal weight to the classification process. Identifying significant attributes may further improve the prediction accuracy. Secondly, multi-source and multi-label techniques can be applied into Wineinformatics. Our testing included using one source, Wine Spectator Magazine wine reviews. There is more than one reliable source for wine reviews. A single wine can be reviewed by more than one source. We suggest using a dataset that includes for each wine multiple sources for its review data. Last but not least, we will test more classification algorithms on different bigger datasets.

References

1. Kantardzic, M.: Data Mining: Concepts, Models, Methods, and Algorithms. Wiley, Hoboken (2011)
2. International Organization of Wine and Vine. http://www.oiv.int/oiv/cms/index?lang=en
3. Foods & Wines from Spain. Wine. http://www.winesfromspain.com/icex/cda/controller/pageGen/0,3346,1549487_6763472_6778161_0,00.html
4. Sun, L.-X., Danzer, K., Thiel, G.: Classification of wine samples by means of artificial neural networks and discrimination analytical methods. Fresen. J. Anal. Chem. **359**(2), 143–149 (1997)
5. Yang, N.: Quality differentiation in wine markets. Washington State University (2010)
6. Masset, P., Weisskopf, J.P., Cossutta, M.: Wine tasters, ratings, and en primeur prices. J. Wine Econ. **10**(01), 75–107 (2015)
7. Storchmann, K.: Introduction to the issue. J. Wine Econ. **10**(01), 1–3
8. Bodington, J.C.: Evaluating wine-tasting results and randomness with a mixture of rank preference models. J. Wine Econ. **10**(01), 31–46 (2015)
9. Stuen, E.T., Miller, J.R., Stone, R.W.: An analysis of wine critic consensus: a study of Washington and California wines. J. Wine Econ. **10**(01), 47–61 (2015)
10. Lewis, D.D.: Naïve (Bayes) at forty: the independence assumption in information retrieval. In: Nédellec, C., Rouveirol, C. (eds.) ECML 1998. LNCS, vol. 1398, pp. 4–15. Springer, Heidelberg (1998)
11. Magerman, D.M.: Statistical decision-tree models for parsing. In: Proceedings of the 33rd Annual Meeting on Association for Computational Linguistics, pp. 276–283. Association for Computational Linguistics (1995)
12. Arias - Bolzmann, L., Orkun, S., Andres, M., Len, L.: Emerald insight. Int. J. Wine Mark. 7 Apr 2014
13. De Villiers, A., Alberts, P., Tredoux, A.G.J., Nieuwoudt, H.H.: Analytical techniques for wine analysis: an African perspective; a review. Analytica Chimica Acta **730**, 2–23 (2012)
14. Chen, B.: Wine Attributes. http://www.cs.gsu.edu/∼cscbecx/Wine%20Informatics.htm. File Wine_Wheel_01242014.dat

15. Wine Spectator. http://www.winespectator.com
16. Wine Spectator (2011) Top 100. http://www.winespectator.com/display/show?id=45906
17. eRobertParker.com. A glossary of Wine Terms. http://www.erobertparker.com/info/glossary.asp
18. Sutton, O.: Introduction to k nearest neighbour classification and condensed nearest neighbour data reduction (2012)
19. Altman, N.S.: An introduction to kernel and nearest-neighbor nonparametric regression. Am. Stat. **46** (1992)
20. Kamber, M., Han, J.: Data Mining: Concepts and Techniques, 2nd edn. Morgan Kaufmann, San Francisco (2005)
21. Hsu, C.-W., Chang, C.-C., Lin, C.-J.: A practical guide to support vector classification. Taipei, 19 March 2015
22. Friedl, M.A., Brodley, C.E.: Decision tree classification of land cover from remotely sensed data. Remote Sens. Environ. **61**(3), 399–409 (1997)
23. Chen, B., Rhodes, C., Crawford, A., Hambuchen, L.: Wine informatics: applying data mining on wine sensory. Accepted by 2014 Workshop on Domain Driven Data Mining (DDDM 2014) (2014)

A Novel Sparsity Based Classification Framework to Exploit Clusters in Data

Sudarshan Babu[✉]

SVCE, Anna University, Chennai, Tamil Nadu, India
sudarshan.warft@gmail.com

Abstract. A huge recent advance in machine learning has been the usage of sparsity as a guiding principle to perform classification. Traditionally sparsity has been used to exploit a property of high dimensional vectors–which is that, vectors of the same class lie on or near the same low dimensional subspace within an ambient high dimensional space–this is seen in algorithms like Basis Pursuit, and Sparse Representation classifier. In this paper we use sparsity to exploit a different property of data, which is that data points belonging to the same class constitute a cluster. Here classification is done by determining which cluster's vectors can best convexly approximate the given test vector. So if the vectors of cluster 'A' best approximate or realise the given test vector, then label 'A' is assigned as its class. The problem of finding the best approximate is framed as a ℓ_1 norm minimization problem with convex constraints. The optimization framework of the proposed algorithm is convex in nature making the classification algorithm tractable. The proposed algorithm is evaluated by comparing its accuracy with the accuracy of other popular machine learning algorithms on a diverse collection of real datasets. The proposed algorithm on an average provides a 10 % improvement in accuracy over certain standard machine learning algorithms.

1 Introduction

When similar objects are represented by a set of parameters, these parameters are numerically very close to each other; for dissimilar objects these parameters are numerically far off. Hence, similar objects are placed near each other in euclidean space, and dissimilar objects are placed away from each other. This reason leads to the formation of clusters in data. This structure exhibited by data has enabled several works to successfully extract semantics out of data. Algorithms such as SVM and k-means have underlined the importance of exploiting the clustering property to perform data analysis [13,14]. This class of classification schemes generally relies only on proximity to define if a given vector belongs to a cluster or not. Owing to the availability of several mathematical measures of defining proximity, in euclidean space, these classifiers are characterised with simplicity and low classification time. kNN–a classifier that relies on proximity–is very effective, yet is simple in its construction, resulting in low run times [12].

Recently there has been a surge in employing sparsity to exploit the subspace property [3,4]. Here subspace property refers to a particular structure of data, wherein datapoints of the same class lie in the same subspace. Sparsity acts as a heuristic to

© Springer International Publishing Switzerland 2016
P. Perner (Ed.): ICDM 2016, LNAI 9728, pp. 253–265, 2016.
DOI: 10.1007/978-3-319-41561-1_19

select the subspace that the test vector is the lies in, or is the nearest to. The optimization framework is framed as a ℓ_1 norm minimization problem. The optimization framework is convex, and hence makes the classification scheme tractable. The superlative results achieved by tractable sparsity based classifiers make this set of classifiers ideal for classification.

It can be summarized that both, sparsity based classifiers, and clustering exploiting classifiers have exhibited superlative accuracy and computational tractability; as a result of which these types of classifiers have been used extensively in machine learning and data analysis. The attributes exhibited by these classifiers have motivated us to design an unifying mathematical framework that uses both sparsity and clustering exploitation (for classification). Towards this end, we propose a classification algorithm called Sparsity Based Clustering Exploiting classifier. The algorithm is named so because it exploits clusters in data by using sparsity. It does not perform any clustering; it uses clusters in data to perform classification.

Now that the motivation for our work is provided, the contributions of our work are discussed to underline the novelty of the proposed method.

- The paper proposes a novel classification algorithm called Sparsity Based Clustering Exploiting Classifier (SCEC). An optimization framework, using sparsity as guiding principle has been designed to extract semantics from clusters in data. It is also to be noted that, by deploying sparsity to exploit clustering property, this is one of the first works to suggest an alternative avenue for using sparsity, which otherwise generally, is only used to exploit subspace property and its variants.
- [15] uses Representative Vector Machines(RVMs) to perform classification: the basic idea of RVMs is to assign the class label of a test example according to its nearest representative vector. It further states that classical machine learning algorithms such as Support Vector Machines, Nearest Neighbour, and Sparse Representation Classifier are all a special case of RVMs. The above mentioned algorithms differ only in how the Representative Vector is chosen. In this sense the proposed optimization formulation can also be seen as a new method of choosing the Representative Vector.
- On the applied front, we find that the proposed algorithm on an average provides a 10% improvement in performance over popular classification algorithms such as, Sparse Representation Classifier (SRC), kNN, C4. 5, and Naive Bayes. We make such a claim based on the head to head comparisons we performed on a few datasets that were obtained from varies fields. Further to add more depth to the discussion, we graphically show as to why the algorithm classified, or misclassified certain test cases.

1.1 Topology of the Paper

This paper is broken down as follows, Sect. 2, "Related works", discusses other independent works that are related to the proposed work. Section 3, "Sparsity Based Clustering Exploiting Classification Algorithm" presents the proposed classification algorithm. Section 4, "Results and Analysis" empirically presents the different aspects of the effectiveness of the proposed algorithm. Finally the paper ends with Sect. 5, "conclusion" which underlines the significance of the proposed work.

2   Related Work

In this section we discuss other independent works that either employ sparsity, or exploit clustering, to perform machine learning tasks. Every paragraph begins by discussing a related work, and ends by highlighting the difference between the proposed method and that related work.

[5] seeks to encode the class of test vectors using sparse non negative vectors. The optimization framework enforces the required sparsity and non-negativity. This method strives to extract the benefits of non negative representation mentioned as in [11]. The main disparity between the proposed method and [5] is that, our method tries to use representations that are convexly closed, while their method seems to use only non negative representations.

[4] is one of the first works to use sparsity in classification. It proposes using sparsity to perform face recognition. The algorithm works under the assumption that vectors of the same class span the same subspace. The optimization framework, which is a ℓ_1 norm minimization problem, is designed to exploit this subspace property in data using sparsity as a guide for inference. The work achieved a good measure of success in solving the problem. There are two key differences between [4], and the proposed work: first difference is the initial assumptions of the two algorithms, [4] assumes that objects of the same class lie on the same subspace, while the proposed work assumes that objects of the same class lie within the same cluster; second difference is the optimization framework, [4] has no convexity constraints in its optimization framework, while the proposed method's optimization framework has convexity constraints present in it.

[16, 17] are seasoned and well respected methods in data mining. They use–similar to the proposed work–sparse coefficients to make their predictions; but the addition of convexity constraints in the optimization formulation employed in the current work, sets it apart from the optimization formulation in [16, 17].

[1] proposes group sparsity method as a viable solution for the problem of automatic image annotation. The optimization framework discussed in the paper uses regularized ℓ_1 and ℓ_2 norms in the objective function, sparsity and clustering property in tandem to solve the problem. The favourable results achieved by the paper suggests the effectiveness of the combined usage of both sparsity and clustering property in classification. The difference between our work and [1] is that, our optimization framework allows the columns of the data matrix to combine only convexly, while their optimization framework does not impose any such restrictions.

[2] altered the optimization framework of the standard sparse representation method, by introducing a new constraint to the standard formulation, thereby developed a novel cluster based sparse representation method for denoising of images. This work is one of the first works to combine sparsity and clustering under a unified mathematical framework. Our work is a classifying algorithm, while [2] is a denoising algorithm, this essentially sets the two works different from each other. Apart from this difference, the constraints of the optimization framework are different from each other.

3   Sparsity Based Clustering Exploiting Classification Algorithm

We begin our discussion by providing an intuitive overview of the proposed classification scheme. This is followed by describing the construction of the dictionary matrix D. Finally, the working of the proposed optimization framework is described.

3.1   An Intuitive Overview

The algorithm expects datapoints of the same class to be a part of the same cluster (this is referred to as the clustering property of data). This is shown in the Fig. 1. In a k class classification problem, the algorithm expects k clusters, where each cluster contains data points belonging to one class only. So, when a test vector is given to the algorithm, it simply tries to determine the cluster that is nearest to the given test vector; consequently the test vector is given the label of the vectors of that cluster. So suppose, the test vector was the nearest to cluster formed by training vectors of class 'A', then the algorithm would classify the test vector as class 'A'. This is the–simple, yet effective (as evidenced by our results)–intuition behind the classification algorithm. Classifying test vectors based on the cluster the test vector is the nearest to has been tried before; how we do it using sparsity is one of the novelties of the proposed work.

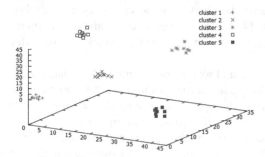

Fig. 1. The figure provides an ideal case representation of how datapoints are distributed in the euclidean space.

3.2   Framing the Dictionary Matrix

Let us consider the problem of framing a dictionary matrix for a dataset that has the following description,

- k-class dataset.
- L training vectors per class.
- Each training vector has m attributes

From the given training data, we first form local dictionary matrices. A local dictionary matrix is formed by horizontally concatenating training vectors belonging to one particular class alone. This is given by Eq. 1.

$$D_i = [v_1, v_2, ..., v_j, ..., v_L] \tag{1}$$

Here D_i is the i_{th} local dictionary matrix, and v_j is the j_{th} training vector of class i. Since there are only L training vectors in each class, there are L columns in each of these k local dictionary matrices.

The dictionary matrix is formed by horizontally concatenating all these k local dictionary matrices as given in Eq. 2.

$$D = [D_1, D_2, ..., D_k] \tag{2}$$

The dictionary matrix is in $\mathbb{R}^{m \times n}$, where m is the number of features, and n (equal to k $\times$ L) is the number of training vectors across all classes.

3.3 Optimization Framework

The Eq. 3 gives the optimization framework of the proposed classification scheme.

$$\underset{x}{\text{minimize}} \|x\|_1$$

$$\text{such that,} \tag{3}$$

$$\|Dx - Y\|_2 = \varepsilon$$

$$\|x\|_2 = 1$$

$$x_i \geq 0$$

$$\text{where } 1 \leq i \leq n$$

The Eq. 3 can be treated as a function, which when provided with the dictionary matrix D, the regularization constant ε, and the unclassified test vector Y, outputs a sparse x vector that encodes the class of the given test vector Y. To encode the class of Y, we want the x vector to give us information as to which cluster the given test vector is a part of, or is the nearest to.

One way to check if a particular vector is a part of a cluster, is to check if that vector can be realised as a convex combination of vectors belonging to that cluster. This checking process is done by the constraints of the optimization problem in Eq. 3. The constraint $\|Dx - Y\|_2 = \varepsilon$, ensures that a combination of vectors in the dictionary matrix D can realize Y with a tolerance of ε. The constraints $\|x\|_2$ and $x_i \geq 0$ (where i satisfies $1 \leq i \leq n$), ensure that the combination is convex. The addition of convexity constraints $\|x\|_2$, and $x_i \geq 0$, is a deliberate effort to force the algorithm to choose convex combinations over linear combinations. This is essential because only convex combinations can ensure if a test vector lies in a particular cluster; linear combinations cannot ensure this.

One possible way of finding which cluster the test vector belongs to can be done iteratively. Here we can try to convexly realise the test vector by first using the training vectors from one class alone, and then, the subsequent residual could be noted. We could repeat the process k times, and then pick the class which has the least residual. But this method will result in a slow classification process as we have to iterate exhaustively through the training vectors of all k classes.

This is where sparsity plays an important role by avoiding the algorithm to iterate through the many training vectors from the several classes. The minimize $\|x\|_1$ part in 3, forces the x vector to be sparse. Thereby this allows only a very few training vectors which are very closely placed to the test vector, to take part in the combination. Owing to the proposed optimization formulation, the classification scene becomes similar to the one as given by Fig. 2. It is seen that, there are only a few sparse and convex combinations that can realise the given test vector. This funnels out most clusters, and lets only a few neighbouring clusters to remain in contention.

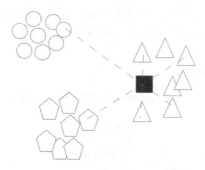

Fig. 2. The figure shows a possible classification scene. Each dashed line shows a convex combination that can realise the given test vector. The data points, that these dashed lines exist between are the points involved in the convex combination.

Now, of the few possible convex combinations that realise the test vector, we need to pick the combination that best lets us know as to which cluster the given test vector belongs to or is the nearest to. One important observation to make is that, only one combination involves training vectors of the same class, while other combinations involve training vectors from different classes. The objective function of the formulation–minimize $\|x\|_1$ in Eq. 3–selects the combination involving data points of the same class (blue dashed line), over other combinations that involve data points from different classes (red dashed lines); this is because $\|x\|_1$ is generally the smallest, when combinations involve data points from the same class. This choosing of a combination involving data points from the same class is very essential. This is so because, only if training vectors of the same class are involved in the combination, will the solution vector 'x' be able to encode the cluster the test vector belongs to, and thereby give us its class. To make it more clear an example is provided let us assume that the first L columns[1] of the dictionary matrix D alone belong to class 'A'. In this case if the test vector belongs to class 'A', then the solution vector x of the optimization problem in Eq. 3 will be as given in Eq. 4

$$x^T = [\alpha_1, \alpha_2, 0, 0, \alpha_5, 0, ..., 0, \alpha_L, 0, ..., 0.., 0, 0] \tag{4}$$

Here, α_1 to α_L are non zero coefficients. The presence of the non zero coefficients are seen only in some of the first L rows of x (or some of the first L columns of x^T), this suggests that some of the first L columns of the dictionary matrix alone have combined

---

[1] L is the number of training vectors per class.

Table 1. Table comparing the performance of the proposed method with the a few commonly used machine learning algorithms.

Classification algorithm	E-coli dataset	Tamil OCR dataset	Tagme dataset	Abalone dataset	Mean accuracy
The proposed method	81	**78**	**60**	19	**59.5**
kNN	**86**	60	**60**	3.5	52.375
C4. 5	78.4	53.33	50.6	**21.5**	50.95
Sparse representation classifier	73	55	59	16	50.75
Naive Bayes	80	NA	40	11.3	43.7

convexly to realise the test vector, which further suggests that the test vector is a part of the same cluster as the first L training vectors of the dictionary matrix. Since we work under the assumption that vectors of the same cluster belong to the same class, we assign the test vector the class of the training vectors that were involved in the convex combination, which in this case is label 'A'. To summarise the working of the proposed method, Algorithm 1 is provided.

Algorithm 1. Sparsity Based Clustering Exploiting Classification Algorithm

---

1: Objective: To determine the class of the test vector Y.
2: Input: Dictionary matrix D, the regularization constant ε , and a test vector Y.
3: Normalize the columns of the dictionary matrix D to unit ℓ_2 norm
4: Solve the ℓ_1 norm minimization problem

$$\underset{x}{\text{minimize}} \|x\|_1$$

$$\text{such that,}$$

$$\|Dx - Y\|_2 = \varepsilon$$

$$\|x\|_2 = 1$$

$$x_i \geq 0$$

$$\text{where } 1 \leq i \leq n$$

5: From x, find the class that has the most non-zero coefficients. Assign that class as the class of the given test vector Y.

---

4   Results and Analysis

In this section three essential aspects of the proposed algorithm are analysed: first, the algorithm's consistency; second, efficiency of the mathematical design of the algorithm; and finally, the statistical significance of ε. But, before we proceed with the analysis we discuss the experimenting methodology, and the backgrounds of the datasets used.

4.1 Experimenting Methodology

Standard machine learning practices are carried out to estimate the accuracy of an algorithm for a given dataset. First, the algorithms are trained with samples whose classes are known. Then the test vector is provided to the trained classifier. The output labels from the classifier are compared with the ground truth table to determine the number of test cases that have been rightly classified, from this, the different accuracies of the algorithm for different datasets are determined.

It is to be noted that this project along with all its associated experiments was run on a fourth generation i5 processor. The code for the project was written in MATLAB, and different optimization problems were solved during the course of the project with the help of cvx software [10].

4.2 About the Datasets

In order to support the idea, that the presence of clustering property can be assumed safely without any loss of generality, a diverse set of datasets has been chosen. There are four datasets that have been chosen, and a brief summary is given about each of these datasets.

Tagme Dataset. This dataset is a five class image classification problem [9]. Here the learning algorithm is trained with a hundred images of buildings, shoes, cars, humans, and flowers, in total five hundred images. These are the five classes involved in the classification problem. Here classification is performed on fifty images, ten from each of the classes.

Ecoli Dataset. This is a dataset that is freely available in the UCI college repository for datasets [7]. This is a 8 class classification problem, with 8 attributes for each observation. The dataset is typically used in protein localization sites determining projects. Based on a set of 8 attributes one of the 8 different protein localization sites such as, cp (cytoplasm) im (inner membrane without signal sequence), pp (perisplasm), imU (inner membrane, uncleavable signal sequence), om (outer membrane), omL (outer membrane lipoprotein), imL (inner membrane lipoprotein), imS (inner membrane, cleavable signal sequence) has to be predicted as the localization site for Ecoli bacteria. The 8 different protein localization sites are the 8 classes. There are 252 training and 84 testing samples in this dataset.

Tamil OCR Dataset. This dataset is obtained from hp's Tamil character dataset [8]. The dataset in total has more than three hundred class, but only sixteen classes are chosen. The sampled dataset made from the original hp dataset consists of one hundred and twenty eight training vectors and thirty two test vectors across 16 classes.

Abalone Dataset. This dataset is again obtained from the UCI college repository for datasets [7]. This is a 29 class problem, with each class having eight attributes. The classifier is used to predict the age of an abalone using the eight attributes. Here four thousand training samples and hundred and seventy seven testing samples are used.

4.3 Analysis of Consistency

From Table 1, it can be seen that when tested with datasets from various fields, the proposed method in terms of mean accuracy leads the columns by slightly more than 7%. This signifies the consistency of algorithm, as well as its robustness to noise arising from various datasets. Another factor that is testimony to the consistency of the proposed method is that, the accuracy of the proposed method is always above the average accuracy for any dataset. For example, the average accuracy achieved by all five classification algorithms on the abalone dataset (column 5 Table 1) is 14.26%, though the proposed method has not achieved the best accuracy for this dataset (c4. 5 has 21.5%), it has achieved 19%, which is a healthy 5% over the average of 14.26%. However, kNN's accuracy drops down to a meager 3.5%, no such sudden drops in accuracy can be seen in the case of the proposed algorithm. This bolsters our claim on the consistency exhibited by the proposed algorithm.

4.4 Analysis of the Efficiency of the Algorithm

In this section we show that the algorithm does what it is exactly designed to do, and that, it breaks down only when the clustering property is not exhibited; In other words we try to show that the algorithm breaks down only when the given test vector is placed far off from the cluster formed by the training vectors of its class. This is shown by the graphs given by Figs. 3, 4, 5 and 6. The x-axis is the test vector id. The y axis is the distance between the test vector, and the cluster formed by training vectors of its class. So if a vector has $(10, 0.2)$ as its coordinates, then it means that, the 10th test vector is 0.2 units of length away from the centroid of the cluster formed by the training vectors of its class. A purple '+' sign means it has been correctly classified, while a green 'x' sign means it has been + wrongly classified. A common trend that can be seen is that, the misclassified test vectors are generally placed atop the correctly classified test vectors in these graphs. This suggests that, the misclassified test vectors are generally placed far off from their corresponding clusters, while the correctly classified ones are near their clusters. This corroborates with our claim that, the algorithm generally breaksdown only when test vectors are away from the cluster that they should be in.

In Fig. 3 shows how the test vectors are positioned in the Ecoli dataset. It can be seen that mostly all test vectors that have been correctly classified are below the 0.4 mark in the y-axis. This implies that these test vectors are near the cluster formed by the training vectors of their class. It can also be seen that the wrongly classified ones are above the 0.40 mark in the y-axis, suggesting that they are far off from their class' cluster. However, near the 40, 50, 80 mark in the x-axis three deviants could be seen. They are classified wrongly even though those test vectors were found to be fairly close to their corresponding clusters. On further investigation, we found out that these vectors were classified wrongly because, they were closer to a cluster formed by the training vectors of another class, than they were to the cluster formed by training vectors of their class.

The Fig. 4 shows for the Tamil OCR dataset, whether test vectors were classified correctly or wrongly, based on how near or far, they were to their class' cluster. The same said for the Ecoli dataset applies here. The same general trend is found here,

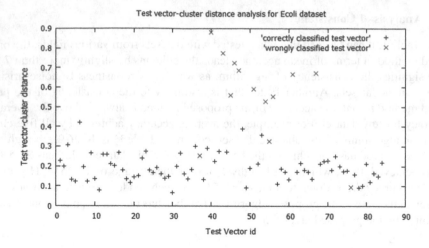

Fig. 3. Analysis of test vector-cluster distance for the Ecoli dataset.

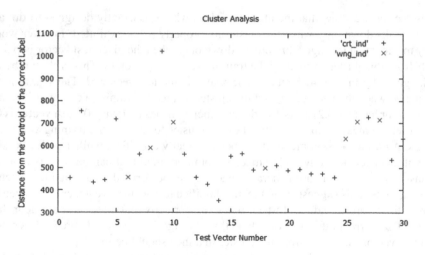

Fig. 4. Analysis of test vector-cluster distance for the Tamil OCR dataset.

where correctly classified test vectors are seen lying near the bottom, while the wrongly classified ones are seen atop them, with a few deviants.

The Fig. 5, for the Tagme data, relates the positioning of test vectors, to the how they were classified. This graph might seem to have a bit more deviants than the other ones. But the general trend of having the correctly classified test vectors at the bottom, while having the wrongly classified ones atop them is still prevalent here.

The abalone dataset exhibits this trend very vividly even though we have had 1033 testing vectors. This is given by Fig. 6. Here it is seen that almost all correctly classified vectors are below the 1 mark in the y-axis, while the wrong ones are generally above it.

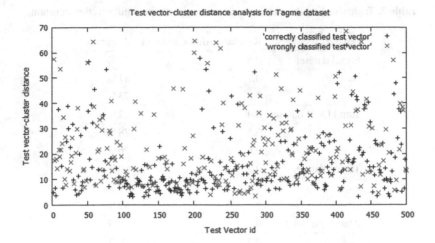

Fig. 5. Analysis of test vector-cluster distance for the Tagme dataset

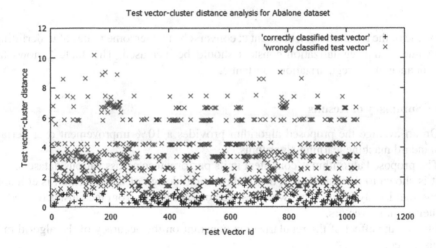

Fig. 6. Analysis of test vector-cluster distance for the Abalone dataset

These graphs support our claim that the proposed algorithm misclassifies only when the test vector is far off from the cluster formed the training vectors of its class. Thereby stands testimony to the efficiency of the design.

4.5 Statistical Significance of ε

The constant ε in Eq. 3 has more purpose than just adding tolerance to the realization. It is the regularization constant in the optimization formulation presented by Eq. 3. It is used to find a trade-off between the variance and bias of the model. It can be used to overcome issues of overfitting, and underfitting. We can tackle overfitting by lowering

Table 2. Table showing the variation in accuracy with the regularization constant.

Dataset	Regularization constant ε	Accuracy
E-coli dataset	0.6	77 %
	0.7	81 %
	0.8	73 %
Tamil OCR dataset	0.4	62 %
	0.5	78 %
	0.6	76 %
Tagme dataset	0.5	60 %
	1	58 %
	2	59 %
Abalone dataset	0.1	17 %
	0.2	19 %
	0.3	16 %

the value of the regularization constant ε; conversely, to overcome issues of underfitting, the value of the regularization constant should be increased. The Table 2 shows the significance of the regularization constant ε.

4.6 Summary of Results

- On an average the proposed algorithm provides a 10 % improvement over certain standard machine learning algorithms.
- The proposed algorithm is consistent in its performance across various datasets.
- It is shown to the readers that the mathematical design of the algorithm is efficient, and that it only breaksdown when test vectors are far off from the cluster formed by their training vectors.
- Finally, the effect of the regularization constant on the accuracy of the algorithm is discussed.

5 Conclusion

The proposed algorithm, to the authors knowledge is the first dictionary learning method to exploit clustering property in data. The motivation, construction, and working of the algorithm have been discussed in depth. Further essential aspects of the algorithm such as, efficacy, consistency, efficiency of the algorithm's design, and effect of regularization constant have been sufficiently analysed. In Subsect. 4.4 the break down point of the algorithm has also been discussed. Thereby we feel that a wholesome discussion of the work has been provided.

Since the proposed algorithm on an average gives an 10 % improvement over certain standard machine learning algorithms, we strongly feel that the proposed algorithm could supplant some of algorithms discussed in Table 1, especially in applications where the clustering property is more pronounced. We finally end with a profound hope that this work could inspire a profuse amount of work towards employing sparsity in a novel way.

References

1. Zhang, S., et al.: Automatic image annotation using group sparsity. In: 2010 IEEE Conference on Computer Vision and Pattern Recognition (CVPR). IEEE (2010)
2. Dong, W., et al.: Sparsity-based image denoising via dictionary learning and structural clustering. In: 2011 IEEE Conference on Computer Vision and Pattern Recognition (CVPR). IEEE (2011)
3. Ramirez, I., Sprechmann, P., Sapiro, G.: Classification and clustering via dictionary learning with structured incoherence and shared features. In: 2010 IEEE Conference on Computer Vision and Pattern Recognition (CVPR). IEEE (2010)
4. Wright, J., et al.: Robust face recognition via sparse representation. IEEE Trans. Pattern Anal. Mach. Intell. **31**(2), 210–227 (2009)
5. Hoyer, P.O.: Non-negative sparse coding. In: Proceedings of 2002 12th IEEE Workshop on Neural Networks for Signal Processing. IEEE (2002)
6. Elhamifar, E., Vidal, R.: Clustering disjoint subspaces via sparse representation. In: 2010 IEEE International Conference on Acoustics Speech and Signal Processing (ICASSP). IEEE (2010)
7. Lichman, M.: UCI machine Learning Repository. School of Information and Computer Sciences, University of California, Irvine. http://archive.ics.uci.edu/ml
8. HP-labs: Isolated Handwritten Tamil Character Dataset Developed by HP India Along with IISc (2006). http://lipitk.sourceforge.net/datasets/tamilchardata.htm. Accessed 30 Sept 2010
9. Tagme dataset. http://events.csa.iisc.ernet.in/opendays2014/events/MLEvent/index.php. Accessed 26 Mar 2014
10. Grant, M., Boyd, S.: Graph implementations for nonsmooth convex programs. In: Blondel, V., Boyd, S., Kimura, H. (eds.) Recent Advances in Learning and Control. LNCS, vol. 371, pp. 95–110. Springer, Heidelberg (2008)
11. Lee, D.D., Seung, H.S.: Learning the parts of objects by non-negative matrix factorization. Nature **401**(6755), 788–791 (1999)
12. Cover, T.M., Hart, P.E.: Nearest neighbor pattern classification. IEEE Trans. Inf. Theor. **13**(1), 21–27 (1967)
13. Hartigan, J.A., Wong, M.A.: Algorithm AS 136: a k-means clustering algorithm. Appl. Stat. **28**, 100–108 (1979)
14. Osuna, E., Freund, R., Girosi, F.: Training support vector machines: an application to face detection. In: Proceedings of 1997 IEEE Computer Society Conference on Computer Vision and Pattern Recognition. IEEE (1997)
15. Gui, J., et al.: Representative vector machines: a unified framework for classical classifiers. (2015)
16. Tibshirani, R.: Regression shrinkage and selection via the Lasso. J. Roy. Stat. Soc. Ser. B (Methodol.) **58**, 267–288 (1996)
17. Chen, S.S., Donoho, D.L., Saunders, M.A.: Atomic decomposition by basis pursuit. SIAM J. Sci. Comput. **20**(1), 33–61 (1998)

Mining Event Sequences from Social Media for Election Prediction

Kuan-Chieh Tung[1], En Tzu Wang[2], and Arbee L.P. Chen[3(✉)]

[1] Institute of Information Systems and Applications,
National Tsing Hua University, Hsinchu, Taiwan
jackytong8085@hotmail.com
[2] Computational Intelligence Technology Center,
Industrial Technology Research Institute, Hsinchu, Taiwan
m9221009@em92.ndhu.edu.tw
[3] Department of Computer Science and Information Engineering,
Asia University, Taichung, Taiwan
arbee@asia.edu.tw

Abstract. Predicting election results is a challenging task for big data analytics. Simple approaches count the number of tweets mentioning candidates or parties to do the prediction. In fact, many other factors may cause the candidates to win or lose in an election, such as their political opinions, social issues, and scandals. In this paper, we mine rules of event sequences from social media to predict election results. An example rule for a candidate can be as follows: "(big event, positive) → (small event, negative) → (big event, positive)" implies a victory to this candidate. We detect events and decide event types to generate event sequences and then apply the rule-based classifier to build the prediction model. A series of experiments are performed to evaluate our approaches and the experiment results reveal that the accuracy of our approaches on predicting election results is over 80 % in most of the cases.

Keywords: Election prediction · Social media · Event detection · Sentiment analysis · Sequential classification rule mining

1 Introduction

With the rise of social media, such as Facebook, Instagram, Twitter, and Google plus, many people use them to express their opinions and share their mood and life. Since social media contain much personal information, such as preference, interesting stuffs, living place, job, family members, and so on, they become a good source for research on various subjects. Moreover, social media can be regarded as a sensor for detecting earthquake [1], influenza epidemics [2], burst events [3, 4], etc.

Election related research such as election prediction and political preference identification has become a popular field in recent years [5–10]. Existing approaches focus on counting the number of tweets mentioning candidates or political parties to predict election results or analyzing sentiment of tweets to identify user political preferences. In reality, many factors can cause a candidate to win or lose an election.

© Springer International Publishing Switzerland 2016
P. Perner (Ed.): ICDM 2016, LNAI 9728, pp. 266–281, 2016.
DOI: 10.1007/978-3-319-41561-1_20

In this paper, we propose a novel framework to detect sequences of political events relevant to candidates from social media and analyze the corresponding comments about the events to derive rules for predicting election results. One example rule is shown in Fig. 1. It says in the days before the election day, if the following event sequence happens "(big event, positive) → (small event, negative) → (big event, positive)," the target candidate will win the election. Here, (big event, positive) means it is a big event and has a positive influence to the candidate.

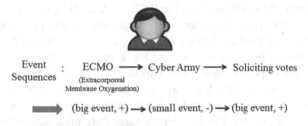

Event   :   ECMO ——→ Cyber Army ——→ Soliciting votes
Sequences   (Extracorporeal
             Membrane Oxygenation)

➡➡ (big event, +) ——→ (small event, -) ——→ (big event, +)

Fig. 1. A winning rule for a candidate

Post-ID	Date	User-ID	Article
35440	08/03	Sunpaper	[News] Sean Lien donate....
35441	08/03	Qn1234	[Ask]Who will win ?
35442	08/03	MicroMM	Re:[News] Sean Lien donate..
35443	08/03	mayamen	[Scoop]Legislator scandal!!
35444	08/03	Ipad3	[Announcement]policy revise
35445	08/03	XXXXDog	Re:[Announcement]policy
35446	08/03	cpblguu	Re:[Scoop]Legislator scandal

Fig. 2. Entering board screen

In this paper, we focus on Taiwan mayor election held on November 29th, 2014, and consider 155,921 articles from PTT Bulletin Board System, published during August 1st to November 28th, 2014 to do the prediction. PTT is the most popular BBS in Taiwan, established in 1997. Major assemblies such as the parade on the death of Chung-chiu Hung (洪仲丘事件, 2013) and Sunflower Student Movement (太陽花學運, 2014) recruited people through PTT, which led to 250 thousand people and 500 thousand people to be on the street, respectively. The influence of PTT in Taiwan can be huge and therefore we use the data of PTT to analyze the election in Taiwan.

On PTT, we can post articles or messages without length limitation. Figures 2 and 3 show the entering board screen and the entering article screen, respectively. In the following, we describe five important interactive features of PTT, including *Post*, *Repost*, *Comments to agree the post*, *Comments to disagree the post*, and *Comments without sentiments to the post*. In this paper, an article can be a post or a repost.

Post. Each PTT user can post on the board.

Repost. If the users want to express their opinions on a specific post, they can write a repost.

Comments to agree the post (like tag). If the users agree or like a specific post, they can give a like tag and write short comments to the post.

Comments to disagree the post (dislike tag). If the users disagree or dislike a specific post, they can give a dislike tag and write short comments to the post.

Comments without sentiments to the post (arrow tag). If the users have no sentiments on the post but want to express their thoughts, they can give an arrow tag and write short comments to the post.

In our framework, we (1) use the LDA (Latent Dirichlet Allocation) model [11] to discover topics related to election from the PTT articles, (2) identify events and decide the event type for an event, (3) generate event sequences, and (4) build a rule-based classifier and use it to predict election results. The experiment results show that the prediction accuracy of our approach is over 80 % in most of the cases.

The rest of this paper is organized as follows. In Sect. 2, the related works on election prediction, event detection, and sequential classification rule mining are reviewed. Section 3 states four methods for event type determination and describes the election prediction algorithm. The experiment results are presented in Sect. 4. Finally, Sect. 5 concludes this work.

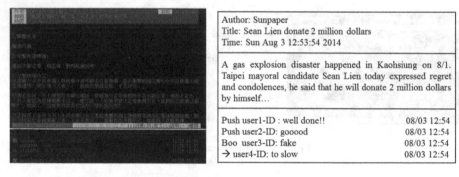

| Author: Sunpaper |
| Title: Sean Lien donate 2 million dollars |
| Time: Sun Aug 3 12:53:54 2014 |

A gas explosion disaster happened in Kaohsiung on 8/1. Taipei mayoral candidate Sean Lien today expressed regret and condolences, he said that he will donate 2 million dollars by himself…

Push user1-ID : well done!!	08/03 12:54
Push user2-ID: gooood	08/03 12:54
Boo user3-ID: fake	08/03 12:54
→ user4-ID: to slow	08/03 12:54

Fig. 3. Entering article screen

2   Related Works

Election prediction such as predicting the election results or political preferences has become popular in recent years. Many approaches have been proposed. Sprenger et al. [9] collects the tweets with the names of the six parties in the German parliament, and counts the number of tweets mentioning a party to predict the election. Gaurav et al. [6] uses the same concept in prediction by considering the candidates' name, alias and nickname to collect messages. O'Banion et al. [8] proposes a method to predict the voting behavior. At first, the candidates' name, financial status, and job are used to collect messages. Then, positive/negative words in each message are used to compute the sentiment score for each candidate. The results are close to the election polls. Several approaches predict political preferences of people. [5] count the number of political keywords mentioning a party to predict political preferences of people. [7] collects history logs of Twitter users by using political relevant hashtags. Then, for each user, the number of positive words to the corresponding party in the history logs minus the number of negative words is used to decide the political preferences of the users. Unankard1 et al. [10] collects tweets of the users to compute their sentiment scores, and considers the popularity degree of the events mentioned in the tweets. The political preferences of the users can therefore be decided and the voting percentage of the candidates be predicted.

Event detection is important in many research fields. [1, 2] focus on specific topics related to influenza epidemics and earthquake, respectively. They use related keywords of the specific topics to detect events. If the related keywords increase more than usual, some events are considered to have happened. The experiments show that detecting events from the search engines or social media can be more efficient and faster than the official releases. In [3, 4], event detection is not limited to specific topics. Liu et al. [3] assumes that if an event happens, some relevant words become more active than usual. An approach is proposed to detect the burst words based on a time window. The messages containing the burst words are then clustered to find interesting topics. Xie et al. [4] proposes a detection framework by computing the acceleration of three quantities including the total number of tweets, the occurrence of each word, and the occurrence of each word pair.

Approaches on integrating classification and association rule mining have been proposed and applied to health care mining, intrusion detection, transaction mining, etc. Liu et al. [12] proposes a solution by mining candidate rules, and selecting rules covering all training data to build a classification model. When classifying a new data object, the best rule (i.e. the rule with the highest confidence) is applied. Li et al. [13] uses a similar way to generate and select rules. The major difference is that multiple rules are used to do the classification. Finally, the major difference of [14, 12] is its selecting the best k rules for each class to predict the labels of the new data objects.

3   The Proposed Approach

There are five parts in the system, including *topic determination, event detection, sentiment analysis, event type determination* and *prediction model building*. The topic determination discovers the topics the collected articles belong to. The event detection discovers highly relevant events of the corresponding candidates. The sentiment analysis identifies the influence of the events to the corresponding candidates. The event type determination decides the event type by considering the influence and size of the events. The prediction model building generates rules to predict election results.

There are 22 cities involved in the 2014 Taiwan mayor election. We focused on the six major cities (Taipei, New Taipei, Taoyuan, Taichung, Tainan, and Kaohsiung), to which over 95 % of the collected articles refer. Each major city election has two major candidates.

In addition to collecting the PTT articles, a list of 98 politicians (among them, 48 are from the political party KMT, 37 from DPP, and 15 from others) is also constructed. These articles are preprocessed by the CKIP Chinese word segmentation system [15]. The punctuation and Chinese stop words that carry less important meanings are removed from the articles.

3.1   Topic Determination

After preprocessing the articles, we use the LDA model [11] to discover topics of election. We then use the topics to identify events. We decide an appropriate number for the topics manually.

Moreover, the LDA model also computes the topic distribution for each article. For example, an article may belong to topic 1, topic 2 and topic 3 with weights 30 %, 20 % and 50 %, respectively. In order to simplify the processing, we assign the topic with the most weight to each article if the entropy of the topic distribution is smaller than a predefined threshold. And we abandon an article if the entropy of the topic distribution is higher than the predefined threshold.

Observation: Fig. 4 shows one of the topics of the Taipei election, which contains keywords of doctor, organ, ECMO, and surgical during the period of 8/1/2014-8/31/2014. Two peaks of the number of articles belonging to the topic are identified, which represent two different events.

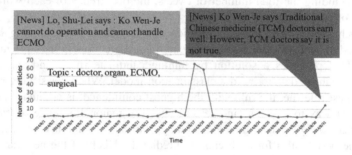

Fig. 4. From topic to events

3.2   Event Detection

[3] claims that if an event happens, the number of relevant words of the event will be higher than usual; [4] indicates that when an event occurs, there is a sudden surge of its popularity. We use the same concept, and consider the number of articles and the number of comments on the articles to detect events. The details are described as follows.

After finishing the topic determination, we have a total of K topics. Given a time window w and period T, we have a total of m time points, where $m = T/w$. The number of articles (or comments) at the time point i, $i = 1, \ldots, m$ is denoted S_i.

Definition 1. The average number of articles (or comments) for a topic at a time point is denoted μ and is equal to

$$\mu = \frac{\sum_{i=1}^{m} S_i}{m \times K} \tag{1}$$

Definition 2. If the difference between S_i and S_{i-1}, divided by μ is larger than a predefined threshold α, we say an event occurs at time point i.

$$\frac{S_i - S_{i-1}}{\mu} > \alpha \qquad (2)$$

Definition 3. If an event continuously occurs from the time point i to the time point j, we say the event is occurring from time point i to time point j.

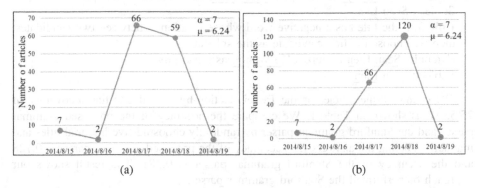

Fig. 5. Event detection example

Figure 5 shows an example of event detection for the Taipei election. We use the number of articles to detect events. The time period is from 8/1/2014 to 8/31/2014. The total number of the articles is 4836, and the number of topics is 25. The time window is 1 day, and the predefined threshold is 7.

The X axis represents date and the Y axis represents the number of articles. In Fig. 5(a), the difference between the number of articles on 8/17 and that of 8/16 divided by the mean is 10.25, which is larger than the predefined parameter α ($\frac{66-2}{6.24} = 10.25 > 7$). We say an event occurs on 8/17. In Fig. 5(b), the difference between the number of articles on 8/18 and that of 8/17 divided by the mean is 8.65, which is larger than the predefined parameter α ($\frac{120-66}{6.24} = 8.65 > 7$). We say an event is occurring from 8/17 to 8/18.

3.3 Sentiment Analysis

There are three parts, i.e. title, content, and comments in an article. In this section, an ad hoc grammar parser is designed to decide the subject of an article (the candidate this article is talking about) from its title. From the comments of the article, we can then compute a *support score* to the corresponding candidate.

The Stanford grammar parser [16] can be used to identify the subject of the article. However, since the title is often shorthanded, it does not work well. We therefore designed an ad hoc approach with a politician dictionary to identify the subject as follows (the candidates in the following examples are underlined).

Rule 1: if the title contains only one candidate, then the candidate is the subject.
Example: Sean Lien becomes a part-time worker at a drink shop to experience the life of citizens
Subject: Sean Lien
Rule 2: if the title has a colon (:), the subject is designated to the candidate whose location is closest to the colon.
Example: Ko Wen-je scolds again Sean Lien: he just likes to complain
Subject: Sean Lien
Rule 3: if the title has a negative word whose location is between two candidates, then we choose the first candidate as the subject.
Example: Sean Lien criticizes Ko Wen-je as a bad guy
Subject: Sean Lien

We examined the articles on the Taipei election by our rules to find a coverage of 75.55 % as shown in Table 1. We compare the accuracy of the proposed grammar parser and the Stanford grammar parser by randomly choosing five hundred titles and manually labeling the answers. The accuracy of the proposed grammar parser is 0.728 and the accuracy of the Stanford grammar parser is 0.381. The result shows our approach outperformed the Stanford grammar parser.

Table 1. Rule coverage (Taipei election)

Title	Number	Coverage
Rule1 + Rule2 + Rule3	5267	75.55 %
No candidates	1215	17.43 %
Have candidates but cannot judge	489	7 %

After identifying the subject of the articles, we compute the support scores for the corresponding candidates by using the number of like tags and dislike tags.

Assume the subject of an article is candidate A, the support score SS is the number of like tags m minus the number of dislike tags n.

$$SS_{C_A} \text{ from an article} = m - n \qquad (3)$$

where SS_{C_A} denotes the support score for candidate A.

We do not analyze the content of the comments because in political articles people like to use ironic words and informal sentences in the comments, which makes it difficult to analyze. To show the accuracy of our approach of using the like tags and dislike tags to compute support scores, we randomly chose 9154 comments and manually labeled the answers. We found out that 8055 comments were correctly decided and 1099 comments were incorrectly decided using the like tags and dislike tags. That is, an accuracy of 0.88 is achieved by using the like tags and dislike tags to reflect the support for candidates.

3.4   Event Type Determination

Let $E = \{e_0, \ldots, e_{m-1}\}$ be a set of m events. Let n_i be the number of articles (or comments) of e_i. Each event e_i has three attributes, including *occurrence time*, *event size* and *event influence*. We consider "big" and "small" for the value of event size and "positive," "negative," and "useless" for the event influence. We only consider four event types, that is, (big, positive), (big, negative), (small, positive), and (small, negative). Take one of the four types for example: (big, positive) means it is a big event and has positive influence to the candidate in consideration. Moreover, based on the occurrence time the events form an event sequence. The event size and event influence are decided as follows:

Definition 4 (Big Event and Small Event). Event e_i is a big event if the n_i of e_i is larger than the average number of articles (or comments) of all the events.

$$\text{Type of } e_i \text{ is big if } n_i > \frac{\sum_{i=0}^{m-1} n_i}{|E|} \tag{4}$$

Otherwise, it is a small event.

In the following, we describe how to compute the support scores from an event to the corresponding candidates.

Assume there are n articles in event e_i, and candidate A is the subject of k articles, where $0 \leqq k \leqq n$, the support score SS for candidate A from event e_i is the summation of the support scores for candidate A from the k articles.

$$SS_{C_A} \text{ from } e_i = \sum_{i=1}^{k} (SS_{C_A} \text{ from } article \, i) \tag{5}$$

where SS_{C_A} denotes the support score for candidate A.

Definition 5 (Positive Event and Negative Event). For an event e_i, if the difference of the support scores of the two competing candidates C_1 and C_2 is larger than a pre-defined impact factor f, and the score of C_1 is larger than the score of C_2, then the event e_i is defined to be positive to C_1 and negative to C_2, vice versa.

$$\text{Type of event: positive to } C_1 \text{ and negative to } C_2 \text{ if } \frac{|SS_{C_1} - SS_{C_2}|}{|SS_{C_1}| + |SS_{C_2}|} \geq f, SS_{C_1} > SS_{C_2} \tag{6}$$

$$\text{Type of event: negative to } C_1 \text{ and negative to } C_2 \text{ if } \frac{|SS_{C_1} - SS_{C_2}|}{|SS_{C_1}| + |SS_{C_2}|} \geq f, SS_{C_1} < SS_{C_2} \tag{7}$$

where SS_{C_1} denotes the support score of candidate C_1 and SS_{C_2} denotes the support score of candidate C_2.

Definition 6 (Useless Event). For an event e_i, if the difference of the support scores between the two competing candidates C_1 and C_2 is smaller than the predefined impact factor f, then event e_i is a useless event.

$$\text{Type of event is useless if } \frac{|SS_{C_1} - SS_{C_2}|}{|SS_{C_1}| + |SS_{C_2}|} < f \qquad (8)$$

where SS_{C_1} denotes the support score of candidate C_1 and SS_{C_2} denotes the support score of candidate C_2.

After the event types are determined, the event sequences can be generated by Algorithm 1.

3.5   Prediction Model Building

After generating the event sequences, we apply a rule-based model to predict the election results. An approach named Event Sequence Classification (ESC) is proposed to mine sequential classification rules. It is adopted from [12] with the following three steps: classification rule mining, prediction model construction, and class label prediction.

Our training dataset $T = \{tr_0, \cdots, tr_{n-1}\}$ consists of n records. Each record tr_i consists of an event sequence and a class label, denoted $e_0, \ldots, e_{m-1} \rightarrow c_i$. As mentioned before, each event is of one of the four types: (big, positive) denoted b^+, (big, negative) denoted b^-, (small, positive) denoted s^+, and (small, negative) denoted s^-. Each class label has two types: "win" denoted W and "lose" denoted L. An example training dataset is shown in Table 2. Notice that events in the same parenthesis mean that these events occur in the same time window.

Table 2. An example training dataset

Sequence	Label
$(s^+s^+)b^+b^-(b^+s^+)s^-b^-$	W
$b^+b^+s^+s^-$	W
$(b^+b^+s^+)s^+b^+s^+(b^-b^-s^+)$	W
$(s^-s^-)b^-b^+(b^-s^-)s^+b^+$	L
$b^-b^-s^-s^+$	L
$(b^-b^-s^-)s^-b^-s^-(b^+b^+s^-)$	L

Step 1: Classification rule mining. In this step, we apply PrefixSpan algorithm [17] and a minimum support threshold to find sequential patterns. Then, we compute the confidence of each sequential pattern. If its confidence is larger than a minimum confidence threshold, it is kept in a rule set.

Step 2: Prediction model building. After finishing classification rule mining, the rules kept in the rule set are sorted in the order defined as follows [12]:

Given two rules, r_i and r_j, $r_i \succ r_j$ (called r_i precedes r_j) if

1. the confidence of r_i is larger than that of r_j, or
2. they have the same confidences, but the support of r_i is larger than that of r_j, or
3. both confidence and support of r_i and r_j are equal, but r_i is generated earlier than r_j.

Step 3: Class label prediction. The number of the mined rules is always even. That is, if we have a rule such as (big, positive), (small, negative) $\rightarrow W$, there must be a corresponding rule: (big, negative), (small, positive) $\rightarrow L$. This is because the articles often influence positively to one candidate, yet negatively to the other. In addition, both the confidences and supports of the two contrary rules are same. Given two contrary rules r_i and r_j with the corresponding class c_i and c_j, if

1. the new record only matches the precedent of r_i, then c_i is assigned and vice versa; or
2. the new record matches the precedent of both r_i and r_j, then c_i is assigned to the new record which matches in more places of the precedent of r_i and vice versa.

Algorithm 2 presents our event sequence classification method.

4 Experiments

A series of experiments were performed to evaluate our approach on (1) generating sequences, (2) building the model, and (3) predicting the election results. In Subsect. 4.1, the experiment settings are described. In Subsect. 4.2, we evaluate the four methods we used to generate sequences. In Subsect. 4.3, we discuss two parameters for event detection. In Subsect. 4.4, we present the predicting results by varying related parameters.

4.1 Experimental Settings

We use the dataset collected from the PTT Gossiping board, published during August 1st to November 28th, 2014. The dataset contains 155,921 articles and 5,532,824 comments from 106,551 users. We focus on the elections in six major cities in Taiwan, each of the elections has two major candidates. We generate event sequences by using the PTT data between two consecutive polls and use the results of the newer poll to be the labels of our training data. As a result, we have a total of 130 event sequences as the training data. Then, we generate event sequences by using the PTT data between the last poll and the day before the election to do the prediction.

The default values of the parameters including the event occurrence threshold α, the impact factor f, the minimum support $msup$, and the minimum confidence $mconf$ are set to 7, 0.2, 0.3, and 0.5, respectively.

4.2 Sequence Generation Evaluation

In this section, we compare four sequence generation methods, including the Proposed Grammar Parser–Article (PGP-A), Stanford Grammar Parser–Article (SGP-A), the

Proposed Grammar Parser–Comment (PGP-C), and Stanford Grammar Parser–Comment (SGP-C).

The numbers of events detected in sequences are shown in Fig. 6. We have the following observations for the result. First, no matter how we vary α and f, PGP-A and PGP-C generate more events than SGP-A and SGP-C. Second, the number of events remains constant no matter what f is. It means that people give one-side opinions to candidates in most events.

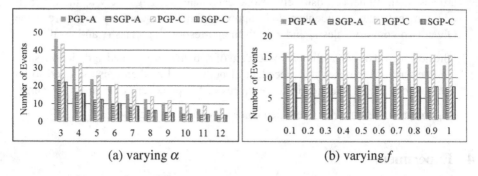

(a) varying α (b) varying f

Fig. 6. Comparing numbers of events

Moreover, we evaluate the *event quality* of the four methods, which is defined as the proportion of the events qualified to form the event sequence, as shown in Fig. 7. We see that the event quality of PGP-A and PGP-C is always better than the SGP-A and SGP-C. Therefore, we choose PGP-A or PGP-C to generate event sequences.

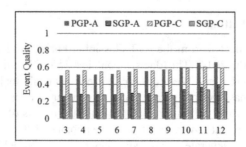

Fig. 7. Event quality comparison (varying α)

4.3 Event Detection Parameters

The two event detection parameters, i.e. number of articles and number of comments were compared. Using the Taipei election during August 1st to October 31th for example, 18 events were detected using the article parameter while 16 were detected using the comment parameter, and there are 14 common events. We conclude that the two event detection parameters are highly correlated.

4.4 Prediction Result Evaluation

As shown in Fig. 7, we see that the event quality of PGP-A and PGP-C is almost the same. We choose PGP-A to generate sequences and use our ESC method to predict the election results. In addition, we compare the ESC method with the well-known CBA [12] and CMAR [13] methods on accuracy and number of rules.

In the ESC method, we consider all rules. When we predict the label of the test data, the best rule satisfied is chosen for prediction. The comparison of the three methods is shown in Fig. 8. We consider four parameters, including α, f, $msup$ and $mconf$. In the following, we discuss the impact of setting these parameters on the performance of the three methods.

Setting of α. As the experiment results show in Fig. 8(a), we see that if we set α too low, too many events may be identified while if we set α too high, too few events may be identified.

Setting of f. As the experiment results show in Fig. 8(b), we see that no matter how we vary f, the accuracy remains about the same. The reason is the same as the case shown in Fig. 6(b). We conclude that people give one-side opinions to candidates in most events. Moreover, the accuracy of ESC is better than CBA and CMAR.

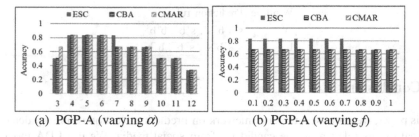

(a) PGP-A (varying α)   (b) PGP-A (varying f)

Fig. 8. Accuracy comparison

4.5 Summary on 2014 Taiwan Mayor Election

We show the best case on predicting election results is that only the election of New Taipei city is incorrectly predicted. In order to find out the reasons, we compute the support scores to each candidate as shown in Table 3. We can see that most of the candidates who are not in KMT party get higher approval rate than the candidates who are in KMT party. We conclude that people give one-side approval to the candidates who are not in KMT party and that is the reason why most of these candidates win the election.

We find common rules on predicting election results as shown in Table 4. We can see that if a prediction rule contains more positive influence events, the candidate will have a higher chance to win the election. Conversely, if it contains more negative influence events, the candidate will have a higher chance to lose the election.

Table 3. Support score to each candidate

City	Candidate (party)	Positive	Negative	None
Taipei	Sean Lien (KMT)	38.4 %	43.6 %	18 %
	Ko Wen-je (None)	70 %	14.5 %	15.5 %
New Taipei	Chu Li-luan (KMT)	48.7 %	28.5 %	22.7 %
	Yu Shyi-kun (DPP)	66.7 %	12.6 %	19.7 %
Taoyuan	Wu Chih-yang (KMT)	55.8 %	23 %	21.1 %
	Cheng Wen-tsan (DPP)	56.2 %	15.5 %	28.3 %
Taichung	Hu Chih-chiang (KMT)	39.3 %	37.7 %	22.8 %
	Lin Chia-ling (DPP)	56.5 %	19.5 %	23.9 %
Tainan	Huang Shiou-shuang (KMT)	54.8 %	21.4 %	23.7 %
	Lai Ching-te (DPP)	68.8 %	12.1 %	19 %
Kaohsiung	Yang Chiu-hsing (KMT)	34.6 %	43.6 %	21.7 %
	Chen Chu (DPP)	55.8 %	25.6 %	18.5 %
Total	KMT	38.9 %	42.5 %	18.4 %
	Not KMT	68.2 %	15.5 %	16.2 %

Table 4. Common rules

Win/lose	Prediction rules
W	$s^+s^+b^-s^+$, s^+b^+, b^+b^+
L	$s^-s^-b^+s^-$, s^-b^-, b^-b^-

5 Conclusion

In this paper, we propose a novel framework on predicting election results by detecting political events relevant to the candidates from social media. We use LDA model to discover topics of articles and then use the numbers of articles (or comments) to detect events. Four methods are proposed to generate event sequences. In PGP-A and PGP-C, the subject identification is based on our proposed grammar parser. In SGP-A and SGP-C, the Stanford grammar parser is applied to identify the subject of the articles. The experiments show that the event quality of PGP-A and PGP-C is better than that of SGP-A and SGP-C. Moreover, we propose the Event Sequence Classification method to predict election results. We use the best rule to do the prediction. The accuracy of ESC is better than CBA and CMAR. The highest accuracy of ESC is 83 %.

In the near future, we will consider political preferences of people to do sentiment analysis. Once we know the political preference of people, the correctness of sentiment analysis can be improved.

Appendix

Algorithm 1 (Event Sequence Generator)

Input: city, start time, end time, A collection of P
       topics, observed average µ of the four months
       data, observed parameter (article or comment),
       predefined event occur threshold α, predefined
       impact factor f.

Output: Candidate 1 event sequence and Candidate 2
event sequence.

1. **For** p = 0 to Topic P-1
2.   **While** time is between start time and end time
3.   Event detection parameter can be article
         or comment.
4.   **If** event detected by Definition 1-3
5.   Add event to the event set.
6. p = p + 1
7. **For** each event in the event set
8.   Decide event size attribute by Definition 4.
9.   **For** each article in event
         Compute support scores for each candidate.
10.   End For
11.   Decide event influence attribute by Definition 5-
   6
12.   End For
13.   Sort the event by event occurrence time.
14.   Return candidate 1 event sequence and candidate 2
      event sequence.

Algorithm 2 (Event Sequence Classification)

Input: Training data set TR, Test data TS, minimum support msup, and minimum confidence mconf.
Output: Class Label for Test data
Classification Rule Mining(TR, msup, mconf)
1. Apply PrefixSpan algorithm, using msup to find the frequent sequential pattern in TR
2. **For** each sequential pattern
3.   Compute confidence
4.   **If** the confidence of sequential pattern $\geq$ mconf
5.   Add to Rule set
6. End for
Build Predict Model(Rule set)
1. Sort the rule set by following rules, given two rules, r_i and r_j, $r_i > r_j$ (also called r_i precedes r_j)
2. **If** the confidence of r_i is greater than of r_j
3. **Else if** their confidence are same, but the support of r_i is greater than that of r_j
4. **Else** both the confidences and supports of r_i and r_j are the same, but r_i is generated earlier than r_j
5.   Add to Predict model.
Predict Class Label (TS, Predict Model)
1. **While** have not finished checking all the rule in Predict Model
2. Check two contrary rule r_i and r_j with corresponding class c_1 and c_2
3. **If** TS contains only r_i, then label c_i to new record, vice versa
4. **Else if** TS contains r_i and r_j, if the number of times of r_i matching the TS is more than r_j's, then label c_i to new record, vice versa
5. **Else** go on to check next two contrary rules

References

1. Sakaki, T., Okazaki, M., Matsuo, Y.: Earthquake shakes twitter users: real-time event detection by social sensors. In: Proceedings of the 19th International Conference on World Wide Web, WWW 2010, Raleigh, North Carolina, USA, pp. 851–860 (2010)
2. Ginsberg, J., Mohebbi, M.H., Patel, R.S., Brammer, L., Smolinski, M.S., Brilliant, L.: Detecting influenza epidemics using search engine query data. Proc. Nat. **457**, 1012–1014 (2009)
3. Liu, C., Xu, R., Gui, L.: Burst events detection on micro-blogging. In: Proceedings of the International Conference on Machine Learning and Cybernetics, ICMLC 2013, pp. 1921–1924 (2013)
4. Xie, W., Zhu, F., Jiang, J., Lim, E.-P., Wang, K.: Topicsketch: real-time bursty topic detection from twitter. In: Proceedings of the 13th International Conference on Data Mining, ICDM 2013, Dallas, TX, USA, pp. 837–846 (2013)

5. Fink, C., Bos, N., Perrone, A., Liu, E., Kopcky, J.: Twitter, public opinion, and the 2011 Nigerian Presidential election. In: Proceedings of the IEEE Conference on Social Computing, SocialCom 2013, Washington, DC, USA, pp. 311–320 (2013)
6. Gaurav, M., Srivastava, A., Kumar, A., Miller, S.: Leveraging candidate popularity on twitter to predict election outcome. In: Proceedings of the 7th Workshop on Social Network Mining and Analysis, SNAKDD 2013, Chicago, Illinois, USA, pp. 1–7 (2013)
7. Makazhanov, A., Rafiel, D.: Predicting political preference of twitter users. In: Proceedings of the International Conference on Advances in Social Network Analysis and Mining, ASONAM 2013, Niagara, Ontario, Canada, pp. 298–305 (2013)
8. O'Banion, S., Birnbaum, L.: Using explicit linguistic expressions of preference in social media to predict voting behavior. In: Proceedings of the International Conference on Advances in Social Network Analysis and Mining, ASONAM 2013, Niagara, Ontario, Canada, pp. 207–214 (2013)
9. Sprenger, A.T.O., Sandner, P.G. Welpe, I.M.: Predicting elections with twitter: what 140 characters reveal about political sentiment. In: Proceedings of the 4th International Conference on Weblogs and Social Media, ICWSM 2010, Washington, DC, USA, pp. 178–185 (2010)
10. Unankard, S., Li, X., Sharaf, M.A., Zhong, J., Li, X.: Predicting elections from social net works based on sub-event detection and sentiment analysis. In: Proceedings of the 15th International Conference on Web Information System Engineering, WISE 2014, Thessaloniki, Greece, pp. 1–16 (2014)
11. Blei, D.-M., Ng, A.-Y., Jordan, M.-I.: Latent Dirichlet allocation. Proc. J. Mach. Learn. Res. JMLR 2003 3, 993–1022 (2003)
12. Liu, B., Hsu, W., Ma, Y.: Integrating classification and association rule mining. In: Proceedings of the 4th International Conference on Knowledge Discovery and Data Mining, KDD 1998, New York, USA, pp. 80–86 (1998)
13. Li, W., Han, J., Pei, J.: CMAR: accurate and efficient classification based on multiple class-association rules. In: Proceedings of the 2001 International Conference on Data Mining, ICDM 2001, San Jose, California, USA, pp. 369–376 (2001)
14. Yin, X., Han, J.: CPAR: classification based on predictive association rule. In: Proceedings of the 3rd SIAM International Conference on Data Mining, SDM 2003, San Francisco, CA, USA, pp. 331–335 (2003)
15. Ma, W.-Y., Chen, K.-J: Introduction to CKIP Chinese word segmentation system for the first international Chinese word segmentation bakeoff. In: Proceedings of the Second SIGHAN Workshop on Chinese Language Processing, SIGHAN 2003, Sapporo, Japan, JPA (2003)
16. Levy, R., Manning, C.-D.: Is it harder to parse Chinese, or the Chinese Treebank? In: Proceedings of the International Conference on Association for Computational Linguistics, ACL 2003, Sapporo, Japan, pp. 439–446 (2003)
17. Pei, J., Han, J., Mortazavi-Asl, B., Pinto, H., Chen, Q., Dayal, U., Hsu, M.: PrefixSpan: mining sequential patterns by prefix-projected growth. In: Proceedings of the 17th International Conference on Data Engineering, ICDE 2001, Heidelberg, Germany, pp. 215–224 (2001)

Multi-exponential Lifetime Extraction
in Time-Logarithmic Scale

Andrew V. Knyazev[✉], Qun Gao, and Koon Hoo Teo

Mitsubishi Electric Research Labs (MERL),
201 Broadway, Cambridge, MA 02139, USA
Andrew.Knyazev@merl.com
http://www.merl.com/people/knyazev

Abstract. Methods are proposed for estimating real lifetimes and corresponding coefficients from real-valued measurement data in logarithmic scale, where the data are multi-exponential, i.e. represented by linear combinations of decaying exponential functions with various lifetimes. Initial approximations of lifetimes are obtained as peaks of the first derivative of the data, where the first derivative can, e.g., be calculated in the spectral domain using the cosine Fourier transform. The coefficients corresponding to lifetimes are then estimated using the linear least squares fitting. Finally, all the coefficients and the lifetimes are optimized using the values previously obtained as initial approximations in the non-linear least squares fitting. We can fit both the data curve and its first derivative and allow simultaneous analysis of multiple curves.

Keywords: Lifetime extraction · Exponential · Multi-exponential · Least squares · Fitting · Numerical differentiation · Dynamic ON-resistance · Semiconductor

1 Introduction

Separation of exponentials, i.e. lifetime extraction from multi-exponential measurement data represented by linear combinations of decaying exponential functions with various lifetimes, is a classical area of research, related to the inverse Laplace transform and the problem of weighted moments, with numerous applications and a vast literature spreading over a century; see, e.g., [11, Chap.IV] and references there. We are interested in a particular case, where the data, the lifetimes, and the corresponding coefficients in the linear combinations are all real-valued, in contrast to a typical scenario of multiple signal classification for frequency estimation and emitter location. One additional assumption is that the lifetimes are widely spread, by orders of magnitude, practically requiring a time-logarithmic scale to represent the measurement data, which rules out traditional methods utilizing uniform time grid. Our particular application relates to analyzing trapping and detrapping of charge carriers, i.e. electrons and electronic holes, in electronic devices, as in [7,8]. Trapping an electron means that

---

Accepted to ICDM 2016. An extended preliminary version posted at arXiv.

P. Perner (Ed.): ICDM 2016, LNAI 9728, pp. 282–296, 2016.
DOI: 10.1007/978-3-319-41561-1_21

the electron is captured from the conduction band, while detrapping an electron conversely means capturing a hole from the valence band; see [10, Chap. 11].

In semiconductor devices, traps pertain to impurities or dislocations that capture the charge carriers, and keep the carriers strongly localized; see, e.g., [1,13,15,16]. Effects of the traps on performance of the semiconductor devices are temporal and eventually decay over time, i.e. the behavior of the device stabilizes and the measured quantity, e.g., the ON resistance, approaches a constant. Each trap can be assumed to behave exponentially decaying in time t with a specific lifetime τ, such as a purely exponential process in time t described by a function $c_\tau e^{-t/\tau}$, in a process reversed to a charge carrier avalanche. The parameter τ is commonly called the lifetime, and is inversely proportional to the decay rate, since $d(c_\tau e^{-t/\tau})/dt = -c_\tau e^{-t/\tau}/\tau$. The coefficient (also called the magnitude) c_τ expresses the magnitudes of the purely exponential process $c_\tau e^{-t/\tau}$, and represents the initial, at the time $t = 0$, contribution of the corresponding process to the measured data. For semiconductor traps, the magnitude c_τ may be related to the total initial, at the time $t = 0$, number of charges participating in the trap process for the fixed lifetime τ, described by the exponential process $c_\tau e^{-t/\tau}$. Depending on the sign of the charge of the carrier and the measured quantity, a single trapping process may lead to an increase with $c_\tau < 0$, called detrapping, or a decrease with $c_\tau > 0$ of the measured quantity.

Collectively, multiple traps are assumed to independently influence the operation of the device in an additive fashion with possibly various lifetimes τ and initial concentrations c_τ of the carriers in every trap. Therefore, the measurement data can be assumed to be multi-exponential in time,

$$I_{data}(t) \approx \sum_{i=1}^{n} c_i e^{-t/\tau_i} + I_\infty, \tag{1}$$

where one needs to determine the number $n > 0$ of terms in the sum, the positive lifetimes τ_i and the corresponding nonzero coefficients c_i, which can be positive in trapping and negative in detrapping processes. In our application, device recovery from the traps can take nano-seconds, minutes, or even days, therefore both t and τ need to be presented in the time-logarithmic scale, e.g., in base 10 we substitute $t = 10^s$ and $\tau = 10^\sigma$. We thus need to determine the values of σ_i and the corresponding nonzero coefficients c_i such that

$$I_{data}(s) \approx \sum_{i=1}^{n} c_i e^{-10^{s-\sigma_i}} + I_\infty, \tag{2}$$

e.g., by minimizing the least squares fit.

Since the traps may act at the nano-scale and the measurements are performed at the macro-scale, the number n of terms in the sum, if every single trap is given its own index, is practically infinite, i.e. the sum mathematically turns into the integral

$$I_{data}(t) \approx \int_{\tau_{min}}^{\tau_{max}} c_\tau e^{-t/\tau} d\tau + I_\infty, \tag{3}$$

where one needs to determine the interval of the lifetime values $0 < \tau_{\min} < \tau_{\max}$ and the function c_τ. Formula (3) clearly reminds us of the Laplace and the inverse Laplace transforms, leading to some exponentially ill-posed numerical problems; see, e.g., [4]. Switching to the time-logarithmic scale, one can make a formal substitution $t = 10^s$ and $\tau = 10^\sigma$ in integral (3), or consider the integral analog of the sum appearing in (2) as follows

$$I_{data}(c) \approx \int_{\sigma_{\min}}^{\sigma_{\max}} c_\sigma e^{-10^{s-\sigma}} d\sigma + I_\infty, \tag{4}$$

where one needs to determine the function c_τ, while the interval of the lifetime logarithm values $[\sigma_{\min}, \sigma_{\max}]$ can in practice be often selected a priori.

The analysis of the recovery is extremely important as the traps severely degrade the performance and reliability of semiconductor devices. Trap analysis is also important for characterizing the formation and behavior of the traps so that the devices can be modeled, designed, and manufactured with improved performance and reliability. The lifetimes are affected by material temperatures and activation energies of the traps. The captured or released coefficient could be a function of the initial number of the traps to be filled or number of carriers in the traps to be released, respectively. Methods that can extract the lifetimes of the trapping and detrapping processes from the measurement data allow detecting and analyzing the traps. The measured data, e.g., the ON resistances, are undoubtedly noisy. The noise distribution function over time is unclear, except that measuring in a short time range is practically difficult, so one may expect larger measurement errors, compared to a long time range.

Having the transient data, measured as a function $I_{data}(t)$ of time t, one goal can be to determine the constant I_∞, the total number n of purely exponential components present, and numerical values of the lifetimes τ_i and the corresponding magnitudes c_i for every exponential component in (1), wherein the value n is chosen as small as possible while fitting the data within a given tolerance. In the time-logarithmic scale, we determine σ_i in (2), having the data measured for some values of s. Such a problem can be called a "discrete lifetime extraction," where the extracted lifetimes represent the dominating lifetimes, averaged over all actual physical traps at the nano-scale.

An alternative goal can be to determine the constant I_∞, the interval of the lifetime values $[\tau_{\min}, \tau_{\max}]$, and the function c_τ, describing the distribution of the lifetimes in (3). In the time-logarithmic scale, instead of c_τ we determine the function c_σ in (4) describing the distribution of the logarithms of the lifetimes, having the data measured as a function of s. Such problems can be called "continuous lifetime extraction." Computationally, continuous lifetime extraction problem (4) can be approached by introducing some grids for the s and the σ in the time-logarithmic scale, approximating the integral (4) using a quadrature rule, and solving for c_σ on the grid for the σ.

We note that the problem of multi-exponential extraction is key in many applications, e.g., fluorescence imaging and magnetic resonance tomography, but common ranges of the exponents and measurement times may not be nearly as large as in our application of analyzing traps in semiconductors that requires using the logarithmic scale, both for the exponents and the measurement times.

2 Differentiating Exponential Functions in the Time-Logarithmic Scale

– I differentiate you!
– I am e^t...

What possibly exciting would one find by looking at the first derivative of exponential functions? The answer is surprising, if the time-logarithmic scale is used. In Fig. 1, several basis functions from the linear combination (1) with $\tau_i = 1, 10, 100, 1000$ and their corresponding derivatives are plotted in the time-logarithmic scale.

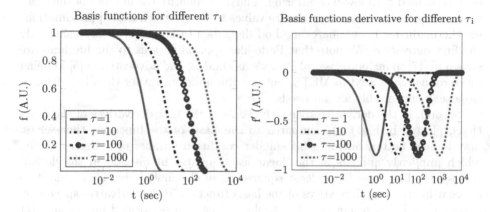

Fig. 1. The exponential basis functions and their analytic derivatives in the time-logarithmic scale (Color figure online)

We notice in Fig. 1(b) that the peaks of the absolute values of the first derivatives of the basis function show us the locations of the lifetime constants τ! This can also be proved analytically using the following expression for the first derivative the time-logarithmic scale

$$\frac{d\left(e^{-t/\tau}\right)}{ds} = -\log(10)\,10^{s-\sigma}e^{-t/\tau}, \text{ where } t = 10^s \text{ and } \tau = 10^\sigma \qquad (5)$$

by direct calculation of one more derivative, which we leave as an exercise for the reader. Formula (5) gets simplified to $d\left(e^{-t/\tau}\right)/ds = -e^{s-\sigma}e^{-t/\tau}$, if $t = e^s$ and $\tau = e^\sigma$, but we use base 10 to follow the original format of the measurement data. Therefore, for a purely exponential, i.e. with $n = 1$, data curve (1), its only lifetime τ_1 can be easily and quickly determined by locating the point of the maximum of the first derivative.

We also observe in Fig. 1(b) that the first derivative of every basis function is reasonably well localized around the corresponding value of the lifetime constant τ, quite quickly vanishing away from τ, especially to the right. The vanishing can be explained analytically by the behaviors of the multiplier $10^{s-\sigma}$ if $t = 10^s < \tau = 10^\sigma$ and the multiplier $e^{-t/\tau}$ if $t = 10^s > \tau = 10^\sigma$ in formula (5) of the first derivative. This crucial observation has three important implications for designing the lifetime extraction methods in the time-logarithmic scale.

First, in the multi-exponential case $n > 1$ in (1), assuming discrete lifetimes, as in Sect. 5, the local picks of the absolute of the first derivative of the data curve can reveal the total number n of dominant lifetimes and their approximate locations. Multi-exponential transient spectroscopy (METS) in [12, 14] is based on multiple differentiation in the time-logarithmic scale to approximate the dominant lifetimes. However, already even the second order numerical derivative is found to be too sensitive to measurement noise and inexact computer arithmetic in our numerical tests and unable to provide reliable lifetime estimates, even combined with low-pass filtering. Thus, we identify the number of the dominant lifetimes and their approximate values to be used as initial approximations for the nonlinear least squares method described in Sect. 5 typically using only the first derivative. We note that Pade-like approximations to the lifetimes are shown in [17] to include classical Prony's method [3] and advocated in [5] as being more robust compared to METS, but use the uniform, rather than logarithmic, time scale, unsuitable for our needs.

Second, in the derivative space, the basis of the first derivatives of the functions, plotted in Fig. 1(b), compared to the basis of the functions themselves, may behave better, having a bit smaller condition number of a Gram matrix, which implicitly appears in the linear least squares fit method in the derivative space. Thus, the linear least squares fit of the first derivative of the data curve using the first derivatives of the basis function in the derivative space may expected to be more numerically stable and can be performed on a computer with the standard double precision floating point arithmetic with larger n, in contrast to the linear least squares fit of the data curve itself using the original basis function, descried in Sect. 3 following [7–9]. The continuous lifetime extraction in the logarithmic scale using dense, e.g., uniform, grids in the logarithmic scale is somewhat more numerically feasible in the first derivative space, as we mention in Sect. 4. We can also take advantage of the first derivatives of the basis functions by adding a weighted quadratic term containing the first derivatives to the function minimized by the nonlinear least squares fitting.

3   Continuous Lifetime Extraction on a Uniform Lifetime Grid in the Time-Logarithmic Scale by Del Alamo et al.

The methodology of [7–9] to extract time constants of the dominant traps is essentially based on an assumption of a continuous distribution of lifetimes, approximating the integral in (4) by a sum in (2) with a large number n using uniformly spaced grid on a given interval $[\sigma_{min}, \sigma_{max}]$ of the logarithms σ_i of the lifetimes τ_i with unknown coefficients c_i. The trapping and detrapping transient data, $I_{data}(t)$, is analyzed by fitting the data to a weighted sum of pure exponentials in (1) using the linear least squares method. The fitting is performed to minimize the sum of $|I_{data} - I_{fitted}|^2$ at the measured points, where the magnitudes c_i are the fitting parameters to be computed, whereas the lifetimes τ_i are predefined. For the fitting functions in [7–9], $n = 100$ exponentials are typically used with the lifetimes τ_i that are equally spaced logarithmically in time. Positive (negative) values of c_i correspond to the trapping (detrapping) processes and each c_i represents the magnitude of a trapping (detrapping) process with respect to the lifetime τ_i. The extracted values of the magnitudes c_i can be used to construct a time-constants spectrum by plotting them as a function of τ. The vertical axis is in terms of arbitrary units (A.U.).

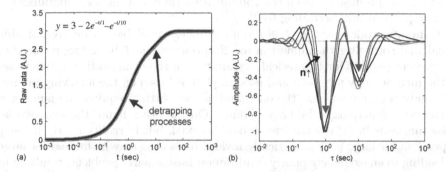

Fig. 2. Examples from [9]. (a) Time-domain signal of a synthetic current transient (red:data, blue:fitted curve). (b) Time constant spectrum extracted from the fitting of the time-domain signal with various numbers of exponentials ($n = 20, 50, 100, 500$) (Color figure online).

In [9], the synthetic data curve $y = 3 - 2e^{-t/1} - e^{-t/10}$ is used to demonstrate the methodology as shown in Fig. 2(a). The time domain signal in this example is composed of two pure exponential detrapping components with lifetime constants $\tau_1 = 1$ and $\tau_2 = 10$. The corresponding coefficient for the lifetime constants are -2 and -1, correspondingly. Various numbers of exponentials, $n = 20, 50, 100$, and 500, are tested to perform the linear least squares fitting. The calculated amplitudes are plotted in the time-constant spectrum in [9, Fig. 2] and reproduced here in Fig. 2(b). The time-constant spectrum lifetime distribution reveals the two exponential components and their relative amplitudes. With $n = 20$, the maximum errors in the lifetime and amplitude are 11 % and 5 %, correspondingly. As the number n of the grid points for τ increases, the time constants and amplitudes may become more accurate, but at the cost of more computing time for the fitting process, according to [9].

Locating the actual values $\tau_1 = 1$ and $\tau_2 = 10$ by checking the peaks of the computed time-constant spectrum lifetime distribution in Fig. 2(b) requires some educated guessing and multiple calculations with various n, since a few evidently erroneous distortions clearly appear in the time-constant spectrum in Fig. 2(b). These errors are explained in reference [9, p. 134] by the following two reasons. First, the number n of exponential components is used for the fitting may be not large enough. Second, the basis functions used in the sum (1) are not orthogonal to each other. The least squares fitting as formulated in [7,9] is a simple quadratic minimization problem. However, in order to prevent over-fitting that makes the time-constant τ spectrum numerically unstable, some constraints, such as lower and upper bounds or smoothness in the spectrum, have been added. Due to such constraints, the numerical minimization takes especially long time to be performed as the dimensionality n of the minimization problem increases. The typically chosen $n = 100$ number of terms in (1) represents a practical compromise between the computation time and the meaningfulness of the computed result, according to [9].

Computer codes used in [7–9] are not publicly available, so we are not able to calculate ourselves the numerical results presented in [7–9] and reproduced in the present paper. Our theoretical explanation of the computational difficulties of the methodology proposed and used in [7–9] is based on the following numerical analysis arguments. On the one hand, a good fitting requires the number n of the basis functions to be large enough. On the other hand, the exponential basis functions in (1) are not just non-orthogonal, but in reality nearly linearly dependent for large n. A numerically invalid basis thus may be formed for large n, leading to an extremely poorly conditioned least squares problem, resulting in erroneous lifetime computations due to unavoidable computer round-off errors and the measurement noise in the data. Applying the proposed constraints to numerically stabilize the least squares minimization may dramatically increase the computational time, while still cannot satisfactory resolve the inaccuracy in the lifetime calculations. In the next section, we summarize our own experience extracting the continuous lifetimes on a dense grid using the least squares, without any constraints.

4   Continuous Lifetime Extraction in the Time-Logarithmic Scale Revisited and Regularized

The case of the time-logarithmic scale is not common in the literature, so in this section we present our heuristic experience making a naive version of the continuous lifetime extraction in the time-logarithmic scale to work using the linear least squares applied to (2) without any constraints for reasonably large values of n for several representative examples of synthetic data, using standard double precision computer arithmetic and off-the-shelf software libraries. It is important to realize that solving the integral equation of the first kind, given by (4), using its approximation (2) with a large number n on a dense grid on an interval $[\sigma_{\min}, \sigma_{\max}]$ of the logarithms σ_i of the lifetimes τ_i with unknown coefficients c_i and utilizing a dense grid of the logarithms s of the times t of data measurements, is clearly an ill-posed problem, possibly extremely sensitive to details of its setup and numerical procedures.

The first choice that needs to be made is related to an experiment design. We conjecture that an optimal grid for the measurements is a uniform grid of the logarithms s of the times t of data measurements on an interval $[s_{\min}, s_{\max}]$ that needs not only include all the values of σ_i corresponding to significantly nonzero coefficients c_i in the synthetic data, but in fact also have some more room, so that at the both end points of the interval $[s_{\min}, s_{\max}]$ the measurement data $I_{data}(s)$ behaves like a constant, having a small absolute value of the first derivative $I'_{data}(s)$ with respect to s. For example, if the synthetic data curve is generated by a Gaussian function c_σ centered at the mean point 0 with the standard deviation 1, a good smallest interval $[s_{\min}, s_{\max}]$ can be $[-4, 4]$ for the regularized (described later) linear least squares or $[-5, 5]$ for the standard linear least squares, to be reasonably numerically stable.

The second choice is the type of the grid for σ, where we advocate again using a uniform grid of the logarithms σ of the lifetimes τ on an $[\sigma_{\min}, \sigma_{\max}]$. Interestingly, the interval $[\sigma_{\min}, \sigma_{\max}]$ cannot be chosen seemingly naturally to be the same as the already decided interval $[s_{\min}, s_{\max}]$. The reason becomes clear from Fig. 1. The basis function $e^{-10^{s-\sigma}}$ in Fig. 1(a) and its derivative in Fig. 1(b) can get chopped being restricted to the interval $[s_{\min}, s_{\max}]$, no longer representing a desired shape. If the σ is chosen to be within a smaller interval, approximately $[\sigma_{\min}, \sigma_{\max}] = [s_{\min} + 2, s_{\max} - 1]$, for most of the shape of the basis function $e^{-10^{s-\sigma}}$ to fit the interval $[s_{\min}, s_{\max}]$, then we may have not enough range in σ to approximate the actual values of the logarithms σ of the lifetimes τ present in the data. Alternatively, we can choose a larger interval, approximately, e.g., $[\sigma_{\min}, \sigma_{\max}] = [s_{\min} - 1, s_{\max} + 1]$, but then we run into a different trouble of having nearly constant restrictions of the basis function $e^{-10^{s-\sigma}}$ and its first derivative. The latter trouble, however, can be detected and the enlarged interval $[\sigma_{\min}, \sigma_{\max}]$ gets the final tuning by chopping at both ends. In our tests, the tuned enlargement of the interval $[\sigma_{\min}, \sigma_{\max}]$ appears to be more numerically stable, compared to making the interval smaller.

Finally, we need to decide the grid size, which can be different for s and σ. The balanced choice, i.e. the same step lengths of the grids for s and σ, appears to be most numerically stable. Since the suggested interval $[\sigma_{\min}, \sigma_{\max}]$ is only a bit longer, if enlarged, compared to the interval $[s_{\min}, s_{\max}]$, this translates in about the same number of grid points for s and σ. For example, for $[s_{\min}, s_{\max}] = [-5, 5]$ we numerically get reasonably robust results for up to values 200.

However, if the measurements have been already made, it is the number of the measurements that can determine the number of grid points for s on the interval $[s_{\min}, s_{\max}]$. If this number is greater than the largest possible balanced value 200, the number of the grid points for σ has to be correspondingly reduced, to keep the computations numerically stable. Interpolating the measured data to a uniform grid with a smaller number of grid points may create distortions.

To that end, let us describe the actually used standard and regularized linear least squares computational procedures. Using the already decided grids S for s and Σ for σ, correspondingly, two matrices are calculated,

$$F_0 = e^{-10^{S-\Sigma}} \text{ and } F_1 = -\log(10)10^{S-\Sigma} e^{-10^{S-\Sigma}}. \tag{6}$$

The numerical sanity of calculations is checked by computing the condition numbers of the matrices F_0 and F_1. The column vector C, formed by the coefficients c_i is determined by the linear least squares fit of $F_0 C$ to $I_{data}(s)$ (assuming for simplicity of presentation that $I_\infty = 0$) or by fitting $F_1 C$ to $I'_{data}(s)$.

The matrices F_0 and F_1 have a special structure of a Toeplitz (diagonal-constant) matrix. Thus, the linear least squares fit can be performed using specialized methods for Toeplitz matrices. Numerous known fast methods for Toeplitz matrices are often unsuitable for ill-conditioned systems like ours. Since the sizes of Toeplitz matrices of practical importance for our application cannot be that large, the considerations of performance are secondary for us, compared to accuracy. We have implemented an SVD-based Tikhonov regularization and smooth thresh-holding as in, e.g., [6], although it does not actually take advantage of the Toeplitz structure. The choice of the regularization parameter can be done by hand (especially well for synthetic data, where the answer is known!). The Tikhonov regularization stabilizes the computation and typically may accurately determine the general shape of the distribution c_τ, but often gives inaccurate amplitudes, smoothing down the peaks.

Overall, under the assumptions and with the suggestions described in this section, we are able to use our linear least squares computational procedures to solve reasonably challenging synthetic data problems in negligible computer time with good accuracy in the "eye-ball norm". The computational results using the derivative-based fit matrix F_1 may be more numerically stable for noiseless synthetic data curves, compared to original data based matrix F_0. The code performs especially well for smooth distributions c_τ, but can also handle purely discrete cases although resulting in some artifacts near the discontinuity points of the function c_τ. However, the accuracy of the solution in not that good on our actual practical data sets, resulting in noisy without the regularization or strongly depending on the choice of the regularization parameter lifetime distributions, without an evident protocol on choosing the regularization parameter.

One possibly most important apparent reason for such a drop in accuracy for practical measurements is that the range for the measurement time in logarithmic scale $[s_{min}, s_{max}]$ is not large enough for the measured quantity to become stabilized on both ends of the interval and to include the range of the present lifetimes somewhat extended, as required for numerical stability of calculations. On the right end s_{max} of the interval, if the data curve has already approached a constant and no lifetimes are anticipated with $\sigma > s_{max}$ it is easy to extend the interval $[s_{min}, s_{max}]$ to the right, substituting a larger value for s_{max} and extending the data curve using the constant at s_{max}. This approach works well on synthetic data, but does not improve the situation much with the practical data available to us. The main difficulty is with the data at the left end s_{min}

of the interval, where the data curve apparently cannot approach a constant for physical reasons in our application. Indeed, our measurement data represent the influence of the traps in semiconductors. The value of $s_{\min}$ is determined by limitations of existing sensor technologies, the smallest being approximately 10^{-7} in our data. There are no physical reasons to expect an absence of the traps with lifetimes $\tau < 10^{s_{\min}} \approx 10^{-7}$. However, as we have discussed above, due to numerical stability issues of the basis of the exponents, there are limitations on how small $\sigma_{\min}$ can be, given $s_{\min}$. Traps with $\log_{10} \tau$ in the interval $(s_{\min} - 2, \sigma_{\min})$, if present, affect the measurement data strong enough to make the outcome of the exact linear least squares fit wrong.

Our main contribution is changing the paradigm of the continuous lifetime extraction to the discrete one, eliminating the difficulties of the linear least squares fit, described in this section, as well as faced in [7–9]. Next, we present our two independent key ideas reformulating the fitting problem to make the numerical solution quick and reasonably accurate for synthetic and practical data.

5 Discrete Lifetime Extraction: Brief Description and Advantages

Using non-orthogonal basis functions (1) would be numerically more stable in the least squares fitting if the lifetimes τ_i in (1) were *discrete*, i.e. if n was small and the lifetime values τ_i were sparsely distributed separated enough from each other. Both these requirements are violated in the methodology of [9], enforcing the uniform placement of the lifetimes τ_i, essentially assuming a continuous distribution of lifetimes τ in the integral in (3) approximated by a sum in (1). In contrast, for the purpose of computational stability and efficiency, let us assume the discrete distribution of the lifetimes τ_i in (1), and consider possible approaches to the computational lifetime extraction.

We note that the discrete lifetime extraction problem is a particular case of a standard parameter estimation problem, with vast literature. The classical Prony method and its numerous extensions require a constant time sampling interval leading to an equidistant grid in time t, which is impractical for our application, where the time t can range from 10^{-7} to 10^4 seconds and the measurements are given in the time-logarithmic scale. Another classical approach, the exponential peeling method, does not technically require the measurement points in time to be equidistant and can extract the lifetimes one after the other, having thus a disadvantage of an increasing and unrecoverable error after every performed extraction. Many methods, especially in audio-related applications, such as multiple signal classification for frequency estimation and emitter location, are specifically aimed at the case, where the data and all parameters in (1) are complex-valued, not taking advantage of our purely real case. Some recent approaches, popular in compressed sensing, minimize the sparsity of the lifetimes at the costs of heavy computations, allowing an optimal resolution determining closely located discrete lifetimes, which is irrelevant in our application.

We choose a generic minimization software, e.g., available in Octave and MATLAB, performing the iterative nonlinear least squares fit of the data to the sum (1), varying all parameters in (1) using real computer arithmetic. Speed up convergence of the iterations of the minimizer, we explicitly derive the Jacobian of the minimized least squares fit function and provide it to the minimization software. Any iterative nonlinear least squares method is more computationally difficult, compared to the linear least squares, and typically requires reasonable initial approximations for all the parameters in (1) for fast and accurate computations. We discuss our choice of the initial approximations in the next section.

In Fig. 3, we demonstrate an application of our fitting algorithm to the same synthetic data curve as in Fig. 2 to illustrate our methodology. The comparison between the fitting curve and the synthetic data is shown in Fig. 3(a); the error between them is also shown to be bounded by 7×10^{-6}. The time-constant spectrum from our algorithm are shown in Fig. 3(b). Our time-constant spectrum clearly and accurately demonstrate the two exponential components (1 and 10) with their absolute values of corresponding coefficient (-2 and -1). There are no undesirable false lifetimes in our time-constant spectrum. Therefore, no extra effort is required to further process the time-constant spectrum, which might be needed in [9]. The maximum error in our time constant and corresponding coefficient fitting is 6×10^{-6} in this example.

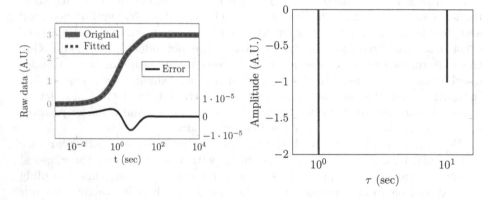

Fig. 3. Our discrete time-constant analysis: (a) Time domain signal of the same synthetic current transient (Blue:data, red:fitted curve, black:error). (b) Time constant spectrum extracted by our method (Color figure online).

The advantages of our methodology to extract the discrete lifetimes in this example, as seen in Fig. 3, are as follows.

o The time-constant spectrum of our method clearly and accurately demonstrates the two lifetimes and their corresponding coefficients as appears in the synthetic function, in contrast to the time-constant spectrum of [9] as shown in Fig. 2. It is difficult to determine whether the small peak as shown in Fig. 2(b) is an actual lifetime constant or an error.

o The magnitudes corresponding to lifetimes are the actual coefficients' values whereas the magnitudes in [9] only reflect the ratio between the actual values of coefficients.

o Our method is numerically stable and no constraint is applied during the optimization. In contrast, constraints such as lower and upper bounds or smoothness in the spectrum have been added in [9] in order to prevent over-fitting. Determining reasonable constraints may be difficult. Constrained minimization can be slow.

o Our method is highly computationally efficient. It takes less than a second to calculate the time-constant spectrum for the synthetic function as shown in Fig. 2(a) while the code in [9] takes hours.

6   Discrete Lifetime Extraction: Detailed Description

The details of our algorithm are as follows.

1. In Fig. 4 101, interpolate the data using a dense uniform logarithmic grid in time domain with the grid size of 2^n. The measurement data are usually not measured uniformly in log scale. This step is a preparation for the following steps because discrete Fourier transform will be used to handle the measurement noise and a uniform grid is needed. Grid size of 2^n will improve the FFT from $O(N^2)$ to $O(NlogN)$ by utilizing the well-known radix-2 Cooley-Tukey algorithm [2].

2. In Fig. 4 102, filter out the noise in the measurement data. There are two alternative approaches to accomplish this in our algorithm. First one, perform a cosine Fourier transform to obtain the cosine Fourier spectrum. By analyzing the cosine Fourier spectrum and applying a proper filter in frequency spectrum, the noise in the measurement data can be filtered. Second one, apply smoothing to the data curve with measurement noise to obtain a smooth data curve.

3. In Fig. 4 103, obtain the derivative of the data curve. We implement two alternative approaches to accomplish this. First one, we transform the data into Frequency domain utilizing the Fast Fourier Transform (FFT), obtain the derivative using the Fourier spectral method, and perform inverse FFT to find the derivative in real time domain. Second one, numerically calculate the derivative using the finite difference method.

4. In Fig. 4 104, identify the nearby positions of the dominant trap lifetimes τ_i in time-constant spectrum by finding the maxima of the absolution value of derivative. Each maximum in the absolute value of derivative corresponds to an exponential component with the lifetime constant near that location. This step will give good initial approximation for the values of trap lifetimes (not only the locations, but also the number of trap lifetimes).

5. In Fig. 4 105, use linear least squares method to find the values of the coefficients c_i corresponding to the trap lifetime. This step will give a good initial approximation for the coefficients c_i.

6. In Fig. 4 106, use all the above lifetimes and coefficients as the initial approximation and employ the non-linear least squares method to optimize both c_i and τ_i simultaneously for all $i = 1, \dots, n$. This step will improve our results in the previous few steps for τ_i and c_i.

The flowchart of our algorithm is shown in Fig. 4.

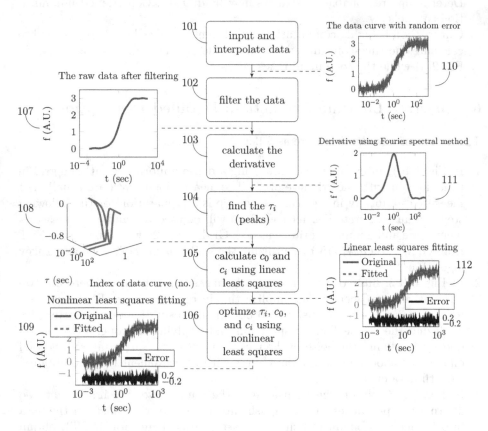

Fig. 4. The flowchart of our algorithm (Color figure online)

We can also perform simultaneous analysis of multiple data curves, with the same lifetimes, but various coefficients, which is beneficial, e.g., if the same device is tested multiple times under slightly different conditions. The derivative of every data curve separately is used to determine potential lifetimes, that are then consolidated into a single set for all curves, using thresh-holding and averaging of nearby values. The remaining steps perform the least squares fits for all curves together, increasing robustness of calculations.

7   Application on Experimental Data Representing Dynamic ON-resistance

We finally test our algorithm on experimental data from [8, Fig. 2]. We obtain in Fig. 5 a similar data fit and the envelop shape of our discrete time-constant spectrum agrees with the continuous one obtained in [8, Fig. 2].

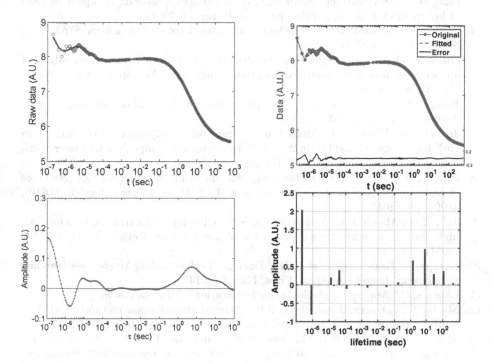

Fig. 5. The original data, the fitted curve, and error (top) and time-constant spectrum (bottom) using continuous [8] (left) vs. our discrete (right) fit (Color figure online).

8   Conclusion

We extract trap lifetimes and corresponding coefficients from experimental data, in time logarithmic scale, of dynamic ON-resistance for semiconductors. The proposed methods are numerically stable, extremely computationally efficient, and result in high quality fitting of noisy synthetic and experimental data.

Acknowledgments. We thank Donghyun Jin and Jesus del Alamo for providing us the raw data for their [8, Fig. 2] that we use in Sect. 7, and the corresponding results of the fit by their method, displayed in Fig. 5.

References

1. Binari, S.C., Klein, P.B., Kazior, T.E.: Trapping effects in GaN and SiC microwave fets. Proc. IEEE **90**(6), 1048–1058 (2002)
2. Cooley, J.W., Tukey, J.W.: An algorithm for the machine calculation of complex fourier series. Math. Comput. **19**(90), 297–301 (1965)
3. de Prony, G.C.F.M.R.: Essai expérimental et analytique sur les lois de la Dilatabilité des fluides élastiques et sur celles de la Force expansive de la vapeur de l'eau et la vapeur de l'alkool, à différentes températures (1795)
4. Epstein, C.L., Schotland, J.: The bad truth about Laplace's transform. SIAM Rev. **50**(3), 504–520 (2008)
5. Gutierrez-Osuna, R., Nagle, H.T., Schiffman, S.S.: Transient response analysis of an electronic nose using multi-exponential models. Sens. Actuators B Chem. **61**(1–3), 170–182 (1999)
6. Hansen, P.C.: Deconvolution and regularization with Toeplitz matrices. Numer. Algorithms **29**(4), 323–378 (2002)
7. Jin, D., del Alamo, J.A.: Mechanisms responsible for dynamic ON-resistance in GaN high-voltage HEMTs. In: 2012 24th International Symposium on Power Semiconductor Devices and ICs (ISPSD), pp. 333–336 (2012)
8. Jin, D., del Alamo, J.A.: Methodology for the study of dynamic ON-resistance in high-voltage GaN field-effect transistors. IEEE Trans. Electron Devices **60**(10), 3190–3196 (2013)
9. Joh, J., del Alamo, J.A.: A current-transient methodology for trap analysis for GaN high electron mobility transistors. IEEE Trans. Electron Devices **58**(1), 132–140 (2011)
10. Kompa, G.: Basic Properties of III-V Devices - Understanding Mysterious Trapping Phenomena. Kassel University Press, Kassel (2014)
11. Lanczos, C.: Applied Analysis. Dover Publications, Mineola (1988)
12. Marco, S., Samitier, J., Morante, J.R.: A novel time-domain method to analyze multicomponent exponential transients. Meas. Sci. Technol. **6**(2), 135 (1995)
13. Meneghesso, G., Verzellesi, G., Danesin, F., Rampazzo, F., Zanon, F., Tazzoli, A., Meneghini, M., Zanoni, E.: Reliability of GaN high-electron-mobility transistors: state of the art and perspectives. IEEE Trans. Device Mater. Reliab. **8**(2), 332–343 (2008)
14. Samitier, J., Lopez-Villegas, J.M., Marco, S., Camara, L., Pardo, A., Ruiz, O., Morante, J.R.: A new method to analyse signal transients in chemical sensors. Sens. Actuators B Chem. **18**(1), 308–312 (1994)
15. Sozza, A., Dua, C., Morvan, E., diForte Poisson, M.A., Delage, S., Rampazzo, F., Tazzoli, A., Danesin, F., Meneghesso, G., Zanoni, E., Curutchet, A., Malbert, N., Labat, N., Grimbert, B., De Jaeger, J.-C.: Evidence of traps creation in GaN/AlGaN/GaN HEMTs after a 3000 hour on-state and off-state hot-electron stress. In: IEEE International Electron Devices Meeting Technical Digest, IEDM 4p., p. 593 (2005)
16. Stoklas, R., Gregusova, D., Novak, J., Vescan, A., Kordos, P.: Investigation of trapping effects in AlGaN/GaN/Si field-effect transistors by frequency dependent capacitance and conductance analysis. Appl. Phys. Lett. **93**(12), 124103 (2008)
17. Weiss, L., McDonough, R.N.: Prony's method, Z-transforms, and Pade approximation. SIAM Rev. **5**(2), 145–149 (1963)

Fuel Pipeline Thermal Conductivity
in Automatic Wet Stock Reconciliation Systems

Pawel Foszner[1,2](✉), Aleksandra Gruca[2], and Jakub Bularz[1]

[1] AIUT Sp. z o.o., ul.Wyczkowskiego 113, 44-109 Gliwice, Poland
Pawel.Foszner@aiut.com
[2] Institute of Informatics, Silesian University of Technology, Gliwice, Poland
http://www.aiut.com.pl/

Abstract. In the fuel industry, as in any other, it is important to have control over the resources that directly generate profit. In the case of a petrol station this is a gasoline that passes a very complicated path from the terminal, through underground tank up to the tank inside the car. At most stages of its journey, the system can control the volume, height, temperature, the physical processes which is subjected to and based on this, predict its state. This is done by analysing the telemetry data and various types of flow models. In this paper we want to concentrate on one of the most difficult parts of this system. Difficult, because usually undocumented and invisible - a system of piping between the tank and the fuel distributor. Fuel that is located there, is usually beyond the observation of telemetry system and its volume changes over time. We want to focus on the nature of these changes and predict its impact on the balance of the fuel at the station.

Keywords: AWSR · Thermal conductivity · Reconciliation

1 Introduction

In general Automatic Wet Stock Reconciliation Systems (AWSR systems) are about controlling fuel flow on petrol/gas stations. Inspection focuses on monitoring the flow of fuel at any stage of its delivery and sales, and rapid detection and elimination of any anomalies. An efficient and reliable AWSR system should:

1. Correctly determine the fuel level in the tank.
2. Accurately detect the amount of fuel supplied from the outside (deliveries).
3. Accurately determine the timestamps and volume of fuel sold.
4. Detect all anomalies (leaks, surpluses, etc.).

In this work we want to focus on the task 3. In the simple station model the amount of fuel sold out (measured by dispenser) should always be consistent with the amount of fuel that decreases from a tank during the transaction, assuming that we have correctly calibrated dispensers, which true for the most of tanks. But in case of rarely used distributors (for example one transaction for a day), the model should consider additional factors, which normally could be neglected.

P. Perner (Ed.): ICDM 2016, LNAI 9728, pp. 297–310, 2016.
DOI: 10.1007/978-3-319-41561-1_22

One of such factors is the change of the volume of fuel outside the observed system (piping between the tank and the distributor). Usually, the monitoring system do not include piping and information about fuel volume changes there can be lost. In this paper we want to fill this gap and describe numerically processes that take place there. In particular, we want to focus on the volume changes while waiting for a transaction. Fuel, just like any other liquid, is subjected to a basic principle of thermodynamics of fluids, which means that it changes its volume due to temperature. The nature of these changes depends on the reference density (which is for the fuel 15 °C).

Fuel is supplied to the distributor using system of pumps and pipes. During refueling of a vehicle, the fuel is first delivered from the piping, and then, from the underground tank. For a reconciliation process it is important to compare the volume of the fuel sold (measured by a distributor) with the volume of the fuel descended from the reservoir. In the latter case, to asses the amount of fuel, the difference of probe height before and after transaction is measured. Such approach provides the correct results, as long as the volume of fuel in the piping does not change between transactions.

For a constant sale, that is a sale for which the pauses between transactions are not too long, the assumption that the temperature of the fuel in the tank is equal to the temperature of the fuel dispensed is almost correct (although affected by minimum error). Fuel temperature in the tank is calculated as the average temperature of all the sensors immersed in the liquid and in most cases this is the only measure of the fuel temperature at the station. For the majority of cases it is sufficient, but we believe that for a small amount of little used tanks, the fuel volume change in piping system can significantly affect the reconciliation process. In this paper we present the thermal model which is a part of AWSR system. Proposed model incorporates into the reconciliation analysis the process of fuel volume change in the piping due to temperature difference. We analyse the model performance on the data from the real gas station and we show that its application significantly reduces reconciliation error.

2   Related Work

Companies use AWSR systems usually for economic or environmental reasons. Also the reason may be the need to meet certain legal requirements. In case of legal or governmental regulations most widespread publications are US Environmental Protection Agency (EPA) booklets. US EPA regulations are released as a series of brochures and standards related to environmental issues. In terms of AWSR systems most important are instructions about inventory control in underground tanks [11], test procedures for leak detection [10] and introduction into statistical reconciliation [13]. The most important in these documents is the fact that EPA sets thresholds, goals and procedures for fuel reconciliation systems or simple inventory control.

Very often publications on complex systems appear in the form of patents [5,6,8,9]. Research in this area were carried out in the early 70s, where W. F. Rogers published a statistical analysis methods that were used at a later time [1,7]. Many literature references focuses only on the detection of leakage of fuel [14] or on statistical methods useful in terms of reconciliation [3,12]

According to the best authors knowledge there is no literature reference to the problem of volume changes in piping in the context to temperature variations in AWSR systems. Existing solutions implemented on AWSR systems focus only on the aspects directly targeted by telemetry.

3   Heat Model

3.1   Cumulative Variance

Reconciliation Error. The basics underlying in AWSR analysis is to determine the quality of fuel reconciliation and the most common measure is the difference between inventory of fuel and the volume of fuel sold. In case the dispenser nozzles are correctly calibrated and there are no anomalies such as theft or leak, both volumes should be equal. Reconciliation error is the difference between the amount fuel measured by distributor and a decrease of fuel in the tank measured by probe height difference, and it is computed as follows:

$$E(t_1, t_2) = V_s(t_1, t_2) - V_o(t_1, t_2), \tag{1}$$

where t_1, t_2 are timestamps between which measurements are taken, V_s is a fuel sold between t_1, t_2 and V_o is fuel outcome calculated as difference of fuel stocks:

$$V_o(t_1, t_2) = V(t_1) - V(t_2). \tag{2}$$

In addition, to meet all possible cases, there is a need to include fuel volume that was delivered externally (represented by $V_d(t_1, t_2)$ component):

$$E(t_1, t_2) = V_s(t_1, t_2) - V_o(t_1, t_2) - V_d(t_1, t_2). \tag{3}$$

Cumulative Variance. The reconciliation error provides information about only a single time interval. It is also prone to measurement noise such as random error. Therefore, to see the whole picture and tank trends, there is a need to cumulate this error over the time domain:

$$CV(t_n) = \sum_{i=0}^{n-1} E(t_i, t_{i+1}), \forall n > 0 \tag{4}$$

Such defined measure allows analysing the tank trends and react much faster in case of anomalies. Exemplary cumulative variance plot is presented in the Fig. 1.

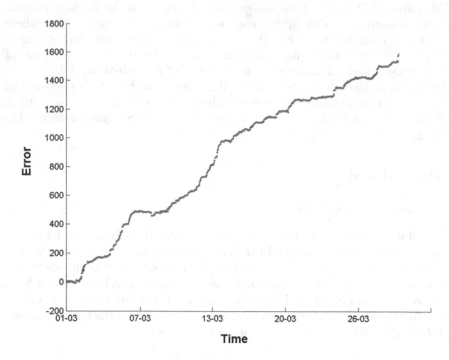

Fig. 1. Sample cumulative variance plot with increasing reconciliation error over the time.

Presented plot contains one month of data. Horizontal axis represents time data (in dd-mm format) and vertical axis represents cumulated error in liters. On the graph were placed around 2500 data points. Each i-th point represents value defined by the equation (3), and average distance between timestamps t_1 and t_2 is equal to 15 min. Cumulative sold over this period was equal to 327,47 m^3. Errors on presented plot has strong increasing trend, however, as a percentage in relation to the sale is equal to 0,51 %. It can be read as permanent surplus or, more likely, wrong suited calibration table.

3.2   Background

The gas station model which is analysed by AWSR system typically consists of a fuel tank and a fuel distributor. The fuel tank can be monitored for volume, height, pressure and temperature. The second element of the telemetry is in the fuel dispensers and such model ignores the changes in the volume of fuel in the path between tank and the distributor due to the temperature differences. Usually fuel remains in the piping too short to its volume has changed, therefore the influence of the outside temperature can be omitted in analysis. However there remains, a small group of premium fuel tanks for which these changes can have a significant influence for the reconciliation outcome.

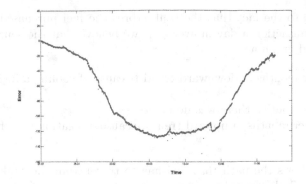

(a) Cumulative variance of very rarely purchased premium fuel.

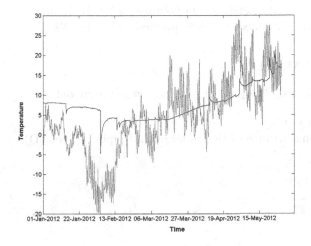

(b) Comparison of fuel (black) and ambient (gray) temperature.

Fig. 2. Data from the rarely used tank and the effect of temperature on fuel in it

The Fig. 2a represents example of very rarely sold fuel. The illustration in Fig. 2b shows differences between fuel temperature inside underground tank and ambient temperature at the station. It is clear that it has an impact on fuel temperature and one can distinguish 3 periods:

- winter months with very low ambient temperature,
- neutral months,
- summer months with very high ambient temperature.

These periods clearly affect cumulative variance presented in the Fig. 2b. However, before stating our hypothesis we need to ascertain that: (1) the tank is well calibrated and has the correct conversion table, and (2) the sale is counted accurately. Due to the fact that the tank stores the fuel purchased very infrequently (one transaction a day in average), we believe that the source of errors on the presented graph are:

- In the winter months, a downward trend because of cooling a fuel residual in piping.
- In neutral in months, there was no errors.
- In the summer months, a upward trend because of heating up a fuel residual in piping.

The Fig. 3 shows the path that fuel has to travel from the tank (A) to the distributor (B). Fuel is supplied by a system of pipes, pumps, etc. (C), which capacity, location, and construction is not known. Therefore, the whole system is treated as a closed "black box", which has some thermal transmittance (D). The fuel that remains in (C), changes its volume under the influence of the outside temperature (E). This effect (D) is expressed in Watts [W].

3.3   Thermal Model

According to the Fig. 3, unknown parameters in the model are:

- Pipes capacity (C),
- Thermal transmittance of the underground piping system(D),

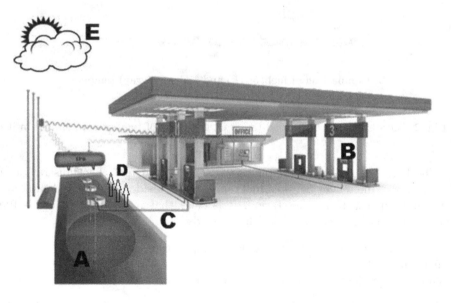

Fig. 3. Simple station model.

And known parameters are:

- Fuel temperature and stocks in tank (A),
- Outside temperature (E),
- Sold volume according to dispensers measurements (B)

In the presented model we try to determine the fuel temperature in piping which is directly related to the output fuel volume.

Theorem 1. *Amount of the energy transferred is dependent on the substance, in proportion to the cross-sectional area of barrier, the temperature difference and the time of heat transfer and inversely proportional to the thickness of the barrier [2]:*

$$Q = \lambda \frac{S * \Delta T * t}{d}[J],\tag{5}$$

which can be also denoted in the following way (after transformation equation for the specific heat):

$$Q = m * c * \Delta h[J],\tag{6}$$

where:

- Q – delivered heat
- λ – thermal transmittance coefficient of system (D) $\left[\frac{W}{m*K}\right]$
- S – cross-sectional area through which the heat flows
- ΔT – difference between fuel temperature in piping and ambient temperature
- t – time (duration [s])
- d – thickness of the barrier
- m – fuel mass
- c – heat capacity of fuel (for each fuel type set to 2100 $\frac{J}{Kg*K}$)
- Δh - fuel temperature difference during time t

Proposition 1. *For the simplicity (the system does not change and we do not know its shape), we assume a constant value $K = \lambda * \frac{S}{d}$. Therefore, a model for the flow of heat is as follows:*

$$Q = K * \Delta T * t[J].\tag{7}$$

Proposition 2. *Fuel temperature in piping can be computed as follows:*

$$\Delta VsTemp = \int_{t_s}^{t_e} \frac{K * \Delta T * t}{m * c} dt,\tag{8}$$

where: $\Delta VsTemp$ is a change of fuel temperature in piping, and timestamps t_s, t_e denote respectively the start time and the end time measurements.

Corrected temperature can be used to determine the actual volume of the fuel that has been taken from the tank, which allows calculating the reconciliation error more precisely.

3.4 Learning Algorithm

Equations for fuel volume in piping from the Sect. 3.3 can be applied to any fuel station. The only factor that may change are parameters: m – fuel mass in piping and K - constant that represent piping shape and thermal connectivity. We propose machine learning algorithm which searches for these parameters. The proposed approach does not require any knowledge about the physical shape of the piping, its depth, etc. Such information is usually unavailable and impossible to obtain. To handle the problem, we propose a heuristic search algorithm based on minimizing reconciliation quality measure (cumulative variance):

$$CV(t_n) = \sum_{i=0}^{n-1}(V_s(t_1,t_2) - V_o'(t_1,t_2) - V_d(t_1,t_2)), \forall n > 0, \qquad (9)$$

where $V_o'(t_1,t_2)$ is the fuel outcome modified by thermal model build on parameters m and K.

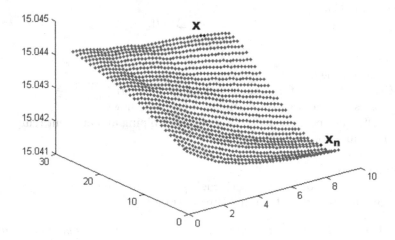

Fig. 4. Visualization of the parameter search algorithm.

The main task of learning algorithm is to find such values of parameters m and K that the function $CV(t_n)$ is minimized. To accomplish this, we propose using the idea of gradient descent method [4], but without computing partial derivatives:

1. Choose in 3D space start point $X = (m, K)$.
2. Around point X compute, in discreet steps, surface S.
3. On the surface S find a new point X_n witch satisfies condition $CV(t_n)^{X_n} < CV(t_n)^{X}$.
4. If $\exists X_n => X = X_n$ go to 2. Otherwise go to 5.
5. Set $X = (m, K)$ as the best parameters for thermal model.

Where $CV(t_n)^X$ is the cumulative variance corrected for thermal model parameters $X = (m, K)$

Figure 4 shows the concept of the search algorithm. An exhaustive search among all subspaces becomes exponentially more complex, which leads to an overwhelming time complexity. However, the space of possible solutions can be significantly narrowed. For example, one can roughly estimate how close to the tank is a distributor and on which depth the piping is placed. Therefore, the total number of possible solutions can narrowed and the parameters can be searched in a finite time.

This proposed solution applies to the situation in which there is only one distributor or many distributors with identical piping connected the tank and all of them are used with the same frequency. The extension of this procedure for many different distributors would significantly increase the dimensionality of the problem. Therefore, in case of D number of distributors, we propose the following procedure, that analyses every distributor independently:

For All D–distribiutors set initial parameters X=(m, k)

for i:=1 to D do
begin
    Find optimal solution for i'th distributor,
    with others left unchanged
end;

4   Results

4.1   Tank Selection

The selection of the test tank used for parameters estimation cannot be random and the tank must meet the following criteria:

1. Must be properly calibrated, that is, the tank calibration curve which maps the relationship between height and volume must correspond to reality.
2. Distributors must have properly calibrated nozzles and correct metering volume.
3. No leakage or any discrepancies may exits.
4. Dispensers should be identically connected to the tank.
5. Distributors should be equally used.
6. There should be large gaps between successive transactions.

Table 1 presents the amount of sale from the gas station where there are 3 different types of fuel available. The presented data was derived from a real station and includes various types of fuel: a simple best-selling diesel, average gasoline and the high-octane and rarely sold gasoline.

Table 1. Sold rate in cubic meters per month on sample tanks

Month/Product	January	February	March	April	May	June	July
Diesel	271,71	270,59	327,47	298,84	316,02	228,92	225,76
Gasoline	42,03	47,25	42,15	50,28	61,16	48,82	52,74
H-O Gas	3,07	3,55	3,05	3,93	3,64	3,34	3,55

Based on the over half year history, we selected the high-octane gasoline tank for the purpose of our analysis, due to the fact that it meets all the above criteria. Its sale is almost one hundred times smaller than the sale of the most frequently chosen fuel. Selected fuel tank has a very rare transactions and the average spacing time between successive transactions is one day. Due to the infrequent refueling, a fuel remains in the piping for a long time, where it is subjected to a process of cooling or heating, and we are not able to control its temperature. Selected tank has a height of 2.4 meters and a maximum volume of 25 cubic meters. Average monthly sales is about 3,000 liters and the calibration curve for the tank is correctly mapped.

4.2 Ambient Temperature

The ambient temperature was downloaded from the site http://polish. wunderground.com. Using the appropriate parameters of the HTTP GET request, the user is able to search the database of measurements at the weather station of his interest. For the selected gas station we use the weather station distant about 4 km. Fuel temperature in the tank was measured every 15 s, while the measurement at the weather station was made every 5 min.

4.3 Cumulative Variance

Starting point for the analysis it was half a year of data which has 120 l of error with a negative sign, and 100 l of error with a positive sign. The original cumulative variance over six month period for the selected tank in presented in the Fig. 5 as a black curve. We can clearly see that in the cold months, when the environmental temperature is significantly lower than the temperature of the fuel in the tank, we note the large negative errors. Conversely, in the warmer months there are positive errors. In the winter months (from the beginning of the year to mid-March) cumulative reconciliation error was about 120 l on minus. This represented approximately 2.6 % of total sales. Starting from spring through summer, when the outside temperature was higher than the temperature in the tank - we recorded an increase in cumulative error at the level of 100 l. This represented approximately 1.3 % of total sales. We believe that this is due to respectively cooling or warming of fuel left in the pipes.

Before starting the work, it was necessary to make sure that the thermal model of piping was the only significant source of reconciliation errors. The first

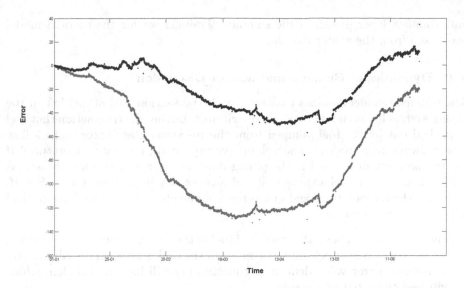

Fig. 5. Results of the thermal model application. Comparison of reconciliation with (black curve) and without (grey curve) thermal model (Color figure online).

step was to assure whether calibration curve correctly maps the height to volume relationship, as it is the most common cause of errors in the fuel reconciliation. To verify the accuracy of the calibration curve, a series of manual measurements has been made, using a specialized dipstick. Another possible cause of the error may be incorrect calculation of sales by the distributor flow-meter. To exclude this issue, nozzle control procedure was conducted. The procedure consisted of multiple pouring a fixed amount of fuel to the measuring flask, and checking if there are any differences between flask stock and distributor total counters volume. Also the tightness of the whole system has been checked as well as the correctness of the outside fuel deliveries.

After eliminating all of the above mentioned possible sources of error, we applied the learning algorithm as presented in the Sect. 3.4, in order to find the parameters of the thermal model. In the Fig. 5 we present the results of our analysis in the form of two cumulative variances for the tank. Grey curve shows the cumulative variance without the use of thermal model, and the black curve presents the temperature correction applied by the model.

Based on the performed analyses, we obtained the following values for a thermal models parameters m and K:

1. Piping with a volume of $m = 80\,l$,
2. Heat transfer coefficient ($K = \lambda * \frac{S}{d}$) equal to 14.

Using estimated parameter values we managed to improve the cumulative variance from −120 to −40 in the winter months and from +100 to +50 during the summer months. Based on the results we estimate that the effect of temperature in the piping on the reconciliation was about $30\,l$ per month in the winter, and about 20 l per month in the summer. Training set for the thermal model was taken from the winter months.

4.4   Dynamics of Changes and Model Discussion

The obtained model assumes that after each transaction, 80 l of fuel left in the piping system between fuel tank and distributor. During the transaction that fuel is pushed out by the fuel pumped from the reservoir. Distributor and its flow meter always measures the gross volume (volume at the current temperature). If the temperature of the fuel in the piping does not change between transactions – its volume does not change as well and has not any impact on transaction. If, however, the temperature, and hence the volume change, then we have to deal with one of two situations:

- The temperature rises – the volume of fuel in the pipe will increase, and during the transaction less fuel will be taken from the tank (less than distributor will indicated). Error value defined in equation (1) will have a plus sign, which will be interpreted as a surplus.
- The temperature drops – the volume of fuel in the pipe decreases, and during the transaction more fuel will be taken from the tank (more than distributor will indicated). Error value defined in equation (1) will have a minus sign, which will be interpreted as a loss.

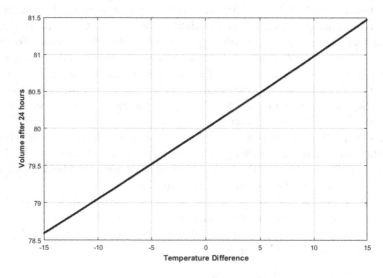

Fig. 6. Dynamics of change in fuel volume depends on difference between ambient and fuel temperature (after 24 hours)

Figure 6 illustrates a fuel volume change after 24 hours depending on the difference in temperature of the fuel and the environment. These theoretical analysis were carried out on the basis of the model parameters developed in the Sect. 4.3. The horizontal axis represents the difference in temperature, for example, −15 means that the ambient temperature is 15 °C lower than the temperature of the fuel in the reservoir (after the last transaction). The vertical axis represents the volume in the pipe after 24 hours of cooling or heating (starting volume is 80 l). The parameters for the thermal model has been copied from the tank presented in the Sect. 4.3.

For the results presented in the Sect. 4.3, the monthly impact on the reconciliation error was up to 30 l. This was one percent (1 %) of sales (3000 l).

5 Conclusions and Future Work

Reconciliation of liquid fuels is a very complicated and complex process. It is affected by many factors, both external (fraud, equipment failure, etc.) and internal (change in temperature, evaporation, etc.). It is very important to understand and model all of them. Also physical phenomena associated with the unobserved (not covered by telemetry) part of the whole system should certainly be considered.

Proposed method was applied to the real data and the results show that the volume changes occurring under the influence of the thermal model inside the piping have a real impact on the reconciliation of the tank. This difference was even 1 %, which is a significant value in terms of the accuracy of the entire system and it should not be omitted.

The proposed models and analysis have been thoroughly tested at the stations with a simple and typical infrastructure (one tank with selected product and a series of identically connected and equally used distributors). Further work in this area will focus on the inclusion into the model different types of infrastructure such as:

- dispensers used with varying frequency,
- many vessels with the same product that share some distributors,
- manifold tanks.

The completion of these works will allow creating an universal model applicable for any type of the gas station.

Acknowledgement. The project was founded by the Polish Council of the National Centre for Research and Development within the DEMONSTRATOR+ program.

References

1. Brown Jr., G.F., Rogers, W.F.: A bayesian approach to demand estimation and inventory provisioning. Technical report DTIC Document (1972)
2. Cengel, Y.: Introduction to Thermodynamics and Heat Transfer+ EES Software. McGraw Hill Higher Education Press, New York (2007)

3. Gadzhiev, C.M.: Detection of and allowance for loss of petroleum products due to leaks and evaporation in tanks. Meas. Tech. **37**(2), 159–161 (1994)

4. Hestenes, M.R.: Multiplier and gradient methods. J. Optim. Theory Appl. **4**(5), 303–320 (1969)

5. Keating, J.P., Dunn, W.W., Dunn, W.D.: Storage tank and line leakage detection and inventory reconciliation method. US Patent 5,297,42 3 (1994)

6. O'connor, P.M.: Automated statistical inventory reconciliation system for convenience stores and auto/truck service stations. US Patent 5,400,253 (1995)

7. Rogers, W.F.: Exact null distributions and asymptotic expansions for rank test statistics. Ph.D. thesis, Department of Statistics, Stanford University (1971)

8. Rogers, W.F., Collins, J.R., Jones, J.B.: Method and apparatus for monitoring operational performance of fluid storage systems. US Patent 5,757,664 (1998)

9. Rogers, W.F., Collins, J.R., Jones, J.B.: Method and apparatus for monitoring operational performance of fluid storage systems. US Patent 6,401,045 (2002)

10. US Environmental Protection Agency: Standard Test Procedures For Evaluating Leak Detection Methods. Automatic Tank Gauging Systems (1990)

11. US Environmental Protection Agency: Doing Inventory Control Right For Underground Storage Tanks (1993)

12. US Environmental Protection Agency: Guidance for Data Quality Assessment. Practical Methods for Data Analysis (1993)

13. US Environmental Protection Agency: Introduction To Statistical Inventory Reconciliation For Underground Storage Tanks (1995)

14. Williams, B.N., Kauffmann, G.A.: Method and apparatus for continuous tank monitoring. US Patent 5,363,093 (1994)

A Novel Multivariate Mapping Method for Analyzing High-Dimensional Numerical Datasets

Edwin Aldana-Bobadilla[1]([✉]) and Alejandro Molina-Villegas[2]

[1] CINVESTAV-Unidad Tamaulipas, Ciudad Victoria, Mexico
ealdana@tamps.cinvestav.mx
[2] The National Commission for Knowledge and Use of Biodiversity,
Mexico City, Mexico
amolina@conabio.gob.mx

Abstract. In modern science, dealing with high dimensional datasets is a very common task due to the increasing availability of data. Multivariate data analysis represents challenges in both theoretical and empirical levels. Until now, several methods for dimensionality reduction like Principal Component Analysis, Low Variance Filter and High Correlated Columns has been proposed. However, sometimes the reduction achieved by existing methods is not accurate enough to analyze datasets where, for practical reasons, more reduction of the original dataset is required. In this paper, we propose a new method to transform high dimensional dataset into a one-dimensional. We show that such transformation preserves the properties of the original dataset and thus, it can be suitable for many applications where a high reduction is required.

Keywords: Feature selection · Dimensionality reduction · Density estimation

1 Introduction

Due to technological advances to massively collect data, the last decades have raised several challenges for data analysis. Data reduction is one of the most important. Typically such reduction can be made in two ways: decrease the number of instances (elements) of the dataset and decrease the number of required attributes to describe each of these instances. These ways are usually called instance selection [1,4,7,9,13] and feature selection [10,12] respectively. In many approaches of Data Mining (DM) and Machine Learning (ML) is compulsory a pre-processing of the dataset based on these ways in order to improve the performance. Particularly to reduce the dimensionality, feature selection methods as Principal Component Analysis (PCA) [11], Low Variance Filter [2] and Highly Correlated Columns [14] are usually applied. These techniques are based on statistical measures that allow us to obtain the set of features with the minimal correlation between them. Since the dataset may have uncorrelated variables

© Springer International Publishing Switzerland 2016
P. Perner (Ed.): ICDM 2016, LNAI 9728, pp. 311–319, 2016.
DOI: 10.1007/978-3-319-41561-1_23

that cannot be eliminated, the number of obtained features might not always be appropriate. A dataset with this characteristic may represent a performance problem when there are constraints of time or space. To avoid this, we propose a method that allows us to represent a n-dimensional instance as a numerical data in one-dimensional space. It is compulsory that such representation preserves the information conveyed by the original data. Our proposal is based on a "discretization" of the data space. We resort to the idea of quantiles which are cutpoints dividing a set of observations or instances into equal sized intervals [6]. Usually the quantiles are defined over one-dimensional space. A set of instances in such space may be grouped by the quantile to which them belong. In this sense, a quantile represents all those instances that are close to each other. Many operations with an instance may be approximated by the range defined by its quantile. For example, assumming many instances $x_i \in \mathbb{R}$ which belongs to the quantile q defined by the interval $[0.25, 0.30)$. The operation $f(x_i) = sin(x_i) \forall x_i \in q$ may be approximated by $f(\underline{q}) = sin(0.25) = 0.004$, $f(\overline{q}) = sin(0.29) = 0.005$ or even $f(q) = f(\underline{q}) + f(\overline{q})/2 \approx 0.004$. We can see that the effectiveness of this approximation depends on the size of the interval that defines to q. Thus an appropriate size value must be determined.

Based on the above, we assume that the instances of a dataset may be represented by a set of quantiles, which preserves the properties of such instances when several operations are applied. As mentioned, typically the quantiles are defined over one-dimensional space. We propose a methodology that extends such definition to n-dimensional space. We want to project every instance of the dataset to its corresponding quantile. Every quantile is identified by an unique numerical code. The projection of an instance to its corresponding quantile allows us to obtain the numerical representation of such instance (encoding instances). The set of encoded instances will be the new dataset, we show experimentally that this set preserves the information and properties of the original dataset when several operations are applied.

The rest of this work is organized as follows: In Sect. 2 we present the main idea about dataset discretization based on quantiles. In Sect. 3 we explain how to map a data instance into one-dimensional space through its corresponding quantile. Then, in Sect. 4, we show the experimental methodology to measure the effectiveness of our method and present the results. Finally, in Sect. 5 we conclude and mention some future work.

2   Data Discretization

Given a dataset Y in a one-dimensional space, we can divide its space into a set of quantiles as it is shown in Fig. 1.

In this case, a quantile q_i is an interval of the form $q_i = [\underline{y_i}, \overline{y_i}]$ where $\underline{y_i}$ and $\overline{y_i}$ are the lower and upper limits of q_i respectively. To determine the values of $\underline{y_i}$ and $\overline{y_i}$ a quantile width δ must be defined. Such value is given by:

$$\delta = \frac{|max(Y) - min(Y)|}{N} \tag{1}$$

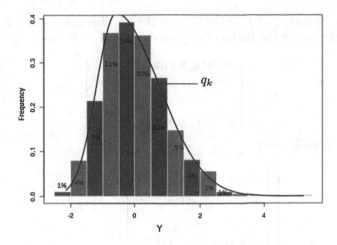

Fig. 1. A possible division of the space of an one-dimensional random variable Y. The quantile q_k contains a proportion of instances of Y.

where N is a prior value of the desired number of quantiles. Based on the above, the first quantile is defined as a half-closed interval of the form:

$$q_1 = [min(Y), min(Y) + \delta) \qquad (2)$$

and the subsequent quantiles are defined as:

$$q_i = \begin{cases} [\overline{y}_{i-1}, \overline{y}_{i-1} + \delta] & \text{if } i = N \\ [\overline{y}_{i-1}, \overline{y}_{i-1} + \delta) & \text{otherwise} \end{cases} \qquad (3)$$

The idea could be extended to higher dimensional data, in which case, a quantile will be a n-dimensional partition of the data space. In this case, given a dataset $Y \in \mathbb{R}^n$ with instances of the form $y = [y_1, y_2, ..., y_n]$, we can divide the data space into a set of n-dimensional quantiles.

Each n-dimensional quantile is composed by a set of intervals that determine the upper and lower limits for each dimension. Such definition is expressed as:

$$q_i = [[\underline{y}_{i1}, \overline{y}_{i1}], [\underline{y}_{i2}, \overline{y}_{i2}], \ldots, [\underline{y}_{in}, \overline{y}_{in}]] \qquad (4)$$

where $\underline{y}_{i,k}$ and $\overline{y}_{i,k}$ are the lower and upper limit of q_i in the kth dimension and the width of each interval is now given by:

$$\delta_k = \frac{|max(Y_k) - min(Y_k)|}{N} \qquad (5)$$

The variable Y_k corresponds to the data in the kth dimension. We can generalize the way to determine the limits of a quantile when $Y \in \mathbb{R}^n$ as:

$$q_1 = \begin{bmatrix} [min(Y_1), min(Y_1) + \delta_1] \\ [min(Y_2), min(Y_2) + \delta_2) \\ \vdots \\ [min(Y_n), min(Y_n) + \delta_n] \end{bmatrix}^T \tag{6}$$

for the first quantile, and:

$$q_i = \begin{cases} \begin{bmatrix} [\overline{y}_{(i-1),1}, \overline{y}_{(i-1),1} + \delta_1] \\ [\overline{y}_{(i-1),2}, \overline{y}_{(i-1),2} + \delta_2] \\ \vdots \\ [\overline{y}_{(i-1),n}, \overline{y}_{(i-1),1} + \delta_n] \end{bmatrix}^T & \text{if } i = N \\ \begin{bmatrix} [\overline{y}_{(i-1),1}, \overline{y}_{(i-1),1} + \delta_1) \\ [\overline{y}_{(i-1),2}, \overline{y}_{(i-1),2} + \delta_2) \\ \vdots \\ [\overline{y}_{(i-1),n}, \overline{y}_{(i-1),1} + \delta_n] \end{bmatrix}^T & \text{otherwise} \end{cases} \tag{7}$$

for subsequent quantiles. Note that in general a partial order is formed corresponding to the left-to-right precedence relation where $q_i < q_j$ if $\exists k$ such that $\overline{y}_{i,k} < \underline{y}_{j,k}$, for $k \leq n$. Thanks to the precedence relation, we can assign to each quantile a numerical code that preserves the partial order. To illustrate the above idea, in Fig. 1 the leftmost quantile can be identified as 1 while the rightmost quantile can be identified as 10. Even any other numerical basis can be used (instead of 10 base), as long as the order is preserved.

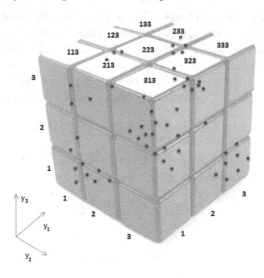

Fig. 2. Possible encoding of three-dimensional quantiles

Now consider Fig. 2 which illustrates a possible encoding of three-dimensional quantiles. The quantile code is formed by combining of the sequence number of the intervals that define q_i in each dimension. It means, that a three-dimensional quantile encoded as 111 is defined by the leftmost intervals per dimension. Likewise, a three-dimensional quantile encoded as 333 is defined by the rightmost intervals per dimension. In such encoding, we can see that the quantile encoded as 113 precedes the quantile encoded as 323. The evident precedence order given in one-dimensional space is preserved in higher order spaces.

3   Mapping Data to One-dimensional Space

So far, we have shown how to divide the data space into a set of n-dimensional quantiles. Now, the idea is to map the dataset instances to the quantile to which they belong. Those instances that belong to q_i can be represented (encoded) through the code assigned to it. Thus, given a set of quantiles Q defined from a n-dimensional dataset Y, a one dimensional dataset Y', composed only by encoded instances, can be obtained. Table 1 shows a set of quantiles Q for an hypothetical dataset $Y \in \mathbb{R}$.

Table 1. Example of quantile encoding for a one-dimensional dataset.

Code	Lower	Upper
01	[−0.63	−0.60)
02	[−0.60	−0.58)
03	[−0.58	−0.55)
04	[−0.55	−0.53)
05	[−0.53	−0.50)
06	[−0.50	−0.48)
05	[−0.48	−0.45)
08	[−0.45	−0.43)
09	[−0.43	−0.40)
10	[−0.40	−0.38]

Possible instances of Y have been encoded, according to the quantile code to which they belong, as it is shown in Table 2.
From this example the encoded dataset Y' is:

$$Y' = \{01, 01, 02, 02, 03, 03, 04, 04, 05, 05, 07, 07, 08, 08, 09, 09, 10, 10, 10, 06\} \quad (8)$$

The above idea can be extended to n-dimensional case. For instance, let Y be a dataset in $\mathbb{R}^3$. Assume that the space of the dataset in the kth dimension (Y_k) is divided into five intervals (for illustrative purposes) where

Table 2. Example of an encoding of a one-dimensional dataset.

Y	Code	Y	Code
−0.63	01	−0.48	07
−0.62	01	−0.46	07
−0.60	02	−0.45	08
−0.59	02	−0.44	08
−0.58	03	−0.43	09
−0.57	03	−0.41	09
−0.55	04	−0.40	10
−0.54	04	−0.39	10
−0.53	05	−0.38	10
−0.52	05	−0.50	06

$min(Y_1) = -1.22$, $max(Y_1) = 1.94$, $min(Y_2) = -1.21$, $max(Y_2) = 2.70$, $min(Y_3) = -1.26$ and $max(Y_3) = 1.12$. Also, assume that $\delta_1 = 0.64$, $\delta_2 = 0.78$ and $\delta_3 = 0.48$. In this way, the leftmost quantile is comprised by the intervals set $[-1.22, -0.58), [-1.21 - 0.43), [-1.26 - 0.78)$, which correspond to the leftmost intervals per dimension. Using a decimal numerical code of two digits to identify every interval, the corresponding quantile can be encoded as $01 - 01 - 01$ or merely 010101. In Table 3, this and other quantiles are illustrated including their corresponding boundaries per dimension.

Table 3. Sample of quantiles for an hypothetical three-dimensional dataset.

Quantile code	Y_1	Y_2	Y_3
010101	$[-1.22 - 0.58)$	$[-1.21 - 0.43)$	$[-1.26 - 0.78)$
010103	$[-1.21 - 0.57)$	$[-1.21 - 0.43)$	$[-0.79 - 0.31)$
010104	$[-1.20 - 0.56)$	$[-1.21 - 0.43)$	$[\,0.17\ 0.65)$
010201	$[-1.21 - 0.57)$	$[-0.430.35)$	$[-1.26 - 0.78)$

Given the set of quantiles Q, assume an instance $y \in Y$ of the form $[-1.15, -0.5, -0.4]$. Since this instance lies into the quantile 010103, it can be represented by the quantile code.

Based on this representation process, we can obtain a one-dimensional dataset denoted as Y'. Note that the above representation method may involve a loss of information allegedly depending on the number of quantiles. This value is implicitly associated to the number of bins in which the space of Y_k (the dataset in the kth dimension) is divided. In this regard, an appropriate selection of this value is compulsory. Typically, Sturges's rule is used [5,6]. Alternative rules, which attempt to improve the performance of Sturges's rule without a normality assumption, are Doane's formula [3] and the Rice rule [8]. In this paper, we prefer the Rice rule, which is to set the number of intervals to twice the cube root of the number of instances. In the case of 1000 instances, the Rice rule yields 20 intervals instead of the 11 recommended by Sturges' rule. One of our main goals in near future is to drive experiments to minimize the loss information.

Having defined the way to represent a n-dimensional data, in the next sections we show the experimental results which allow us to confirm that our proposal is promissory.

4   Experiments and Results

In this section, first we show in Subsect. 4.1 some preliminary results of the effectiveness of our method to preserve the properties of the original dataset in clustering analysis. Subsequently, in Subsect. 4.2, we evaluate the effectiveness through a wide sample of systematic experiments that allow us to generalize the results.

4.1   Preliminary Results: Preserving Clustering Properties

We applied a non supervised classification algorithm over the well known Iris dataset (available at https://archive.ics.uci.edu/ml/datasets/Iris). This dataset (in what follows Y_{iris}) contains 3 classes of 50 instances each, where each class refers to a type of iris plant.

Figure 3 shows the iris data classes obtained by the k-means algorithm for three centers, using Y_{iris} and the encoded dataset Y'_{iris}. The plot was obtained using the *plotcluster* function of *fpc* R module with the default parameters.

Table 4 is the confusion matrix obtained for three classes. We can see that in this case the ratio of well classified instances is 80 %.

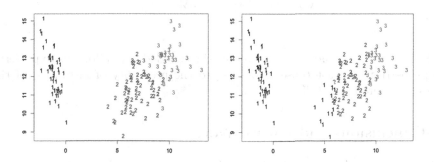

Fig. 3. Classification of k-means using Y_{iris} vs. encoded dataset Y'_{iris}.

The above results show that the encoded dataset achieves to retain the information conveyed by the original data. However, these results are not enough to generalize this observation. In the following subsection, we present an experimental methodology that allows us to generalize the effectiveness of our proposal.

Table 4. Confusion matrix of k-means.

	Reference class (Y_{iris})		
Predicted class Y'_{iris}	Class 1	Class 2	Class 3
Class 1	39	6	0
Class 2	11	35	4
Class 3	0	9	46

4.2   General Results

In order to generalize the above results, we generated a wide set of synthetic datasets (about 5000) in $\mathbb{R}^n$ for $n = 2, 3, ...100$. To generate each dataset, a number of classes was defined a prior. For each dataset an encoded dataset (based on our proposal) was obtained. Subsequently k-means algorithm is applied to both non-encoded (Y) and encoded dataset (Y'). As mention the effectiveness is given by the percentage matching between the class labels found with Y dataset and the class labels found with Y'. The average result is shown in Table 5. For completeness, we show the confidence interval of the results with a -value of 0.05.

Table 5. Average effectiveness

	Matching(%)
Average	0.8895
Standard deviation	0.2181
Lower limit	0.8806
Upper limit	0.8983
Confidence level	95 %

The experiments show that in average the encoded data allows us to preserve the properties (at least those associated to the proximity of instances) of the dataset in more than 80 %.

5   Conclusions and Future Work

In this paper we have described a method to encode multivariate data. Such encoding, could be interpreted as a data reduction method, in a way that using the codes we are able to apply data mining methods and obtain similar results as using the original data. We applied the reduction method over iris data and results show good performance in classification. Also we show that statistically our method achieves to preserve about 80 % of information conveyed by the original data.

However, there are some more hypotheses to prove in order to generalize the reduction method. For instance, we suspect that precision in classification tasks could improve if a finer number of quantiles is used; more experiments are needed in order to explore this idea. Another future research will be the inclusion of non numerical data types.

Acknowledgments. The authors acknowledge the support of Consejo Nacional de Ciencia y tecnología (CONACyT) and Centro de Investigación y Estudios Avanzados-CINVESTAV.

References

1. Aldana-Bobadilla, E., Alfaro-Prez, C.: Finding the optimal sample based on shannon entropy and genetic algorithms. In: Sidorov, G., Galicia-Haro, S.N. (eds.) MICAI 2015. LNCS, vol. 9413, pp. 353–363. Springer, Heidelberg (2015)
2. Cox, K.A., Dante, H.M., Maher, R.J.: Product appearance inspection methods and apparatus employing low variance filter, 17 August 1993. US Patent 5,237,621
3. Doane, D.P.: Aesthetic frequency classifications. Am. Stat. **30**(4), 181–183 (1976)
4. Gowda, K.C., Krishna, G.: The condensed nearest neighbor rule using the concept of mutual nearest neighborhood. IEEE Trans. Inf. Theor. **25**(4), 488–490 (1979)
5. Hyndman, R.J.: The problem with sturges rule for constructing histograms. Monash University (1995)
6. Hyndman, R.J., Fan, Y.: Sample quantiles in statistical packages. Am. Stat. **50**(4), 361–365 (1996)
7. Kalegele, K., Takahashi, H., Sveholm, J., Sasai, K., Kitagata, G., Kinoshita, T.: On-demand data numerosity reduction for learning artifacts. In: 2012 IEEE 26th International Conference on Advanced Information Networking and Applications (AINA), pp. 152–159. IEEE (2012)
8. Lane, D.M.: Online statistics education: an interactive multimedia course of study (2015). http://onlinestatbook.com/2/graphing_distributions/histograms.html. Accessed 03 Dec 2015
9. Liu, H., Motoda, H.: Instance Selection and Construction for Data Mining, vol. 608. Springer, Heidelberg (2013)
10. Reeves, C.R., Bush, D.R.: Using genetic algorithms for training data selection in RBF networks. In: Liu, H., Motoda, H. (eds.) Instance Selection and Construction for Data Mining, vol. 608, pp. 339–356. Springer, Heidelberg (2001)
11. Shlens, J.: A tutorial on principal component analysis (2014). arXiv preprint arXiv:1404.1100
12. Skalak, D.B.: Prototype and feature selection by sampling and random mutation hill climbing algorithms. In: Proceedings of the Eleventh International Conference on Machine Learning, pp. 293–301 (1994)
13. Randall Wilson, D., Martinez, T.R.: Reduction techniques for instance-based learning algorithms. Mach. Learn. **38**(3), 257–286 (2000)
14. Lei, Y., Liu, H.: Feature selection for high-dimensional data: a fast correlation-based filter solution. ICML **3**, 856–863 (2003)

Association Rule Hiding for Privacy Preserving Data Mining

Shyma Mogtaba$^{(\boxtimes)}$ and Eiman Kambal

Department of Computer Science, Faculty of Mathematical Sciences,
University of Khartoum, Khartoum, Sudan
shyma_mogtaba@hotmail.com, Eimankambal@gmail.com

Abstract. Privacy preservation is a big challenge in data mining. The protection of sensitive information becomes a critical issue when releasing data to outside parties. Association rule mining could be very useful in such situations. It could be used to identify all the possible ways by which 'non-confidential' data can reveal 'confidential' data, which is commonly known as 'inference problem'. This issue is solved using Association Rule Hiding (ARH) techniques in Privacy Preserving Data Mining (PPDM). Association rule hiding aims to conceal these association rules so that no sensitive information can be mined from the database.

This paper proposes a model for hiding sensitive association rules. The model is implemented with a Fast Hiding Sensitive Association Rule (FHSAR) algorithm using the java eclipse framework. The implemented algorithm is integrated with a Weka open source data mining tool. Model analysis and evaluation shows its efficiency by balancing the trade-off between utility and privacy preservation in data mining with minimal side effects.

Keywords: Association rule mining · Association Rule Hiding (ARH) · Privacy Preserving Data Mining (PPDM)

1 Introduction

Over the last twenty years, there has been extensive growth in the amount of private data collected about individuals. This data comes from a number of sources, including medical, financial, library, telephone, and shopping records. Rapid advances and development of computing technologies has made it possible to correlate and analyze these massive amounts of data.

Data Mining is the process of extracting knowledge from various available data sources, and may also be referred to as data archaeology, data dredging, knowledge extraction and pattern analysis. The primary goal of data mining is to find relevant data from the data sources using various techniques. Large information repositories, data warehouses and databases contain a wide range of data from which the knowledge is discovered. Various data mining models help to identify the relevant data, which is then used for query processing, decision making and management of available information [1]. In some applications, extensive data mining may be carried out; in this case, it becomes crucial to ensure that information is not exposed to the public. Consequently,

© Springer International Publishing Switzerland 2016
P. Perner (Ed.): ICDM 2016, LNAI 9728, pp. 320–333, 2016.
DOI: 10.1007/978-3-319-41561-1_24

privacy preservation has become important concern with regards to the success of the data mining. Privacy Preserving Data Mining (PPDM) deals with protecting the privacy of individual data or sensitive knowledge without losing its utility. People have become well aware of intrusions on their personal data and are thus very reluctant to share sensitive information. However, privacy preservation may lead to inadvertent results when applying data mining [2].

In order to share data while preserving privacy, the data owner must come up with a solution that preserves privacy while ensuring accurate mining results. One of the main productive approaches for privacy preserving data mining is using association rule mining. Association rules is a powerful and popular tool for discovering relationships hidden in large data sets. These relationships can be represented in the form of frequent itemsets or association rules, which can be used to identify all the possible ways in which 'non-confidential' data can reveal 'confidential' data, i.e., identifying the different paths that can reveal such data which is commonly known as 'Inference problem'. Inference rule hiding aims to hide these association rules so that no sensitive information can be mined from the database [3].

In this paper, our focus is on hiding sensitive rules with minimal compromise of data utility using association rule hiding techniques. The approach for rule hiding is based on selectively modifying the database transactions, and data owners perform this modification before publishing their data. The general framework for hiding sensitive rules approach is shown in Fig. 1.

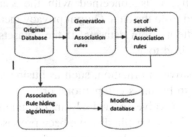

Fig. 1. General framework for hiding sensitive association rule [4]

This paper is organized as follows. Section 2 gives a background for privacy preserving data mining techniques, and presents evaluation measures for privacy preserving algorithms. Section 3 describes the proposed model, presents the design framework for building privacy preserving association rule mining model and experimental data set, as well as explains the strategy used and the experiment methodology. Section 4 provides the experiment results on the privacy preserving data mining model. Section 5 discusses the results on the privacy preserving data mining model. And Sect. 6 present the conclusion and possible future work.

2 Privacy Preserving Data Mining

2.1 Privacy Preserving Data Mining Issues

From a general point of view, privacy issues arising in data mining are classified into three categories: input privacy, output privacy and owner privacy, as outlined below.

2.1.1 Input Privacy

The first category is related to the data before analysis and is known as data hiding or input privacy. Specifically, data hiding perturbs the disclosed data in order to prevent the miner from extracting confidential or private information. Input privacy methods aim at addressing environments where users are unwilling to provide their personal information, or deliberately provide false information, to data recipients because they fear that their privacy may be violated.

The goal of these methods is to guarantee that such personal information can be released to (potentially untrusted) data recipients in a privacy-preserving way that still allows the data recipients to build accurate data mining models from the released data [5, 6].

2.1.2 Output Privacy

The second category concerns the information or knowledge that a data mining method may discover after having analyzed the data, and is known as knowledge hiding or output privacy. Specifically, it is concerned with the sanitization of confidential knowledge patterns derived from the data. Output privacy methods aim to eliminate the disclosure of sensitive patterns from datasets. If the datasets were shared as-is, then such patterns could easily lead to

i. The disclosure of sensitive information, such as business or trade secrets that can provide an advantage to business competitors, or
ii. Discrimination, if they involve individuals in the input data who have certain unique characteristics. Several methods have been proposed to offer output privacy by eliminating sensitive patterns from the released data in a way that minimizes data distortion and side-effects [5, 6].

2.1.3 Owner Privacy

A third line of research involves protocols that enable a group of data owners to collectively mine their data, in a distributed fashion, without allowing any party to reliably learn the data (or sensitive information about the data) that the other owners hold that is, the sources of the data. For this purpose, several cryptographic methods have been recently proposed to facilitate the privacy-preserving distributed mining of data that reside in different data warehouses. These methods assume that the data are either horizontally or vertically partitioned among the different sites, and that any sensitive disclosures should be limited in the data mining process [5, 6].

2.2   Existing Methods in Privacy Preserving Data Mining

This subsection gives a brief description for privacy preserving association rule mining techniques.

2.2.1   Privacy Preserving Association Rule Mining Techniques

Association rules are a powerful and popular tool for discovering relationships hidden in large data sets. The relationships can be represented in the form of frequent itemsets or association rules.

Privacy preserving data mining using association rules refers to the area of data mining that seeks to protect sensitive information from unsolicited or unsanctioned disclosure [7, 8]. Most of the initial works on privacy preservation address concerns related to the input of data mining where private information is revealed directly by inspection of the data without sophisticated analysis. Over the past few years, interest has shifted towards dealing with the problem of hiding sensitive patterns. Our work is more related to the concern over the output of association rule mining methods, where threats are caused by what mining tools can discover. Several methods have been proposed to offer output privacy by eliminating sensitive patterns from the released data in a way that minimizes data distortion and side-effects [4, 6].

The strategies of association rule hiding are classified as follows:

i. *Heuristic based approach*
    This approach hides sensitive association rules using two methods.
    (a) *Data distortion based technique*
        M. Attallah et al. were the first use this technique of hiding association rules by decreasing the support of the rule up to an acceptable level, after which the confidence of the rule is reduced up to certain threshold [4].
        Verykios et al. proposed five different algorithms for hiding association rules, including: increase support of left-hand side (LHS) and decrease support of RHS items, association rule hiding using hidden counters, decrease support of RHS items in rule clusters (DSRRC), modified decrease support of RHS item of rule clusters (MDSRRC), and fast hiding sensitive association rules (FHSAR) [4].
        (1) *ISL (Increase Support of LHS) and DSR (Decrease Support of RHS)*
            Wang and Jafari proposed two algorithms for hiding sensitive predictive association rules, namely ISL and DSR. The ISL algorithm hides the rule by increasing the support of left-hand side items of the rule, while the DSR algorithm hides the rule by decreasing the support of right-hand side items of the rule. [4, 9, 10].
        (2) *Association rule hiding using hidden counters*
            Belwal et al. presented a heuristic based algorithm for addressing the problem of association rule hiding. They proposed a method for hiding sensitive association rules based on ISL, modified the definition of support and confidence and introduced the use of a hidden counter in determining support and confidence [4, 9, 10].

(3) DSRRC *(decrease support of RHS items in rule clusters)*

Modi and Rao et al. proposed an algorithm for hiding sensitive association rules based on a heuristic approach in the form of the DSRRC
(decrease support of RHS items in rule clusters) algorithm, which is based
on DSR. It hides the association rule by making clusters of sensitive
association rules based on right-hand items, then calculates the sensitivity
of each cluster, i.e. the sum of sensitivity of each item present in the
cluster. It then indexes the sensitive transactions for each cluster and sorts
all the clusters in decreasing order of sensitivity. The hiding process
conceals the rule by deleting common RHS items of the rules in cluster
from the sensitive transactions. The advantage of the proposed algorithm
is that it maintains database quality and does not make major changes to
the database. However, its disadvantage is that it can only hide those
sensitive association rules that contain single items in the right-hand side
of the rule [4, 9, 10].

(4) *MDSRRC (modified decrease support of RHS item of rule clusters)*

Domadiya et al. proposed a heuristic based algorithm for hiding sensitive
association rules called the MDSRRC. A modification of the DSRRC
algorithm, it overcomes the limitation of DSRRC, and is able to hide the
sensitive association rules that contain multiple items in the right-hand
side. Its main advantage is that it does not make major changes to the
database and is able to hide rules that contain multiple items in the
right-hand side of the rule [4, 9, 10].

(5) *FHSAR (fast hiding sensitive association rules)*

Weng et al. [11] proposed an algorithm for hiding sensitive association
rules called the FHSAR. It can hide all sensitive association rules without
modifying the given database. In addition, the algorithm can be applied to
any database size and is very time-efficient, scanning the database only
once in order to hide sensitive rules. Given database D, sensitive rules
(SAR), *min_support*, and *min_confidence*, the goal of FHSAR is to
generate a database to be released, 'D', in which the sensitive association
rules are completely hidden and the side effects generated are minimized
[4, 9–11].

(b) *Data blocking-based technique*

Y. Saygin et al. proposed an algorithm for hiding sensitive association rules
based on the data blocking technique, where rules are hidden by changing the
value of some items in the database matrix from 0 or 1 to ? (Unknown). So the
support of certain items goes down to a certain level and thus the rule mining
algorithm is unable to mine sensitive rules [4, 6, 9].

ii. *Border-based approach*

The border based approach hides sensitive association rules by modifying the
border in the lattice of frequent and infrequent itemsets of the original database.
The border consists of itemsets that separate frequent from infrequent itemsets.
Sun and Yu were the first to introduce the concept of border [4, 6, 9].

iii. *Reconstruction-based approach*

This approach is implemented by first perturbing the data and then reconstructing the distributions at an aggregate level in order to perform the association rules mining. This approach has three phases, in which first generates frequent itemsets from the original database and second performs a sanitization algorithm over frequent itemsets by selecting a hiding strategy and identifying sensitive frequent itemsets according to sensitive association rules. The third phase generates a sanitized database by using an inverse frequent itemset mining algorithm, and then releases this database [9].

iv. *Exact approach*

This is a non-heuristic algorithm, which formulates the rule-hiding problem into a constraint satisfaction problem or optimal problem that is solved by integer programming. Divanis and Verykios were the first to use the exact approach for hiding rules. It provides an optimal solution for the problem of association rule hiding [4, 6, 9].

v. *Cryptographic approach*

Cryptography based approaches solve the problem of Secure Multi-Party Computation (SMC). When the database is distributed among several sites, multiple number parties may share their private data without leaking any sensitive information at their end. It is divided into two categories: vertically partitioned distributed data and horizontally partitioned distributed data. In these approaches, instead of distorting the database, it encrypts original database itself for sharing. The communication cost of this approach is very effective but it is very expensive for large amount of datasets; however, it falls short of providing a complete answer to the problem of privacy preserving data mining. It also fails to fully protect output of data mining methods where the threats are caused by what data mining tools can discover [4, 6, 9].

2.2.2   Analysis of Privacy Preserving Association Rule Mining Techniques

Heuristic algorithms may suffer from undesirable side effects that lead them to identify approximate hiding solutions. This is because heuristics always aim at selecting best decisions with respect to the hiding of sensitive knowledge. Heuristic algorithms may cause undesirable side effects to nonsensitive rules, e.g. lost rules and new rules. Border-based approaches provide an improvement over pure heuristic approaches, but they are still reliant on heuristics to decide upon the item modifications that they apply on the original database. As a result, in many cases these methodologies are unable to identify optimal hiding solutions, although such solutions may exist for the problem at hand. Algorithms in exact approaches have very high time complexity due to the time that is taken by the integer programming solver to solve the optimization problem [12].

From the previous analysis, we can conclude that while several available privacy preserving data mining techniques and methods are proposed, the heuristic approach is the most reliable approach in regards to sensitive association rule hiding.

From work done in the heuristic approach, we have analyzed that there are mainly two strategies to hide any association rule. The first strategy hides sensitive rules by inserting a new itemset in selected transactions that will decrease confidence of the rule.

The second strategy is to remove the itemset from selected transactions. It can be done in two ways. For rule we can remove itemset to decrease the confidence of rule or we can reduce the support count of large itemset by removing items from selected transactions.

It is observed that whenever an item inserting strategy is used, it creates more artifactual patterns because it increases the support of some itemsets such that they becomes frequent. It sometimes fails to hide some sensitive association rules due to new patterns created as side effects. On the other hand, if an item removing strategy is used, then some frequent itemsets become infrequent, which affects non-sensitive rules that are hidden as side effects.

Our work has the basis of rule-hiding approach using the item removing strategy, which has fewer side effects in hiding sensitive rules and does not modify the given database.

2.2.3   Evaluation Measures

It is difficult to diagnose proper criteria in evaluation of algorithms and significant privacy preserving tools while meeting all required criteria for an algorithm.

Usually, evaluation measures selected according to the user's needs.

In the area of association rule mining tasks, we evaluate some association rule hiding algorithms mentioned in the previous section. Evaluation of algorithms is usually done based on a number of measures: *lost rules, artificial rules, dissimilarity and hiding failure*.

i. *Lost rule*: Non-sensitive rules that will be lost due to hiding sensitive rules. This is measured as:

$$Lost\ rule = \frac{|R\ (D)| - |R\ (D') \bigcap R(D)|}{R\ (D)|} \tag{1}$$

Where R (D) is a mined rule from a major database rather than a sensitive rule.

ii. *Artificial rule*: those rules that are not mined by support and confidence levels that are defined by users from the main database, but will be mined from a cleaned database after it is secured. This is measured as:

$$Artificial\ rule = \frac{|R\ (D')| - |(D) \bigcap R\ (D')|}{|R\ (D')|} \tag{2}$$

iii. *Dissimilarity*: performed change is between the main database and cleaned database. This is measured as:

$$Dissimilarity = \frac{\sum_{i=1}^{n} fD\ (i) - fD'\ (i)}{fD\ (i)} \tag{3}$$

In this formula, i stands for an item in the main database of D and fD (i) is its frequency in the database.

iv. *Hiding failure*: the degree of sensitive rules that are mined after applying security on a cleaned database is measurable as:

$$Hiding\,failure = \frac{|Rh\,(D')|}{|Rh\,(D)|} \qquad (4)$$

Where Rh (D) will be a sensitive rule and Rh (D') will be a mined rule from the cleaned database [6, 10].

3   The Proposed Model

This section presents the design framework for building the privacy preserving association rule mining model.

3.1   Data Understanding

Real-life dataset, a German credit dataset from UCI Repository [13], is employed. The number of instances of this dataset is 1000 instances on 20 attributes. Each record in the German credit dataset represents an applicant for credit; and each applicant was rated as "good credit" or "bad credit".

3.2   Modeling

Association rule hiding algorithms get strong and efficient performance for protecting confidential data. Association rule hiding aims to hide the sensitive rules. A rule is said to be sensitive if its disclosure risk is above a certain privacy threshold defined by the data owner. The approach for rule hiding is based on selectively modifying the database transactions; this modification is performed by data owners before publishing their data.

The following case study focuses on bank customers:

Suppose that the bank (data owner) wants to release data for association or classification analysis on rating, but does not want others to infer sensitive information about individuals.

The general framework for this model presented in Fig. 2 shows the sequence of the processes followed throughout this work.

The privacy preserving association rule mining model building consists of four phases, as follows:

i. *Phase 1: find frequent itemsets and association rules*
   This phase consists of two stages:

   Stage 1: *find frequent itemsets*
   An "Apriori Algorithm" is used to search for frequent itemsets occurring in the selected database. Apriori is a basic algorithm proposed for mining frequent

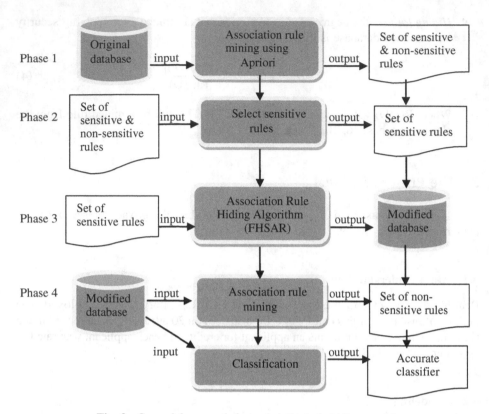

Fig. 2. General framework for association rule hiding model

itemsets for association rules. The name of the algorithm is based on the fact that it uses prior knowledge of frequent itemset properties. Apriori employs an iterative approach known as a level-wise search, where k-itemsets are used to explore (k + 1)-itemsets.

First, the set of frequent 1-itemsets is found by scanning the database to accumulate the count for each item and collecting those items that satisfy minimum support. The resulting set is denoted by L1. Next, L1 is used to find L2, the set of frequent 2-itemsets, which is used to find L3, and so on, until no more frequent k-itemsets can be found. The finding of each Lk requires one full scan of the database. The search for frequent itemsets is based on the concept of searching using passes where only the candidate itemsets are passed on to subsequent passes [1].

Stage 2: *generating association rules from frequent itemsets*

Once the frequent itemsets from transactions in a database D have been found, it is straightforward to generate strong association rules from them. The rules are mined from the commonly occurring itemsets. From the confidence threshold value entered by the user, the set of all possible association rules is mined. The support value is considered for extracting the maximum frequent itemsets, whereas confidence value is a must for mining the rules [1].

ii. *Phase 2: Sensitive rule selection*
The data owner selects a set of sensitive rules resulted from association rule mining in phase 2 according to a defined threshold of disclosure risk.
The output of this phase is a set of selected sensitive rules.
iii. *Phase 3: association rule hiding*
In the third phase, FHSAR is used. This phase consists of three stages:

> Stage 1: FHSAR scans the database once while it collects information about the correlation between each transaction and sensitive rules.
> Stage 2: each transaction is assigned a weight to denote the number of itemsets in sensitive association rule (SAR) that contain the items in the transaction, after which the transaction with the maximal weight is selected.
> Stage 3: selects the item to be deleted from a selected transaction according to a heuristic function for estimating side effects. The function, checking-and-removing item, repeats to modify transactions one-by-one until all rules in SAR have been hidden. The order of the modifications is according to the prior weight associated with each transaction [11].

Fast Hiding Sensitive Association Rule algorithm (FHSAR) implemented in Java code integrated with Weka open source data mining tool. FHSAR algorithm applied to hide the identified sensitive rules resulted from previous phase.
The output of this phase is modified dataset with removed items from maximal weigh transaction.
iv. *Phase 4: model evaluation*
Model evaluation phase consists of two stages:

> Stage 1: an Apriori algorithm is applied on a modified dataset to evaluate the hiding algorithm based on a number of measures, as previously mentioned: lost rules, artificial rules, dissimilarity, and hiding failure.
> Stage 2: a classifier model evaluated in terms of accuracy, precision, recall, and Root Mean Squared Error (RMSE) for the original and modified data set using C4.5 classification decision tree.

4   Experiment Results

The selected algorithm is implemented in java using the eclipse framework, and tested on credit datasets of different size. This dataset is used for the analysis with Minimum Support as 60 % and Minimum Confidence as 90 %.

The implemented algorithm is applied to the dataset and parameters like the number of lost rules, artificial rules, hiding failure and the number of modified items of the implemented algorithm are recorded. Using the data from different parameters, different graphs are constructed.

As an example, suppose that the data owner (bank) selects a set of sensitive rules resulted from Apriori algorithm according to the defined risk, as follows:

credit_amount = 1000 < X<5000 other_parties = none ==> num_dependents = <2
credit_amount = 1000 < X<5000 num_dependents = <2 ==> other_parties = none

existing_credits = <2 foreign_worker = yes ==> num_dependents = <2

As the result of implementation of association rule hiding algorithm is that, only one item is removed from one record.

Table 1 below shows the number of lost rules, artificial rules, hiding failure and number of modified items generated by the existing algorithm using datasets of various sizes for hiding 3 sensitive rules. From the graph in Fig. 3, it is clear that the numbers of lost rules and artificial rules are independent of database size.

Table 2 shows the number of lost rules, artificial rules, hiding failure and number of

Table 1. Association hiding results for German Credit Dataset |SAR| = 3

Modified items	Hiding failure	Artificial rules	Lost rules	Number of instances
1	**0.3**	7	11	8 Instances
2	**0.3**	19	18	10 Instances
1	**0**	5	3	15 Instances

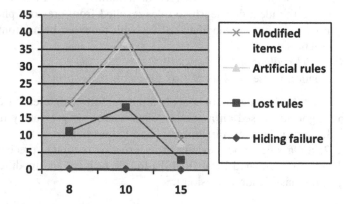

Fig. 3. Bar chart for association hiding results |SAR| = 3

modified items generated by existing algorithm using datasets of various sizes for hiding 6 sensitive rules. From the graph in Fig. 4, it is clear that the numbers of lost rules and artificial rules are independent of database size.

Table 2. Association hiding results for German Credit Dataset |SAR| = 6

Modified items	Hiding failure	Artificial rule	Lost rules	Number of instances
1	**0.3**	5	9	8 Instances
2	**0**	15	20	10 Instances
1	**0**	6	5	15 Instances

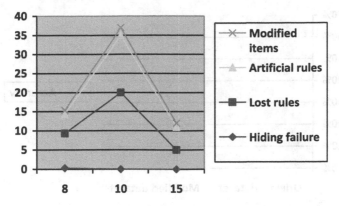

Fig. 4. Bar chart for association hiding results |SAR| = 6

Figure 5 shows the number of lost rules by the implemented algorithm using datasets of various sizes for hiding 3 and 6 sensitive rules. From the graph, it is clear that the numbers of lost rules is independent of the number of sensitive rules.

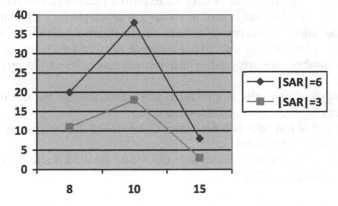

Fig. 5. Bar chart for lost rules for |SAR| = 3 and |SAR| = 6

Table 3 shows the results of classification model accuracy on original and modified datasets. The graph in Fig. 6 indicates that the accuracy of classifier models isn't affected by the modification of data.

Table 3. Classification results for German Credit Dataset

Algorithms	J48			
Measures	Accuracy	Precision	Recall	F-Measure
Original database	66%	0.508	0.6	0.55
Modified database	66%	0.508	0.6	0.55

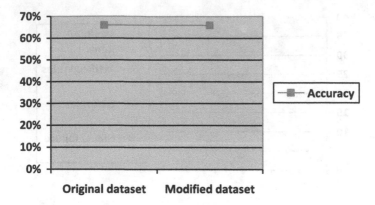

Fig. 6. Bar chart for classification results

5 Discussion

The experimental results can be summarized as follows:

i. As shown, the performance of FHSAR improved because of two reasons:
First, a heuristic function is used to obtain a prior weight for each transaction, by which the order of transactions modified can be efficiently decided. Secondly, the correlations between the sensitive association rules and each transaction in the original database are analyzed, which can effectively select the proper item to modify.

ii. The number of new rules is minimized and independent of the size of database, which can be discovered in Tables 1 and 2.

iii. The number of lost rules is independent of the size of the database, which can be discovered in Tables 1 and 2.

iv. The accuracy of classifier model isn't affected by the modification of data, which can be discovered in Table 3.

6 Conclusion and Future Work

Privacy preserving data mining is a new body of research focusing on the implications originating from the application of data mining algorithms to large databases.

The purpose of the association rule hiding algorithm for privacy preserving data mining is to hide certain crucial information so they cannot be discovered through association rule.

In this paper, the existing association rule hiding algorithm (FHSAR) developed in java code was integrated with Weka in order to balance the trade-off between utility and privacy preservation. This is based on the association rule hiding approach and modifying the database transactions so that the confidence of the association rule can be reduced.

As the results of the experiment show, high performance of implemented algorithms was achieved compared to other association rule hiding algorithms and with minimal side effects in terms of hiding failure, number of new rules and number of lost rules.

Future research can consider testing the proposed evaluation framework for other privacy preservation algorithms, such as data distortion methods and extending Weka data mining tools with association rule hiding algorithms. Furthermore, future research can study the improvement of FHSAR algorithm efficiency and minimizing the side effects generated in most cases.

Acknowledgments. We deeply thank the University of Khartoum and special thanks to the Faculty of Mathematical Sciences for their unwavering support.

References

1. Han, J., Kamber, M., Pei, J.: Data mining: concepts and techniques: concepts and techniques. Elsevier, New York (2011)
2. Parmar, S., Gupta, M.P., Sharma, M.P.: A comparative study and literature survey on privacy preserving data mining techniques. IJCSMC **4**, 480–486 (2015)
3. Giannotti, F., Lakshmanan, L.V., Monreale, A., Pedreschi, D., Wang, H.: Privacy-preserving mining of association rules from outsourced transaction databases. Syst. J. IEEE **7**(3), 385–395 (2013)
4. Upadhyay, N., Tripathi, K., Mishra, A.: A survey of association rule hiding approaches. IJCSITS **5**, 190–195 (2015)
5. Ranjan, J.: Business intelligence: Concepts, components, techniques and benefits. J. Theoret. Appl. Inf. Technol. **9**(1), 60–70 (2009)
6. Aggarwal, C.C., Han, J.: Frequent Pattern Mining. Springer, Switzerland (2014)
7. Chhatrapati, M., Serasiya, S.A.: Research on Privacy Preserving Data Mining Using Heuristic Approach. IJRST **1**, 113–117 (2015)
8. Chopra, J., Satav, S.: Privacy preservation techniques in data mining. IJRET **2**, 537–541 (2013)
9. Kaur, K., Bansal, M.: A review on various techniques of hiding association rules in privacy preservation data mining. IJECS **4**(6), 12947–12951 (2015)
10. Ghalehsefidi, N.J., Dehkordi, M.N.: A survey on privacy preserving association rule mining (2015)
11. Weng, C.-C., Chen, S.-T., Lo, H.-C.: A novel algorithm for completely hiding sensitive association rules. In: IEEE (2008)
12. Shah, K., Thakkar, A., Ganatra, A.: A study on association rule hiding approaches. IJEAT Int. J. Eng. Adv. Technol. **3**, 72–76 (2012)
13. http://archive.ics.uci.edu/ml/datasets.html

Extending Process Monitoring to Simultaneous False Alarm Rejection and Fault Identification (FARFI)

Geert Gins, Sam Wuyts, Sander Van den Zegel, and Jan Van Impe[✉]

Department of Chemical Engineering,
Chemical and Biochemical Process Technology and Control (BioTeC+),
KU Leuven Technology Campus Ghent, Gebroeders De Smetstraat 1,
9000 Ghent, Belgium
{geert.gins,sam.wuyts,jan.vanimpe}@cit.kuleuven.be

Abstract. A new framework for extending Statistical Process Monitoring (SPM) to simultaneous False Alarm Rejection and Fault Identification (FARFI) is presented in this paper. This is motivated by the possibly large negative impact on product quality, process safety, and profitability resulting from incorrect control actions induced by false alarms—especially for batch processes. The presented FARFI approach adapts the classification model already used for fault identification to simultaneously perform false alarm rejection by adding *normal operation* as an extra data class. As no additional models are introduced, the complexity of the overall SPM system is not increased.

Two case studies demonstrate the large potential of the FARFI approach. The best models reject more than 94 % of the false alarms while their fault identification accuracy (> 95 %) is not impacted. However, results also indicate that not all classifier types perform equally well. Care should be taken to employ models that can deal with the added classification challenges originating from the introduction of the false alarm class.

Keywords: Chemometrics · Statistical Process Monitoring (SPM) · Fault Detection & Identification (FDI) · Batch processes

1 Introduction

Modern industry strives towards better-performing processes, resulting in safer, more profitable, and environmentally friendly production. Early detection of deviations from normal operation, as well as the identification of the root causes of these deviations, is of utmost importance [1–4]. Hence, a close monitoring is paramount, and *Fault Detection and Identification/Isolation* (FDI) has received much attention over the last few decades [2–5].

While first-principles models can be used for FDI, this approach is limited to well-known processes of small size [6]. *Statistical Process Monitoring* (SPM)

© Springer International Publishing Switzerland 2016
P. Perner (Ed.): ICDM 2016, LNAI 9728, pp. 334–348, 2016.
DOI: 10.1007/978-3-319-41561-1_25

aims at exploiting already available databases of process measurements for FDI in more complex processes such as (bio)chemical, steel, pulp and paper, or semiconductor industries. Extensive overviews of SPM are available in, i.a., [1–11].

SPM systems typically consist of three steps [12]. First, a model characterizing Normal Operation Conditions (NOC) is established. Next, *fault detection* references current process measurements against this model to detect deviations from NOC. Finally, *fault identification* identifies the root cause of the detected disturbance. Since fault detection in SPM is performed via statistical analysis, the occurrence of false alarms is unavoidable. As they trigger unnecessary— and in some cases incorrect—control actions, which in turn potentially affect process safety, product quality, and profitability, the influence of false alarms on the entire SPM system should be reduced as much as possible. Even so, the influence of these false alarms on fault identification is typically neglected in literature.

Minimizing the influence of false alarms is especially important in the operation of batch processes. In continuous processes, incorrect control actions owing to false alarms impact product quality only for a short amount of time. In batch processes, incorrect control actions can result in the loss of an entire batch. Because batches can take several hours or even days or weeks to complete, and because they are commonly used to manufacture products with high added value (e.g., pharmaceuticals, specialty chemicals, semiconductors, or food products), wasted batches present a substantial economic loss.

This paper addresses the problem of false alarms in SPM. A first possible approach for reducing the number of false alarms is increasing the control limits in SPM (e.g., from 99 % to 99.9 % or even higher). However, this approach substantially lowers detection speed for process upsets, and is therefore not ideal. A second alternative is the use of more accurate fault detection models: a better description of the process yields more accurate monitoring results, in turn reducing false alarms. However, this often requires nonlinear approaches, which present more danger for overfitting and are less robust [13,14]. Furthermore, the more complex mathematics often pose a problem for SPM implementation and maintenance in practice as they turn into *mathemagics* for non-experts [15].

Therefore, this work introduces a methodology which performs a false alarm rejection (i.e., the identification of false alarms) simultaneously with fault identification (i.e., identification of the root cause of a true alarm) by extending an existing FDI system with a False Alarm Rejection and Fault Identification (FARFI) model. In essence, FARFI acts as an *a posteriori* filter to identify and discard false alarms rather than attempting to avoid their generation. This has the advantage that less complex models can be employed for fault detection, benefiting robustness and avoiding *mathemagics*. Any non-linear process aspects not captured by the fault detection model are addressed by the FARFI model. As the fault identification model is typically already non-linear, the overal complexity of the SPM system is not substantially changed. FARFI is presented here for batch processes, but it is equally applicable to continuous processes.

This paper is structured as follows. First, Sect. 2 summarizes current approaches for data-driven FDI. The proposed combined FARFI methodology is presented next, in Sect. 3, followed by a description of the case studies on which its performance is illustrated in Sect. 4. The results of the case studies are discussed in Sect. 5, followed by concluding remarks in Sect. 6.

2 Fault Detection and Identification

Industrial processes are inherently governed by a small number of underlying physical and chemical phenomena, yet they are sampled with numerous different sensors. *Latent variable* approaches such as Principal Component Analysis (PCA)[16] are capable of dealing with the resulting large amounts of correlation present in process data. Even though many complex variants and extensions are available (e.g., [17–21]), linear PCA is the *de facto* reference for data-driven FDI [3]. The SPM methodology for batch process monitoring adopted in this work is presented in more detail in [22]. A brief summary is provided below.

Detection and identification of abnormal process behavior is performed in three steps [12]. First, a mathematical model of the process under NOC is constructed (Sect. 2.1). Next, deviations from NOC are detected by comparing current process operation against the reference NOC model (Sect. 2.2). Finally, any detected disturbance is analyzed to identify its root cause (Sect. 2.3).

2.1 Characterization of Normal Operation Conditions

In a first step, data from I NOC batches is gathered. During each batch, J sensors are sampled K times. The data of each sensor is normalized around their average trajectory over the I batches, removing major nonlinearities and justifying the use of linear PCA [23]. Next, the data is stacked variable-wise in the data matrix $\mathbf{X}$ ($IK \times J$) according to [24].

In a second step, PCA identifies the most important variability in $\mathbf{X}$ via the R leading eigenvectors of the covariance matrix $\mathbf{X}^\top \mathbf{X}$, which are called the *principal components* and stored in the columns of the loadings matrix $\mathbf{P}$ ($J \times R$). The projection of $\mathbf{X}$ on the principal component subspace yields the latent variable (or scores) matrix $\mathbf{T}$ ($IK \times R$) that provide a low-dimensional characterization of the original data. The matrix $\mathbf{E}$ ($IK \times J$) contains the residuals.

$$\mathbf{X} = \mathbf{T}\mathbf{P}^\top + \mathbf{E} \tag{1}$$

$$\mathbf{T} = \mathbf{X}\mathbf{P} \tag{2}$$

Many criteria exist for determining the number of principal components R [16, 25]. All are based on the fraction f_r of the total variation in $\mathbf{X}$ explained by component r (with corresponding eigenvalue λ_r). In this work, an Adapted Wold criterion [25] is applied to the cumulative fraction of explained variance.

$$R = \min \left\{ R \ \middle| \ \frac{\sum_{r=1}^{R+1} \lambda_r}{\sum_{r=1}^{R} \lambda_r} \leq 1.05 \right\} \tag{3}$$

2.2   Detection of Abnormal Events

Two abnormality statistics quantify how much a (normalized) new sample $\mathbf{x}_t$ $(1 \times J)$ at time t deviates from NOC. Hotelling's T^2 monitors the distance from the projection of the new sample on the PCA subspace $\mathbf{t}_t = \mathbf{x}_t\mathbf{P}$ $(1 \times R)$ to the center of the NOC scores at time t to detect extrapolation. The Q statistic monitors the residuals $\mathbf{e}_t = \mathbf{x}_t - \mathbf{t}_t\mathbf{P}^\top$ $(1 \times J)$ to detect model mismatch.

$$T^2(t) = (\mathbf{t}_t - \boldsymbol{\mu}_t) \left(\frac{\mathbf{T}_t^\top \mathbf{T}_t}{I - 1} \right)^{-1} (\mathbf{t}_t - \boldsymbol{\mu}_t)^\top \tag{4}$$

$$Q(t) = \left(\mathbf{x}_t - \mathbf{t}_t\mathbf{P}^\top \right) \left(\mathbf{x}_t - \mathbf{t}_t\mathbf{P}^\top \right)^\top \tag{5}$$

The matrix $\mathbf{T}_t$ $(I \times J)$ is the subset of the NOC scores corresponding to time t, and $\boldsymbol{\mu}_t$ $(1 \times J)$ is the average of $\mathbf{T}_t$. To improve robustness, $\mathbf{T}_t$ and $\boldsymbol{\mu}_t$ are estimated via crossvalidation.

A process upset is identified when either statistic exceeds its upper control limit. As the assumptions underpinning the theoretical derivation of the control limits [23] are not necessarily fulfilled, empirical control limits are employed here. To reduce the number of false alarms, a number—typically 3—of consecutive violations of the control limits are required before an alarm is signaled.

2.3   Fault Identification

Once a process upset is detected, the identification of the root cause is commonly performed via contribution plots [3,26]. When a set of historical process upsets is available, however, fault identification can be performed via classification models. In this context, [22] proposed a classification-based fault diagnosis methodology for batch processes. The classification models can be combined with a local novelty detection framework to identify novel fault classes [27], after which the classification models can be updated to include the new fault types.

Any type of classifier can be used in the presented simultaneous FARFI framework. The FARFI potential is demonstrated in this work using two types of classification model: k-Nearest Neighbors and Least Squares Support Vector Machines. The former is selected for its simple structure yet powerful performance [28,29], the latter for its excellent generalization properties even when only a small number of training samples is available [30].

k-Nearest Neighbors (k-NN) [31] is a simple classification model: a test pattern $\mathbf{z}$ is referenced against a database of N training patterns $\mathbf{z}_n$ $(d \times 1)$ each with known fault label $f_n \in \{1, 2, ..., F\}$ from a set of F known faults. Based on a distance metric (typically the Euclidean distance), $\mathbf{z}$ is assigned the label that is most represented among its k closest neighbors; ties result in no classification.

Least Squares Support Vector Machines. (LSSVM) [32] is a binary classifier based on a database of N training patterns $\mathbf{z}_n$ $(d \times 1)$ each labeled as belonging to one of two classes $y_n \in \{-1, +1\}$. Each pattern is mapped to a higher-dimensional feature space via a nonlinear transformation $\phi(\cdot)$. Next, a linear separation between both classes is defined for a test pattern $\mathbf{z}$.

$$y(\mathbf{z}) = \text{sign}\left(\sum_{n=1}^{N} \alpha_n y_n (\phi(\mathbf{z}) \cdot \phi(\mathbf{z}_n)) + b\right) \tag{6}$$

[33] implicitly define the map ϕ using the kernel function $K(\mathbf{z}, \mathbf{z}_n) = \phi(\mathbf{z}) \cdot \phi(\mathbf{z})$. The model coefficients α_n and bias b are obtained by solving a set of linear equations with inequality contraints [32].

The choice of the kernel function determines whether the LSSVM performs a linear or nonlinear classification. In this paper, linear and RBF kernels (with kernel width σ) are used.

$$K(\mathbf{z}, \mathbf{z}_n) = \mathbf{z}^\top \mathbf{z}_n \qquad\qquad \text{linear kernel} \tag{7}$$

$$K(\mathbf{z}, \mathbf{z}_n) = \exp\left(-\frac{||\mathbf{z} - \mathbf{z}_n||_2^2}{\sigma^2}\right) \qquad\qquad \text{RBF kernel} \tag{8}$$

To achieve multi-class classification in F classes, binarization is required [32].

One-versus-one (OvO) binarization trains a separate LSSVM on each possible pairwise combination of fault classes, resulting in $F(F-1)/2$ LSSVMs. Each LSSVM uses only training patterns from the considered classes. A new pattern $\mathbf{z}$ is assigned to a fault class by majority vote. Ties result in no classification.

One-versus-all (OvA) trains one LSSVM for each of the F fault classes. Each classifier uses the entire set of patterns: the fault class under consideration is labeled as the positive class, while all patterns of the other fault classes are labeled negative. A new pattern $\mathbf{z}$ is assigned to the class for which the class-specific LSSVM predicts a positive label. Again, ties result in no classification.

3   Combined False Alarm Rejection and Fault Identification

In this paper, the concepts of data-driven fault identification presented in Sect. 2.3 are extended to also detect and reject false alarms. Hereto, a small number of samples obtained during NOC is added as an extra *normal operation* class to the reference set of faulty samples and a classifier is trained on this expanded data set. To avoid unbalancedness between the different classes, the number of normal points should be in line with the number of prototypes for the different fault classes. The resulting classification model is capable of distinguishing between normal operation and different types of process faults, hence performing *simultaneous* False Alarm Rejection and Fault Identification (FARFI).

The presented methodology can be interpreted in two ways. In the first interpretation, the fault detection model is the core of the SPM system which distinguishes between good and bad operation. Because the FARFI model filters out NOC samples that are erroneously identified as faulty, it can be considered as an *a posteriori* correction of the fault detection. The second interpretation puts FARFI at the center of the SPM system. The fault detection model is regarded as an *a priori* filter on the incoming data samples, sending only samples that are potentially faulty to the FARFI model. Next, the FARFI model performs the actual distinction between normal operation and the various types of process upsets. Regardless of the interpretation, the net effect of the replacing the FI model with a FARFI model is the overall reduction of the number of false alarms of the entire SPM system.

It is also possible to set up a separate, dedicated false alarm rejection system just prior to analyzing disturbances with a dedicated FI model. However, this approach is more complex as it introduces an additional mathematical model in the entire SPM system. Guided by Occam's razor, this paper therefore investigates the simpler combined FARFI approach introduced above.

4 Case Study

The FARFI procedure proposed in this paper is tested on the widely used PEN-SIM benchmark [34], which is briefly described in Sect. 4.1. Section 4.2 provides more details on the generated data sets. Section 4.3 details the testing procedure.

4.1 PENSIM Benchmark Process

PENSIM is a widely used benchmark for SPM research [34]. It simulates an industrial-scale biochemical batch process for penicillin production.

The fermentation is initially operated in batch mode. During this phase, the biomass uses the substrate initially present in the reactor for growth. Once the substrate concentration drops below 0.3 g/L (after approx. 43 h), a small amount of substrate is continuously fed to the reactor to maintain a low but non-zero concentration. These starvation conditions result in penicillin production. Once a total of 25 L of substrate has been fed (after approx. 460 h), the batch is terminated. During operation, the 11 sensors listed in Table 1 are used to monitor the fermentation. To remove variations in batch duration, indicator variables rather than time are used as progress indicators, as proposed by [34].

Table 1. Online measurements of the PENSIM model and corresponding noise variances.

Sensor measurement	σ_{noise}	Sensor measurement	σ_{noise}
Dissolved oxygen [mmol/L]	0.002	Agitator power [W]	0.5 %
Dissolved CO_2 [mmol/L]	0.060	Substrate feed rate [L/h]	0.5 %
pH [-]	0.010	Coolant flow rate [L/h]	0.5 %
Reactor temperature [K]	0.050	Base flow rate [L/h]	0.5 %
Feed temperature [K]	0.050	Acid flow rate [L/h]	0.5 %
Fermentation volume [L]	0.001	Time [h]	—

Table 2. Simulated fault classes and their magnitude and starting time ranges.

Fault type	Magnitude	Onset time [h]
Agitator power drop	$[-5\%; -30\%]$	$20 - 380$
Aeration rate drop	$[-70\%; -90\%]$	$20 - 380$
DO sensor drift	$\pm[0.50\%/h; 0.75\%/h]$	$20 - 380$
Feed rate drift	$\pm[0.15\%/h; 0.35\%/h]$	$70 - 380$
Feed concentration step	$\pm[1\%; 10\%]$	0
Coolant temperature step	$\pm[1\%; 10\%]$	0

This results in 101 samples for the batch phase, and 501 samples for the fed-batch phase. To retain information on batch speed, time is added as a 12th variable.

The PENSIM model is implemented in the RAYMOND simulator [35] for two case studies. In the first study (*"basic"*), the basic PENSIM model of [34] is used, where measurement noise is only present on the dissolved oxygen (DO) and CO_2 concentrations. In the second case study (*"advanced"*), Gaussian measurement noise with zero mean and standard deviation listed in Table 1 is added to all online measurements. PID controllers are retuned to retain good control performance. Finally, biological variability is introduced as small variations of the maximal specific growth rate of the biomass, as detailed in [22].

4.2 Simulated Data

For each study, 200 NOC batches (without upsets) with varying initial conditions are simulated for the identification of the NOC model for fault detection.

For the training of the classification models, 1000 batches are generated for each of the six process faults specified in Table 2. Magnitude and onset time of each upset is randomly chosen from the listed uniform distributions. An equal number of positive and negative fault amplitudes is used when applicable. An additional 1000 NOC batches are generated for training the FARFI classification model. Hence, the fault identification training set consists of 7000 batches.

Finally, an independent validation set of 700 batches is generated: 100 batches of each fault type, and 100 batches with false alarms.

4.3 Testing Procedure

In each case study, an NOC model for fault detection is first identified. As the process consists of two different phases, a separate model is identified for each. For the basic study, 6 principal components are selected for PCA model of the batch phase model, and 8 components for the model of the fed-batch phase; in the advanced study, 7 and 8 components are retained respectively.

Next, the 7000 training batches and 700 validation batches are monitored with the NOC model. For each faulty batch, the process measurements at the time of alarm are stored and used train the classification models; a random time point is selected for NOC batches. To enable performance comparison, both FI and FARFI classification models are trained and validated. As [36, 37] showed that *fault smearing* inherently present in the computation of contributions can negatively impact fault diagnosis, the (pretreated) process measurements are used as input patterns to the classification models rather than contributions [22, 38].

Previous studies [22, 38] obtained good fault identification with $k = 1$ and $k = 3$ neighbors for k-NN classifiers for similar case studies. The tests also indicated that OvO LSSVM binarization yields better results than OvA binarization. Hence, only OvO is considered in this paper.

The performance of FI models for specific fault types depends on the data pretreatment [22]. A similar influence is expected for FARFI models. Hence, the four different pretreatments of the process data from [22] are also evaluated here. The *"normalized"* pretreatment normalizes the process data around the mean trajectory over all batches to remove the influence of fault onset times. The *"absolute"* pretreatment takes absolute values of the normalized values to increase homogeneity in fault classes with both positive and negative fault amplitudes, but this pretreatment can potentially cause different fault classes to overlap [22]. Finally, *"window 6"* and *"window 11"* pretreatments use short time windows of the absolute, normalized process measurements of respectively 5 and 10 additional time points to better distinguish between drifts and sudden changes. More details on the effect of each pretreatment are provided in [22].

As the number of available faulty batches is limited in practice, the FI and FARFI classifiers are trained with 2–10 training samples per fault class (a total of 12–60 training samples for the FI model and 14–72 training samples for the FARFI model). Validation always occurs on the validation set of 700 batches.

To ensure representative results, this procedure is repeated 100 times, each time selecting new training samples. For the LSSVMs, the training and validation cycle is performed 100 times for each of the selections of training batches (for a total of 10,000 training/validation steps for any given number of training batches per fault class) because the model training contains stochastic elements [32].

5   Results & Discussion

The aim of the FARFI model is the simultaneous rejection of false alarms and identification of process upsets, but correct fault identification is the most important aspect: incorrect rejection or classification of an actual process upset leads to missed alarms or inappropriate control actions and, potentially, to bad quality and/or unsafe batch operation. Hence, two main factors determine the performance of the proposed FARFI methodology: correct rejection of false alarms and—more importantly—correct fault identification.

The *overall* classification rates of the FARFI models are evaluated first. Figure 1 presents the evolution of the correct classification rates as a function of the number of training batches per fault class for the various pretreatments of the tested classifiers for the basic case study. (Similar—in some cases even better—results are obtained for the advanced case study but not presented owing to space limitations.) Fig. 1 indicates that all FARFI classifiers achieve a very good over-all accuracy, especially with the *"window 6"* and *"window 11"* pretreatments. The k-NN classifiers perform best, followed by the LSSVM classifier with RBF kernel. The performance of the LSSVM classifier with linear kernel is slightly worse, indicating that the problem is linearly less well separable.

The classification accuracy *for the fault classes only* of the FARFI models is compared to that of the FI models to assess the impact of false alarm rejection on fault identification. Results (not shown) indicate that the difference in performance between both types of models is less than 0.5 % in most cases for both the basic and the advanced case study. It is concluded that the introduction of false alarm rejection does not negatively influence the fault identification. Only the LSSVM with linear kernel suffers from the introduction of false alarm rejection: performance decreases of more than 3 % and 1 % are observed for the basic and advanced case studies, respectively. This again indicates that the fault identification problem is not well separable linearly.

The accuracy of the false alarm rejection is presented in Fig. 2 for the basic case study. (Again, similar results are obtained for the advanced case study, but not shown owing to space limitations.) A first observation is that k-NN models achieve a very high false alarm rejection, but the variation on the 3-NN model is much lower than that of the 1-NN model. A second observation is that the false alarm rejection rates of both LSSVM models are much lower than the overall classification rate. Both effects are a direct result of the introduction of the false alarm class in the FARFI models. Any fault converges to NOC operation when its amplitude is small enough. This implies that the false alarm class, consisting of NOC samples, is located at the center of all actual fault classes. Hence, some overlap will exist between the false alarm class and faulty samples with relatively small amplitude. Furthermore, the false alarm class is typically more diffuse than the individual fault classes, causing further overlap and makeing the FARFI classification problem more difficult to separate, as indicated by the performance of the LSSVMs. Because k-NN models with larger values for k provide smoother decision boundaries, they are better suited for dealing with class overlap. Hence the better performance of 3-NN compared to 1-NN.

Finally, the fault-specific classification rates are investigated. The results for the 1-NN classifier in the basic case study listed in Table 3 are representative for the other classifiers as well as for the advanced case study. The evolution of the correct classification rates of the fault classes is in line with the results reported in [22, 38]. It is observed in all classifiers that false alarm identification is more accurate with the *"window 6"* and *"window 11"* pretreatments. The reason hereto is that, as the *"normalized"* and *"absolute"* pretreatments employ a single sample point, the entire input pattern $\mathbf{z}$ is affected by a false alarm. In contrast, a single false alarm point influences only part of the input patterns $\mathbf{z}$ for the *"window 6"* and *"window 11"* pretreatments.

These results demonstrate that the proposed methodology is capable of performing simultaneous false alarm rejection and fault identification efficiently. For the presented case studies, the 1-NN and 3-NN models reject more than 94 % of

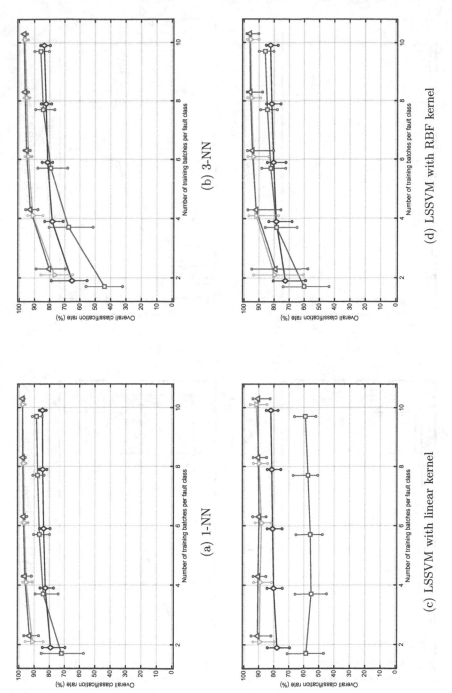

Fig. 1. Correct classification rates over all fault classes *and* false alarms for the tested FARFI models in the basic case study. Main markers indicate median overall classification accuracy; whiskers indicate the $(q_{0.05}-q_{0.95})$ quantile interval. A small horizontal scatter was introduced for visual clarity. (□ Normalized, ◇ Absolute, ▽ Window 6, △ Window 11)

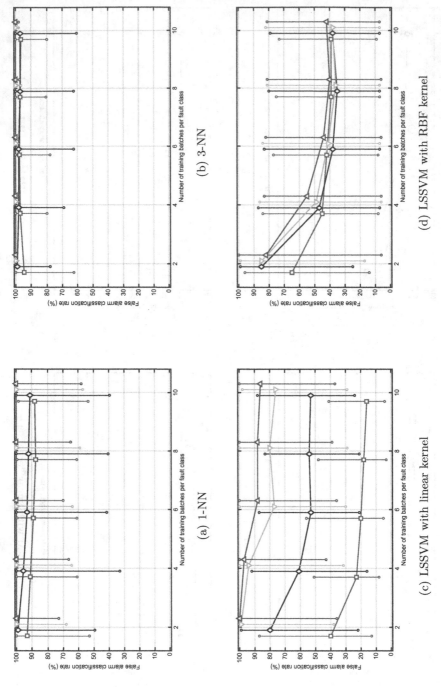

Fig. 2. Correct classification rates for the false alarms only for the tested FARFI models in the basic case study. Main markers indicate median overall classification accuracy; whiskers indicate the ($q_{0.05}$–$q_{0.95}$) quantile interval. A small horizontal scatter was introduced for visual clarity. ($\square$ Normalized, $\lozenge$ Absolute, $\triangledown$ Window 6, $\triangle$ Window 11)

Table 3. Correct classification rates for each fault class using a 1-NN, and a training set size of six faulty batches per fault class for the basic case study.

	Normalized		Absolute		Window 5		Window 10	
	m	$(q_{0.05} - q_{0.95})$	m	$(q_{0.05} - q_{0.95})$	m	$(q_{0.05} - q_{0.95})$	m	$(q_{0.05} - q_{0.95})$
No false alarm rejection, fault identification only (FI)								
Agitator power drop	100.0%	(93.6% – 100.0%)	100.0%	(93.6% – 100.0%)	100.0%	(93.1% – 100.0%)	100.0%	(92.6% – 100.0%)
Aeration rate drop	62.9%	(39.8% – 91.9%)	54.3%	(34.4% – 68.8%)	100.0%	(89.2% – 100.0%)	100.0%	(95.7% - 100.0%)
DO sensor drift	77.3%	(61.9% – 91.5%)	60.2%	(35.2% – 83.0%)	96.6%	(92.6% – 100.0%)	96.6%	(96.6% – 100.0%)
Feed rate drift	100.0%	(50.0% – 100.0%)	100.0%	(98.9% – 100.0%)	100.0%	(100.0% – 100.0%)	100.0%	(100.0% – 100.0%)
Feed conc. step	96.0%	(92.5% – 99.0%)	96.0%	(96.0% – 100.0%)	96.0%	(96.0% – 100.0%)	97.0%	(96.0% – 100.0%)
Coolant temp. step	92.0%	(69.0% – 100.0%)	93.0%	(75.0% – 100.0%)	91.0%	(81.0% – 100.0%)	92.0%	(79.5% – 99.5%)
Combined false alarm rejection and fault identification (FARFI)								
Agitator power drop	100.0%	(92.0% – 100.0%)	100.0%	(92.0% – 100.0%)	100.0%	(91.5% – 100.0%)	100.0%	(91.5% – 100.0%)
Aeration rate drop	69.4%	(41.9% – 95.7%)	54.8%	(31.7% – 68.3%)	97.8%	(88.2% – 100.0%)	98.9%	(93.0% – 100.0%)
DO sensor drift	70.5%	(53.4% – 83.0%)	58.0%	(41.5% – 80.1%)	96.6%	(92.0% – 100.0%)	96.6%	(96.6% – 100.0%)
Feed rate drift	100.0%	(66.0% – 100.0%)	100.0%	(98.9% – 100.0%)	100.0%	(98.9% – 100.0%)	100.0%	(100.0% – 100.0%)
Feed conc. step	96.0%	(95.0% – 97.0%)	96.0%	(96.0% – 98.0%)	96.0%	(96.0% – 97.0%)	96.0%	(96.0% – 97.0%)
Coolant temp. step	89.0%	(71.0% – 97.0%)	93.0%	(75.0% – 99.0%)	91.0%	(81.5% – 98.5%)	92.0%	(81.0% – 97.0%)
False alarm	89.0%	(61.0% – 99.5%)	93.0%	(41.5% – 100.0%)	100.0%	(64.0% – 100.0%)	100.0%	(70.0% – 100.0%)

the false alarms while achieving the same fault identification rates as the basic FI model of more than 95 %. The FARFI LSSVM model with RBF kernel also exhibits fault identification performance similar to its FI counterpart, but it does not reject as many false alarms as the k-NN models. Nevertheless, it still rejects 44 % of the false alarms in the basic study and 80 % in the advanced study. The results of the LSSVM with linear kernel indicate that care should be taken in selecting a model type for FARFI that is capable with dealing with the potential overlap between the false alarm class and the different fault classes.

6 Conclusions and Future Work

In Statistical Process Monitoring (SPM), the occurrence of false alarms should be reduced as much as possible, especially in batch processes. This paper therefore presented a methodology for combined False Alarm Rejection and Fault Identification (FARFI). The presented methodology achieves FARFI by including points of normal operation as an additional class in a fault classification model. By combining false alarm rejection with fault identification in a single step, the overall complexity of the SPM system is not needlessly increased. This approach filters out false alarms rather than trying to prevent their generation. This work focuses on batch process monitoring for FARFI because the impact of false alarms is potentially much larger in batch processes. The proposed FARFI methodology is also applicable to continuous processes, however.

Two case studies demonstrated the large potential of the proposed FARFI approach, with the best models rejecting more than 94 % of the false alarms and correctly identifying more than 95 % of the process faults. While only k-NN and LSSVM classifiers were tested, other types of classifiers can equally well be used for simultaneous FARFI. However, because inclusion of a normal operation class in the classification model can negatively impact fault identification in those situations where the normal operation class partially overlaps with small-amplitude fault samples, care should be taken to select types of classification models that are capable of dealing with this added complexity.

Future research should therefore focus on either of two possible approaches. A first option is to further evaluate the proposed FARFI methodology on other process faults and benchmark processes, with different classification models. This avenue of investigation will yield more insight on the effect of false alarm rejection, and help establish decision guidelines for selecting appropriate FARFI models.

Another option to reduce the impact of false alarm rejection on fault identification could be to assign a higher weight on correct classification of the faults during training of the classifier. The effect would be a FARFI model with a performance similar to that of the LSSVM model with RBF kernel tested in this work: very good fault identification accuracy (> 95 %) combined with lower false alarm rejection rates. As these FARFI models still reject a significant amount of false alarms, they nevertheless improve the overall performance of SPM.

Acknowledgements. Work supported in part by Project PFV/10/002 (OPTEC Optimization in Engineering Center) of the Research Council of the KU Leuven and Project IAP VII/19 (DYSCO Dynamical Systems, Control and Optimization) of the Belgian Program on Interuniversity Poles of Attraction initiated by the Belgian Federal Science Policy Office. The authors assume scientific responsibility.

References

1. Venkatasubramanian, V., Rengaswamy, R., Yin, K., Kavuri, S.: A review of process fault detection and diagnosis–part I: quantitative model-based methods. Comput. Chem. Eng. **27**, 293–311 (2003)
2. Ge, Z., Song, Z., Gao, F.: Review of recent research on data-based process monitoring. Ind. Eng. Chem. Res. **52**, 3543–3562 (2013)
3. Qin, S.: Survey on data-driven industrial process monitoring and diagnosis. Ann. Rev. Control **26**, 220–234 (2012)
4. Van Impe, J.F.M., Gins, G.: An extensive reference dataset for fault detection and identification in batch processes. Chemometr. Intell. Lab. Syst. **148**, 20–31 (2015)
5. Ding, S.: Data-driven design of monitoring and diagnosis systems for dynamic processes: a review of subspace technique based schemes and some recent results. J. Process Control **24**, 431–449 (2014)
6. Yoon, S., MacGregor, J.: Statistical and causal model-based approaches to fault detection and isolation. AIChE J. **46**, 1813–1824 (2000)
7. Aldrich, C., Auret, L.: Unsupervised Process Monitoring and Fault Diagnosis with Machine Learning Methods. Springer, London (2013)
8. Hwang, I., Kim, S.: A survey of fault detection, isolation, and reconfiguration methods. IEEE Trans. Control Syst. Technol. **18**, 636–653 (2010)
9. Kourti, T.: Application of latent variable methods to process control and multivariate statistical process control in industry. Int. J. Adapt. Control Signal Process. **19**, 213–246 (2005)
10. Kourti, T.: The process analytical technology initiative and multivariate process analysis, monitoring and control. Anal. Bioanal. Chem. **384**, 1043–1048 (2006)
11. MacGregor, J., Cinar, A.: Monitoring, fault diagnosis, fault-tolerant control and optimization: data driven methods. Comput. Chem. Eng. **47**, 111–120 (2012)
12. MacGregor, J., Kourti, T.: Statistical process control of multivariate processes. Control Eng. Pract. **3**(3), 403–414 (1995)
13. Bishop, C.: Neural Networks Pattern Recogn. Clarendon Press, Oxford (1995)
14. Hsieh, W.W., Cannon, A.J.: Towards robust nonlinear multivariate analysis by neural network methods. In: Donner, R.V., Barbosa, S.M. (eds.) Nonlinear Time Series Analysis in the Geosciences. Lecture Notes in Earth Sciences, pp. 97–124. Springer, Heidelberg (2008)
15. Van Impe, J.F.M., Augustinus, K., De Prins, M., Gins, G.: Advance warning for loss of separation in an industrial distillation column (2016). Submitted
16. Jolliffe, I.: Principal Component Analysis. Springer, New York (1986)
17. Chen, J., Liu, K.: On-line batch process monitoring using dynamic PCA and dynamic PLS models. Chem. Eng. Sci. **57**, 63–75 (2002)
18. Choi, S., Morris, J., Lee, I.B.: Dynamic model-based batch process monitoring. Chem. Eng. Sci. **63**, 622–636 (2008)
19. Lee, J., Yoo, C., Choi, S., Vanrolleghem, P., Lee, I.B.: Nonlinear process monitoring using kernel principal component analysis. Chem. Eng. Sci. **59**, 223–234 (2004)

20. Lee, J., Yoo, C., Lee, I.B.: Statistical monitoring of dynamic processes based on dynamic independent component analysis. Chem. Eng. Sci. **59**, 2995–3006 (2004)
21. Wang, L., Shi, H.: Multivariate statistical process monitoring using an improved independent component analysis. Chem. Eng. Res. Des. **88**, 403–414 (2010)
22. Gins, G., Van den Kerkhof, P., Vanlaer, J., Van Impe, J.F.M.: Improving classification-based iagnosis of batch processes through data selection and appropriate pretreatment. J. Process Control **26**, 90–101 (2015)
23. Nomikos, P., MacGregor, J.F.: Monitoring batch processes using multiway principal component analysis. AIChE J. **40**, 1361–1375 (1994)
24. Wold, S., Kettaneh, N., Friden, H., Holmberg, A.: Modelling and diagnosis of batch processes and analogous kinetic experiments. Chemometr. Intell. Lab. Syst. **44**, 331–340 (1998)
25. Li, B., Morris, J., Martin, E.: Model selection for partial least squares regression. Chemometr. Intell. Lab. Syst. **64**, 79–89 (2002)
26. Alcala, C., Qin, S.J.: Analysis and generalization of fault diagnosis methods for process monitoring. J. Process Control **21**, 322–330 (2011)
27. Bodesheim, P., Freytag, A., Rodner, E., Denzler, J.: Local novelty detection in multi-class recognition problems. In: Proceedings of the 2015 WACV Conference, pp. 813–820 (2015)
28. King, R., Feng, C., Sutherland, A.: StatLog: a comparison of classification algorithms on large real-world problems. Appl. Artif. Intell. **9**(3), 289–333 (1995)
29. Wu, X., Kumar, V., Quinlan, J., Ghosh, J., Yang, Q., Motoda, H., McLachlan, G., Ng, A., Liu, B., Yu, P., Zhou, Z., Steinbach, M., Hand, D., Steinberg, D.: Top 10 algorithms in data mining. Knowl. Inf. Syst. **14**(1), 1–37 (2008)
30. Abe, S.: Support Vector Machines for Pattern Classification. Springer, Heidelberg (2005)
31. Fix, E., Hodges, J.: Discriminatory analysis, nonparametric discrimination: consistency properties USAF school of aviation medicine, Randolph Field (TX, USA), Technical Report 4 (1951)
32. Suykens, J., Van Gestel, T., De Brabanter, J.: Least Squares Support Vector Machines. World Scientific, Singapore (2002)
33. Aizerman, M.A., Braverman, E.A., Rozonoer, L.: Theoretical foundations of the potential function method in pattern recognition learning. Autom. Remote Control **25**, 821–837 (1964)
34. Birol, G., Undey, C., Cinar, A.: A modular simulation package for fed-batch fermentation: penicillin production. Comput. Chem. Eng. **26**, 1553–1565 (2002)
35. Gins, G., Vanlaer, J., Van den Kerkhof, P., Van Impe, J.F.M.: The RAYMOND simulation package - generating RAYpresentative MONitoring Data to design advanced process monitoring and control algorithms. Comput. Chem. Eng. **69**, 108–118 (2014)
36. Van den Kerkhof, P., Vanlaer, J., Gins, G., Van Impe, J.F.M.: Analysis of smearing-out in contribution plot based fault isolation for statistical process control. Chem. Eng. Sci. **104**, 285–293 (2013)
37. Westerhuis, J.A., Gurden, S.P., Smilde, A.K.: Generalized contribution plots in multivariate statistical process monitoring. Chemometr. Intell. Lab. Syst. **51**, 95–114 (2000)
38. Wuyts, S., Gins, G., Van den Kerkhof, P., Van Impe, J.M.F.: Fault identification in batch processes using process data or contribution plots: a comparative study. Adv. Control Chem. Process. **8**, 1283–1288 (2015)

Parallel and Distributed Data Mining in Cloud

Ivan Kholod[(✉)], Mikhail Kuprianov, and Ilya Petukhov

Saint Petersburg Electrotechnical University "LETI",
ul. Prof. Popova 5, Saint Petersburg, Russia
iiholod@mail.ru, mikhail.kupriyanov@gmail.com,
ioprst@gmail.com

Abstract. The paper describes the approach for a distributed execution of data mining algorithms and using this approach for building a Cloud for Data Mining. The suggested approach allows us to execute data mining algorithms in different parallel and distributed environments. Thus, the created Cloud for Data Mining can be used as an analytic service and a platform for research and debugging parallel and distributed data mining algorithms.

Keywords: Data mining · Big data · Parallel algorithms · Cloud computing · Cloud for Data Mining

1 Introduction

We can observe rapid growth of data volume at present. Data are collected in information systems, generated by different devices, saved as logs of computer system's work, etc. Modern data warehouses provide storage of large amounts of different data. The terms "Big data", "Internet of Things", and "Cloud computing" have recently become very popular. They mean technologies for collecting, storing, and handling the large volumes of data, with a variety of types and a high velocity of generation (Big data). However, all this is worthless if we do not analyze and do not obtain new knowledge from the data.

The technologies like Machine Learning, Data Mining, and Knowledge Discovery are used for discovering new knowledge in data. They use complex mathematical methods and algorithms that need powerful computing resources for analyzing Big data. Cloud and cluster technologies provide unlimited (scalable) resources. Integration of data mining and cloud computing technologies is very important. The result of this integration is a creation of Cloud for Data Mining (CDM). This solution has a number of advantages:

- users always have the latest version of the algorithm;
- algorithms can use all the computational resources available in the "cloud";
- algorithms can be applied to the data stored in the "cloud" and outside of it;
- users user can forget about scaling algorithms.

© Springer International Publishing Switzerland 2016
P. Perner (Ed.): ICDM 2016, LNAI 9728, pp. 349–362, 2016.
DOI: 10.1007/978-3-319-41561-1_26

The CDM should have the following features for a comfortable work of an analyst:

- usability of a multiuser interface;
- execution of a full cycle of analysis;
- access to inner and outer data sources;
- work with different data sources;
- using of all computing resources;
- a wide range of data mining algorithms and others.

Additionally, the CDM must provide the following capabilities for data mining researchers and developers:

- unified API;
- ability to add new data mining algorithms;
- a wide range of parallel and distributed environments.

The paper describes an approach and architecture of the CDM that has these capabilities. The paper is organized as follows. The next section is a review of similar the CDM systems. The Sect. 3 contains the description of a general approach that allows mapping of the algorithm decomposed into blocks on different distributed systems. The Sect. 4 describes the CDM architecture. The last section discusses experiments to compare the developed cloud with the similar solutions.

2   Related Work

Series of data mining and machine learning frameworks have been developed nowadays to provide data analysis services for Big data. They can be used for creation of the CDM. The most famous of them are:

- **Apache Spark Machine Learning Library (MLlib).** [1] is a scalable machine learning library for Apache Spark platform. It consists of common learning algorithms and utilities, including classification, regression, clustering, collaborative filtering, dimensionality reduction, as well as lower-level optimization primitives and higher-level pipeline APIs. It has own implementation of Map reduce paradigm, which uses memory for data storage (versus Apache Hadoop that uses disk storage). It allows us to increase efficiency of algorithm's performance. User can extend a set of machine learning algorithms by his own implementations. However, user should decompose algorithm on map, reduce and other Spark's specific functions. It greatly limits the abilities for parallelization of data mining algorithms.
- **Apache Mahout.** [2] is also a data mining library for Map Reduce paradigm. It can be executed based on Apache Hadoop or Spark based platforms. It contains few data mining algorithms for execution in the distributed environment: collaborative filtering, classification, clustering and dimensionality reduction. User can extend the library by adding new data mining algorithms. The core libraries are highly optimized to allow good performance for non-distributed algorithms as well.
- **Weka4WS.** [3] is an extension of the famous open-source data mining library – Weka (The Waikato Environment for Knowledge Analysis) [4]. The extension implements the framework for support of data mining algorithms execution in

WSRF [5] enabled Grids. Weka4WS allows the execution of all its data mining algorithms on remote Grid nodes. To enable remote invocation, the data mining algorithms provided by the Weka library are exposed as a Web Service, which can be easily deployed on the available Grid nodes. Weka4WS can only handle a dataset contained by a single storage node. This dataset is then transferred to computing nodes to be mined.

The main disadvantages of the described systems are:

- inability to complete the entire cycle of analysis with their help without the need for complex configuration and refinement;
- use of a single platform for distributed analysis (basically MapReduce paradigm and in particular the Apache Hadoop).

The MapReduce paradigm is adapted only for data processing functions, which have the property of a list homomorphisms [6]. Therefore, not all data mining algorithms can be decomposed into *map* and *reduce* functions.

We suggest an architecture of CDM that allows us to execute data mining algorithms on different distributed services.

There are few ready solutions which is the CDM. Last of them is **Azure Machine Learning (Azure ML).** [7] a Software as a Service (SaaS) cloud-based predictive analytics service from Microsoft Inc. It was launched in February, 2015. Azure ML provides paid service which allows us to make a full cycle of data analysis:

- obtaining the data;
- preprocessing data;
- defining features;
- choosing and applying an algorithm;
- evaluating a model;
- publication of a model.

The analysis process is created as a workflow. Each step of the workflow is a module for solution of a single sub-task of analysis (a data reader, a data transformation, an algorithm and other). A module can be executed on a single cluster node. Thus only whole modules can be executed in parallel if the workflow allows it. It greatly limits ability for parallel and distributed execution of data mining algorithms.

Azure ML can import the data from: local files, online source and other projects (experiments) of the cloud. The reader module allows us to load data from outside sources in Internet or other file storages.

The user can apply use only available in Azure ML machine learning algorithms of: classification, regression, Anomaly Detection and clustering. Adding of the algorithms is available from the Machine Learning Marketplace. It contains the additional modules and services that can be integrated through publishing by Azure API.

Chinese Mobile Institute was one of the first who began working in this field. In 2007 it started research and development in the field of cloud computing. In 2009, it officially announced a platform for cloud computing BigCloud, which includes tools for parallel execution of algorithms **Data Mining Big Cloud-Parallel Data Mining (BC-PDM)** [8].

BC-PDM is a SaaS platform based on Apache Hadoop. Users can upload data to the repository (hosted in the cloud) from different sources and apply a variety of applications for data management, data analysis and business applications. The application includes an analysis of parallel applications to perform: ETL processing, social network analysis, analysis of texts (text mining), data analysis (data mining), and statistical analysis.

It includes about ten algorithms and it couldn't be extended by adding new algorithms. This cloud is only used for researching in Chinese Mobile Institute and not available for public use.

These CDM have limited abilities for parallel and distributed execution. The Table 1 contains some features of these systems.

Table 1. The solutions for building data analysis cloud services.

Capabilities	Apache Spark MLib	Apache Mahout	Weka 4WS	Azure ML	BC-PDM
Cloud service model	-	-	-	SaaS	SaaS
User Interface	-	-	Desktop	Web	Web
API Interface	Yes	Yes	Yes	REST	No
Scalable computing	Yes	Yes	No	For single modules	Yes
Data source location	Outside	Outside	Any	Inside and outside cloud	Inside cloud
Distributed computing platform	Apache Spark	Apache Hadoop	WSRF	-	Apache Hadoop
Data pre-processing	ETL	Yes	Yes	Yes	ETL
Analysis cycle	No full	No full	No full	Full	Full
Included data mining algorithms	classification, regression, collaborative filtering, clustering dimensionality reduction, feature extraction	collaborative filtering, classification, clustering, dimensionality reduction	classification, association, regression, clustering	classification, anomaly detection, regression, clustering	classification, clustering, association
Adding new algorithms	Yes	Yes	Yes	From Machine Learning Marketplace	No
Using	Open Source	Open Source	Open Source	Paid	No

3 General Concept

We have used an approach [9] that presents a data mining algorithm as a sequence of functional blocks (based on functional language principles). Classic functions in the functional languages are pure functions. According to Church-Rosser theorem [10], reduction of functional expression of pure function can be fulfilled in any order, also concurrently.

A data mining algorithm can be presented as a sequence of the function calls:

$$\text{dma} = f_n^\circ f_{n-1}^\circ \ldots {}^\circ f_i^\circ \ldots {}^\circ f_1 = f_n(d, f_{n-1}(d, \ldots f_i(d, \ldots f_1(d, \text{ nil}) \ldots) \ldots)), \quad (1)$$

where f_i: is function (function block) of the type $D \to M \to M$, where

- D: is the input data set that is analyzed by the function f_i;
- M: is the mining model that is built by the function f_i.

According to Church-Rosser theorem reduction (execution) of such functional expression (algorithm) can be done concurrently.

One of the main advantages of building algorithms from the function blocks is a possibility of their concurrent execution. For this we need to transform the sequential expression (1) into the form in which the function blocks will be invoked as arguments. For this the high-order *map* function may be used. It allows us to apply some function to the elements of lists. The function can be executed in parallel for different elements. The map function returns the list of results. To reduce the list into a single result the high-order *fold* function can be used. Thus, the function for conversation of sequential express into parallel form has the following view:

$$\text{parallel} :: (D \to M \to M) \to D \to M \to \text{join} \to \text{distrF} \to \text{distrD} \to \text{distrM} \to \text{start} \to \text{get} \to M,$$
$$\text{parallel } m = \text{fold (join, } m, \text{ get, (map(start, distrF(f), distrD(d, m), distrM(d, m))))}$$
$$(2)$$

where

- f (1st argument of the parallel function): a function block is executed concurrently;
- *distrF, distrD* and *distrM* functions for dividing of the functional block f, the input data set d and the mining model m into the lists:

$$\text{distrF} :: (D \to M \to M) \to [(D \to M \to M)]$$
$$\text{distrD} :: D \to M \to [D]$$
$$\text{distrM} :: D \to M \to [M]$$

- *start*: function applying each functional block f from the list [F] to the elements of the lists [D] and [M] and returns a handler h for parallel execution of the parallel function block f:

$$\text{start} :: (D \to M \to M) \to D \to M \to H$$

- *get*: function extracting the mining model from the handler h:

$$get :: H \rightarrow M;$$

- *join:* the function joining the mining models from the list *[M]* and returning the merged mining model *M*:

$$join :: [M] \rightarrow M;$$

- *map*: function applying function start to elements of the lists [F], [D] and [M]:

$$map :: ((D \rightarrow M \rightarrow M) \rightarrow D \rightarrow M \rightarrow H) \rightarrow [(D \rightarrow M \rightarrow M)] \rightarrow [D] \rightarrow [M] \rightarrow [H]$$
$$map\ [h] = list((start(f[0],\ d[0],\ m[0])),\ \ldots,\ (start\ (f[k],\ d[k],\ m[k])))$$

- *fold*: function reducing the list of mining model into single result:

$$fold :: ([M] \rightarrow M) \rightarrow M \rightarrow (H \rightarrow M) \rightarrow [H] \rightarrow M$$
$$fold\ m = join(m,\ get(h[0]),\ get(h[1]),\ \ldots,\ get(h[k])).$$

The list of handlers passing between the map and fold functions is a part of some distributed execution environment. In the general case an execution environment can be presented as a set of handlers:

$$E = \{h_0,\ h_1,\ \ldots,\ h_j,\ \ldots,\ h_n\}, where$$

- h_0: handler to execute sequential function blocks of algorithm;
- h_j: handler to execute parallel function block in the distributed environment.

Each handler must implement the start and get functions for using them in the parallel function. Examples of these handlers are threads, actors, web services and other.

The approach was implemented as a data mining algorithm library – DXelopes [11]. It contains different algorithms and allows us to add new algorithms. The library has adapters for integration with different parallel and distributed environments. Data mining algorithms can be executed in different parallel and distributed environments. We use these features for creating the prototype of the CDM.

4   Architecture of Cloud for Data Mining

Architecture of the CDM is shown in Fig. 1. It can be divided into several levels:

- hardware;
- virtual distributed environment;
- analysis services.

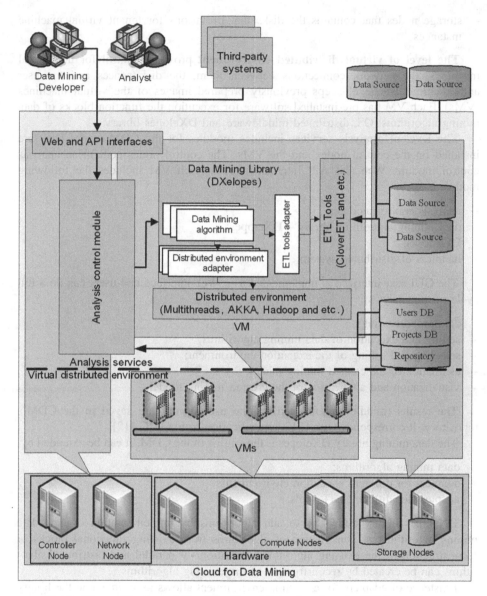

Fig. 1. The architecture of Cloud for Data Mining based on the DXelopes library

The level of hardware includes the pool of computes, storages and networking resources available in the cloud. We distinguish the following nodes:

- control nodes that runs services for managing compute nodes and virtual networks between them;
- compute nodes that runs the hypervisor portion of compute that operates tenant virtual machines or instances;

– storage nodes that contains the disks that provisions for tenant virtual machine instances.

The level of virtual distributed environment provides control for the virtual machines, the network connections between them, the disk spaces and the user authorization. The level keeps previously prepared images of the virtual machines (VMs). Each VM has pre-installed software for execution the function blocks of data mining algorithms: OS, distributed middleware and DXelopes library.

The level of analysis services includes modules for the CDM work. They are installed on the control nodes and the VMs. The control nodes contain an analysis control module, Web and API interfaces for it. Each VM includes the following modules:

– analysis control module;
– data mining algorithms library – DXelopes;
– ETL tools;
– handlers of distributed system.

The GUI user interface is implemented as a Web interface and users can do a full cycle of the analysis:

– data preprocessing;
– setting and execution of data mining algorithms;
– selecting and setting of the execution environment;
– estimation of the created mining model;
– visualization and application of the created mining model.

The results (middle and finished) of the user's work are saved in the CDM's repository. It corresponds with CWM specification from OMG [12].

The data mining library DXelopes is the engine of the CDM. It can be extended by:

– data mining algorithms;
– adapters for different execution environments;
– adapters for different ETL tools.

DXelopes library allows us to add algorithms by implementation of new data mining algorithms or construction of algorithms from existing functional blocks or reconstruction of the existing algorithms. Additionally parallel and distributed algorithms can be created by reconstruction of the existing algorithms.

Existence of adapters for execution environment allows us to integrate the library with the different distributed systems. Now the library is integrated with multithreading and actors model environments (the actors model environment is presented by AKKA framework [13]). Thus, all data mining algorithms implemented in the library can be executed in these environments.

For each environment the CDM contains sets of VMs with installed software. For example, VM of actor model environment has:

– analysis control module;
– DXelopes library;
– configured actor system from the AKKA framework.

The adapters for ETL tools allow us to integrate the library with the different systems which implement data extraction, data transformation and data loading. Currently the CloverETL tool is integrated with the library. It provides loading data into inner storages of the CDM and access to the remote data sources. CloverETL is an extensible software. A user can add new methods for data extraction and data transformation. It increases capabilities of the CDM for data processing.

ETL tools, including CloverETL, can be used to integrate the data from a number of heterogeneous sources. The data mining algorithms of the DXelopes library can process a data stream (continuously incoming data). Thus the integration of the DXelopes library and the ETL tools provides an analysis of Big data in the CDM.

5   Experiments

Series of the experiments were carried out to prove the effectiveness of the CDM implementation. We compared performance of the CDM with:

- Azure ML as a more popular cloud for data analysis;
- Spark MLlib as a more popular distributed platform for data analysis.

The DXelopes library contains a few data mining algorithms: classification, clustering and associations. However we selected clustering algorithm KMeans for experiments because:

- it is simple for understanding;
- it was implemented in all compared systems (DXelopes, MS Azure and Apache Spark Mllib).

We decomposed the KMeans algorithm into the set of functional blocks for distributed execution in the CDM. The KMeans algorithm is presented by the functional expression as:

$$KMeans = findClusters°initClusters, \text{ where}$$

- *initClusters* creates a set of centroids by a random way;
- *findClusters* finds centroids of clusters.

The *findClusters* block is the cycle which calls following function blocks while the cluster's centroids are changed:

- *distributeVectors* computes the distances between vectors (from the data set D) and centroids of the clusters to distribute the vectors between the clusters;
- *updateCentroids* updates centroids of clusters with new sets of vectors.

The last two function blocks can be executed in a distributed way by a few handlers. A handler with the *distributeVectors* function block can handle all the vectors on some nodes and then the handle with the *updateCentroids* function block will recalculate the centroids of the clusters in the central cloud. Thus the KMeans algorithm is mapped on execution environment in the following way:

$$KMeans \rightarrow E = \{(initClusters, h_0), (updateCentroids, h_0),$$
$$(distributeVectors, h_1), \ldots, (distributeVectors, h_k)\}$$

We used two execution environments for our experiments:
- multithreads where the handlers are threads;
- actors model where the handlers are actors.

These environments had been deployed in the CDM on the VMs which were working on high-performance servers supporting hardware virtualization and providing a possibility of performance of the cloud computing systems. For the experiments we used, the prototype of the CDM was installed on the following objects of computing cluster infrastructure:

- IBM FlexSystem × 240 ComputeNode has a model server with the following characteristics:
  - the processor - IntelXeon 2.9 GHz (2 CPU on 6 kernels, performance of calculations in 2 streams on a kernel, only 24 streams on the server);
  - the volume of random access memory - 128 GB;
  - server productivity - 200 GFlops;
  - operating systems: 3 win2012/hyper-v systems;
  - 5 rhel/kvm systems;
  - servers support virtualization and are used for servicing cloud computing.

- IBMFlexSystemp260 ComputeNode has a model server with the following characteristics:
  - the processor - Power7 3.3 GHz (2 CPU on 4 kernels, performance of calculations in 4 streams on a kernel, only 32 streams on the server);
  - the volume of random access memory - 128 GB;
  - server productivity - 400 GFlops;
  - operating systems: 2 AIX/PowerVM systems;
  - servers support virtualization and are used for service of cloud computing.
- two StorageSystemStorwizev700 on 13,6 TiB

This computing cluster infrastructure was also used for work of the Apache Spark. We launched Spark in a standalone deploy mode on 2 CPU on 6 kernels each.

The algorithms of DXelopes and Spark MLLib had executed in distributed environments with a different number of the handlers: 2, 4 and 8. It allowed to compare scalability of these libraries.

The algorithms of Microsoft Azure ML have been launched on cluster itself. For the purpose of a distributed execution we divided data sets on 2, 4 and 8 parts and analyzed them on different nodes of the Azure cluster.

We used the data sets from Azure ML for the experiments to process same data sets as in the Azure ML and as in the CDM. For this we download these data sets from the Azure ML into the CDM. Some parameters of the data sets are presented in Table 2.

Table 2. Experimental data sets

Input data set	Number of vectors	Number of attributes
Iris Two Class Data (ITCD)	100	4
Telescope data (TD)	19 020	10
Breast Cancer Info (BCI)	102 294	5
Movie Ratings (MR)	227 472	4
Flight on-time performance (Raw) (FOTP)	504 397	5
Flight Delays Data (FDD)	2 719 418	5

The experimental results are provided in Table 3. The results are average time (seconds) of the KMeans execution for various data sets and in different clouds. To calculate acceleration of the algorithm's work we measure execution time of sequence algorithm (execution with one handler):

Table 3. Experimental results (s)

Algorithm		ITCD	TD	BCI	MR	FOTP	FDD
DXelopes CDM (Multithreads)	t_1	0.03	1	4.6	14	19.4	132.5
	t_2	0.02	0.4	2.8	7.6	11.5	81.6
	t_4	0.02	0.4	1.7	4.5	6.9	43.4
	t_8	0.02	0.4	1.5	3.7	4.5	27.7
DXelopes CDM (Actors model)	t_1	0.04	1.2	4.3	9.2	12.3	121
	t_2	0.03	0.5	2.6	5.9	6.4	71
	t_4	0.04	0.4	1.5	3.1	3.5	37
	t_8	0.04	0.3	0.9	1.7	2.1	19
Azure ML	t_1	4	5	7	13	26	132
	t_2	4	5	7	11	21	99
	t_4	4	4	5	7	15	63
	t_8	3	3	4	6	11	42
Spark MLib	t_1	3	4.3	4	6.2	10.7	157
	t_2	2	3.6	3.7	5	7	96
	t_4	2.5	3.7	3.3	4.8	6	59
	t_8	3	3.8	3.3	4.6	5	37

$$a = t_1/t_k, \text{ where}$$

- t_1 – time of execution on one handler;
- t_k – time of execution on k handlers.

We compared the proposed approach with the most popular solutions in this area by the following parameters:

- execution time of work algorithms for different data sets (Table 3);
- acceleration of parallel algorithm in environment with maximum number of handlers (8) for different data sets (Fig. 2);

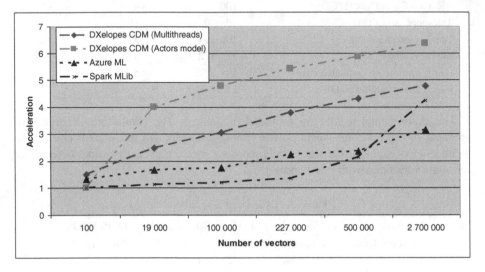

Fig. 2. The acceleration of parallel algorithm execution

- efficiency of parallel algorithm (e = a/k) for different systems and execution environments for the data set (Flight Delays Data) with maximum amount of vectors (Fig. 3)

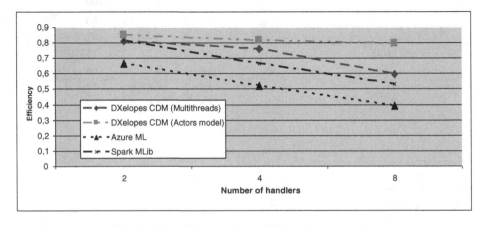

Fig. 3. The efficiency of parallel algorithm execution

The experiments allow us to make the following conclusions:

- The execution time of the algorithm in all environments is almost the same. The actor model environment is a little bit faster.
- Existing solutions have a low acceleration and efficiency for data sets with a few vectors (less 500 000) because they have high overhead on distribution of the data sets between the handlers. The DXelopes library does not divide data set. It restricts an access for different handlers to various parts of data sets.
- The acceleration and efficiency of parallel algorithm execution in the CDM are better then in the Azure ML and the Apache Spark because the DXelopes library allows to distribute the algorithm's blocks between the handlers more flexible. The Apache Spark is restricted by Map-Reduce paradigm and can parallel execute only the map and reduce functions. The Azure ML can only execute whole algorithms on single nodes.

6 Conclusion

The representation of a data mining algorithm as functional expression makes it possible to divide the algorithm into functional blocks. Such a splitting of data mining algorithms into blocks helps to map them to handlers of different distributed environment. It allows us to distribute the functional blocks between the handlers and reaches high acceleration and efficiency of parallel algorithms execution.

We implemented this approach as the DXelopes library. It contains different algorithms and allows us to add new algorithms. The library has adapters for integration with different parallel and distributed environments. It provides execution of the data mining algorithms in these environments.

We built prototype of the CDM based on the DXelopes library. Thus the created prototype has the following key characteristics, which distinguish it from other similar solutions:

- implement Software as a Service and Platform as a Service cloud computing service models;
- scalable with the help of different distributed platforms;
- can be extended by adding new data mining algorithms or modifying their functional blocks;
- provides data processing by different ETL tools.

These capabilities provide using of the CDM as an analytic service and platform for research and debugging parallel and distributed data mining algorithms.

The experiments have confirmed the efficiency of the CDM compared to the existing the most popularly similar solutions. In future we plan to extend the supported distributed platforms and ETL tools and develop release version of the CDM.

Acknowledgments. The work has been performed in Saint Petersburg Electrotechnical University "LETI" within the scope of the contract Board of Education of Russia and science of the Russian Federation under the contract № 02.G25.31.0058 from 12.02.2013. The paper has been prepared within the scope of the state project "Organization of scientific research" of the main part of the state plan of the Board of Education of Russia, the project part of the state plan of the Board of Education of Russia (task 2.136.2014/K) as well as supported by grant of RFBR (projects 16-07-00625).

References

1. Machine Learning Library (MLlib) Guide. http://spark.apache.org/docs/latest/mllib-guide.html. Accessed 05 Apr 2016
2. Grant, I.: Introducing Apache Mahout. http://www.ibm.com/developerworks/java/library/j-mahout/. Accessed 05 Apr 2016
3. Weka4WS. http://gridlab.dimes.unical.it/weka4ws/about/. Accessed 05 Apr 2016
4. Waikato Environment for Knowledge Analysis (Weka). www.cs.waikato.ac.nz/ml/weka/. Accessed 05 Apr 2016
5. The WS-Resource Framework. http://toolkit.globus.org/wsrf/. Accessed 05 Apr 2016
6. Gorlatch, S.: Extracting and implementing list homomorphisms in parallel program development. Sci. Comput. Program. **33**(1), 1–27 (1999)
7. Gronlund, C.J.: Introduction to machine learning on Microsoft Azure. https://azure.microsoft.com/en-gb/documentation/articles/machine-learning-what-is-machine-learning/. Accessed 05 Apr 2016
8. Yu, L., Zheng, J., Shen, W.C., Wu, B., Wang, B., Qian, L., Zhang, B.R.: BC-PDM: data mining, social network analysis and text mining system based on cloud computing. In: Proceedings of the 18th ACM SIGKDD International Conference on Knowledge Discovery and Data Mining, New York, pp. 1496–1499 (2012)
9. Kholod, I., Petukhov, I.: Creation of data mining algorithms as functional expression for parallel and distributed execution. In: Malyshkin, V. (ed.) PaCT 2015. LNCS, vol. 9251, pp. 62–67. Springer, Heidelberg (2015)
10. Alonzo, C., Rosser, B.J.: Some properties of conversion. Trans. AMS **39**, 472–482 (1936)
11. Kholod, I.: Framework for multi threads execution of data mining algorithms. In: Proceeding of 2015 IEEE North West Russia Section Young Researchers in Electrical and Electronic Engineering Conference (2015 ElConRusW), pp. 74–80. IEEE Xplore (2015)
12. Common Warehouse Metamodel (CWM) Specification. http://www.omg.org/spec/CWM/1.1/. Accessed 05 Apr 2016
13. Akka Documentation. http://akka.io/docs/. Accessed 05 Apr 2016

The WoMan Formalism
for Expressing Process Models

Stefano Ferilli[✉]

University of Bari, Bari, Italy
stefano.ferilli@uniba.it

Abstract. Workflow management is fundamental to efficiently, effectively and economically carry out complex processes. In turn, the formalism used for representing workflow models is crucial for effectiveness. The formalism introduced by the WoMan framework for workflow management, based on First-Order Logic, is more expressive than standard formalisms adopted in the literature, and ensures strict adherence to the observed practices. This paper discusses in some details such a formalism, highlighting its most outstanding strengths and comparing it to the current standard formalism (Petri nets), also providing techniques for the translation of workflow models among the two formalisms. The comparison between the two models shows that WoMan is more powerful than standard Petri Nets, and that it can handle naturally and straightforwardly cases that would require complex patterns in Petri Nets.

Keywords: Business process modeling · Process mining · Logic programming

1 Introduction

A *process* consists of a combination of different, inter-related tasks performed by agents [2] (humans or artifacts). A *task* is a generic piece of work to be executed. Boolean preconditions (that must be satisfied in the current state of the world to enable its execution) and postconditions (that must hold after its execution) may be associated to tasks. An *activity* is the actual execution of a task by an agent [17]. Activities spanning some significant period of time are represented by the interval between their start and end events [2]. A process model (or *workflow*) is a formal specification of how a set of tasks can be composed to result in valid processes. Allowed compositional schemes include sequential, concurrent, conditional, or iterative execution [17]. Especially relevant for determining the complexity of a model are [11]: *sequential* vs *concurrent* flow, and *unique* vs *repeated* tasks. Further complexity is introduced by the presence of synchronization among tasks and of invisible or duplicate tasks [4]. A *case* is a particular execution of activities according to a given workflow. Case *traces* are lists of events associated to *steps* (time points). Events of several traces may be collected and interleaved in *logs* [20].

© Springer International Publishing Switzerland 2016
P. Perner (Ed.): ICDM 2016, LNAI 9728, pp. 363–378, 2016.
DOI: 10.1007/978-3-319-41561-1_27

Contemporary society is pervaded by complex processes. A company's success may depend on the proper definition, handling and management of its processes. So, the availability of suitable workflow models is vital. This motivates the development of techniques and tools to model workflows and to supervise and analyze their enactment, but also of suitable formalisms for expressing the models in order to enforce a number of desirable features. Since producing the models is inherently complex, costly and error-prone [11], *Process mining* [21] aims at using a set of sample case traces to infer them automatically [2]. An overview of the current state-of-the-art can be found in [13,18]. Current "hot" topics include the need to consider contextual information [19] and the importance of efficient and declarative approaches [14]. Both are tackled by WoMAN ('WORKflow MANagement') [7,8], a novel framework for workflow learning and management based on First-Order Logic (FOL for short). Incrementality, expressiveness and efficiency are its most relevant features. This paper specifically deals with the formalism used in WoMan to express workflow models, and with its relationships to Petri nets.

After introducing related work in Sects. 2 and 3 reports the details of the representational framework. Then, Sect. 4 discusses some relevant differences with Petri Nets, the standard formalism used in the literature. Section 5 proposes some comments on the utility of the WoMan formalism, and finally Sect. 6 concludes the paper and outlines future work issues.

2   Related Work and Background

Generally speaking, a workflow can be modeled as a directed graph. Nodes are associated to states or activities. Edges represent the potential flow of control among activities; they can be labeled with probabilities and/or boolean conditions on the state of the process, which determine whether they will be traversed or not [1]. Several models have been proposed in the literature for representing processes. In Finite State Machines (FSMs) [3], nodes are associated to states, and edges represent activities. In Hidden Markov Models (HMMs) [9], states represent nodes, and activities correspond to output symbols. Both are unsuitable to model concurrency, which is a serious limitation. Other more specialized formalisms were also attempted, such as the one proposed in [10,12], that distinguishes several types of nodes (Begin, End, Activity, Decision, Split, Join) connected by edges. Activity nodes are associated to tasks and edges can be labeled with probabilities and/or conditions. Subsequent works have mainly focused on *Petri nets*, or on their restriction *WorkFlow nets* (WF-nets) [17], purposely developed to express the control flow in a process. E.g., [17,21] learn models in the form of WF-nets. The α-algorithm family [5,20,22] mines processes in the class of 'sound *Structured WF-nets*'. This class is very limited: it can handle parallelism between pairs of tasks only and does not permit synchronization between tasks. The approach in [6] can learn any Petri net which does not involve either duplicate tasks or more than one place with the same input and output tasks.

Petri/WF nets are the current standard formalism to represent process models. A Petri net is a directed bipartite graph $N = (P \cup T, A)$, where the set of nodes is partitioned into *places* (graphically denoted by circles) P and *transitions* (denoted by squares) T, while *arcs* (denoted by arrows) in A connect either a place to a transition or a transition to a place. Transitions represent tasks, whose causal interdependencies are expressed by arcs. A transition having many output places (called an *AND-split*) starts a concurrent execution. A transition with many input places (called an *AND-join*) reduces different concurrent executions to a single one. A place having many outgoing arcs (called an *OR-split*) determines a choice of which transition is to be executed next. A place with many incoming arcs (called an *OR-join*) indicates a state that can be reached in different alternative ways. Splits of a given kind must be closed by joins of the same kind. Places may contain *tokens* (denoted by dots), to indicate that they are active and may enable the execution of subsequent tasks: a transition is *enabled*, i.e. it may take place, if and only if each of its input places contains at least the number of tokens reported in the corresponding input arc (1 by default). The execution is non-deterministic: given an enabled transition, one cannot say when, or even if, it will take place. A transition that actually takes place (*fires*) represents an activity: it consumes the specified number of tokens from each of its input places and produces in each of its output places the number of tokens specified on the corresponding output arc (1 by default). A configuration of tokens through the net is called a *marking*. WF-nets are Petri nets that involve a single starting place (*source*) and a single terminating one (*sink*) and do not allow 'dangling' nodes. *Sound* WF-nets do not permit deadlocks or live locks (thus guaranteeing termination) or unconsumed tokens in the net after termination and require any task to be on a path from source to sink.

More related to our proposal, also FOL was proposed to describe process models. This setting is called *Declarative Process Mining* and is recognized as being very important when dealing with particularly complex models and domains [15]. Instead of completely specifying a process flow, it imposes only a set of constraints that must be satisfied when executing the process activities.

Additional desirable features for a process model are the capability of handling noise, and the possibility of expressing conditions that determine whether a given activity is to be carried out or not depending on the current status of the execution. The former is usually handled by removing all edges, whose number of occurrences in the training cases is below a given threshold [1,6,10]. The latter is discussed in [1,10,16], as the possibility of learning simple Boolean conditions in the form of decision-tree classifiers.

3 The WoMan Formalism

WoMan uses the Logic Programming formalism (a fragment of FOL). It is based on *Horn clauses*, i.e. implications represented in Prolog style as $l_0 :- l_1, \ldots, l_n$. where l_0 (called the *head*) is the conclusion and $l_1, \ldots, l_n$ (called the *body*) is a conjunction of pre-conditions. Each l_i is an *atom*, i.e. a predicate applied to

```
entry(201509280700, begin_process, motorcycle, 4, start, 1)
entry(201509280700, begin_activity, motorcycle, 4, build_engine, 1)
entry(201509280705, begin_activity, motorcycle, 4, build_frame, 1)
entry(201509280745, end_activity, motorcycle, 4, build_frame, 1)
entry(201509280747, begin_activity, motorcycle, 4, paint_frame, 1)
entry(201509280752, end_activity, motorcycle, 4, paint_frame, 1)
entry(201509280820, end_activity, motorcycle, 4, build_engine, 1)
entry(201509280822, begin_activity, motorcycle, 4, test_engine, 1)
entry(201509280828, end_activity, motorcycle, 4, test_engine, 1)
entry(201509280830, begin_activity, motorcycle, 4, install_engine, 1)
entry(201509280831, begin_activity, motorcycle, 4, install_wheel, 1)
entry(201509280832, begin_activity, motorcycle, 4, install_wheel, 2)
entry(201509280833, end_activity, motorcycle, 4, install_wheel, 1)
entry(201509280834, end_activity, motorcycle, 4, install_wheel, 2)
entry(201509280835, begin_activity, motorcycle, 4, install_seat, 1)
entry(201509280850, end_activity, motorcycle, 4, install_engine, 1)
entry(201509280851, begin_activity, motorcycle, 4, put_fuel, 1)
entry(201509280853, end_activity, motorcycle, 4, install_seat, 1)
entry(201509280855, end_activity, motorcycle, 4, put_fuel, 1)
entry(201509280858, begin_activity, motorcycle, 4, test_engine, 2)
entry(201509280905, end_activity, motorcycle, 4, test_engine, 2)
entry(201509280908, begin_activity, motorcycle, 4, install_radio, 1)
entry(201509280915, end_activity, motorcycle, 4, install_radio, 1)
entry(201509280917, begin_activity, motorcycle, 4, install_plate, 1)
entry(201509280920, end_activity, motorcycle, 4, install_plate, 1)
entry(201509280920, end_process, motorcycle, 4, stop, 1)
```

Fig. 1. Event-based representation of a case

terms as arguments. WoMan works in Datalog, which allows only constants or variables as terms. Clauses having only the head are called *facts*, and represented as just l_0.. Clauses having both the head and the body are called *rules*.

According to foundational literature [1,11] trace elements are 6-tuples, that report information about relevant events for the case they refer to (specifically, task start and end events are needed to properly handle time span and parallelism of tasks [20]). They are represented in WoMan as facts `entry` (T,E,W,P,A,O)., where T is the event timestamp, E is the type of the event (one of **begin_process**, **begin_activity**, **end_activity**, **end_process**), W is the name of the workflow the process refers to, P is a unique identifier for each process execution, A is the name of the activity, and O is the progressive number of occurrence of that activity in that process. An optional field, R, can be added to specify the agent that carries out activity A. Finally, to describe also the context in which the activities take place, WoMan exploits a further kind of event, **context_description**. When $E =$ **context_description**, A is a FOL description of the context at time T, consisting of a set of atoms built on domain-specific predicates.

Figure 1 reports an excerpt of a hypothetical case #4 of a process aimed at manufacturing a motorcycle (events reporting contextual descriptions have been removed). Concurrency is evident in activities that begin when previous activities have not ended yet.

WoMan models are expressed as sets of atoms built on four predicates:

- task(t,C): task t occurred in training cases C.
- transition(I,O,p,C): transition[1] p, occurred in training cases C, is enabled if all input tasks in $I = [t'_1, \ldots, t'_n]$ are active; if fired, after stopping the execution of all tasks in I (in any order), the execution of all output tasks in $O = [t''_1, \ldots, t''_m]$ is started (again, in any order). Transitions represent the allowed connections between activities. If several instances of a task can be active at the same time, I and O are multisets, and application of a transition consists in closing as many instances of active tasks as specified in I and in opening as many activations of new tasks as specified in O.
- task_agent(t,A): task t can be carried out by an agent matching the roles in A.
- transition_agent$([a'_1, \ldots, a'_n], [a''_1, \ldots, a''_m], p, C, q)$: transition p, involving input tasks $I = [t'_1, \ldots, t'_n]$ and output tasks $O = [t''_1, \ldots, t''_m]$, may occur provided that each task $t'_i \in I, i = 1, \ldots, n$ is carried out by an agent matching role a'_i, and that each task $t''_j \in O, j = 1, \ldots, m$ is carried out by an agent matching role a''_j; several combinations can be allowed, numbered by progressive q, each encountered in cases C.

Argument C in these predicates is a multiset because a task or transition may occur several times in the same case. When supervising process executions, it is useful in at least 3 ways:

1. It allows to check that the whole flow of activities that are taking place was encountered in at least one training/sample case. In this way, it is possible to avoid recognizing as valid a new execution that mixes transitions taken from different sample cases.
2. It allows to set limits on the number of repetitions of loops. Indeed, when loops are enacted, one may check that the new execution does not repeat the involved tasks more times than seen in sample/training cases.
3. It allows to compute statistics. Given a model involving overall n sample/training cases, if a task or transition t is associated to cases C_t, then its probability may be approximated by its relative frequency $|C_t|/n$. This allows to handle noise: all tasks or transitions whose frequency does not pass a given threshold may be considered as noise, and be ignored in the model.

Figure 2 shows the possible flows of activities for a hypothetical 'motorcycle' process. It involves many complex features for most process mining systems: short loops ('test_engine'/'fix_engine'), duplicated tasks

---

[1] Note that this is a different meaning than in Petri Nets. In the following, we will distinguish the latter by writing Petri Net transitions.

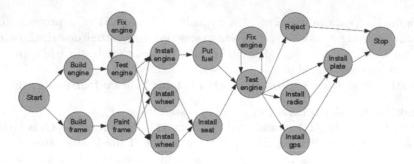

Fig. 2. Activity flow for a motorcycle production process

('test_engine', 'fix_engine', and 'install_wheel'), duplicated sub-processes (the loop 'test_engine'/'fix_engine'), concurrent activities (e.g., 'install_engine' and two concurrent occurrences of 'install_wheel'), concurrent sequences of activities (e.g., the sequence 'build_engine' – 'test_engine'/'fix_engine' with the sequence 'build_frame' – 'paint_frame'). Albeit not apparent in Fig. 2, there are optional tasks, in that 'install_radio' and 'install_gps' may or may not be carried out between 'test_engine' and 'install_plate'. More specifically, let us suppose that in the intended model 'install_gps' may be carried out only in combination with 'install_radio'.

The activity flow component of this model can be easily represented in WoMan formalism as reported in Fig. 3. In Fig. 3 it is associated to 5 sample cases (corresponding to training cases if the model was learned from examples), including case #4 in Fig. 1. Some tasks are carried out in all cases ('build_engine', 'build_frame', 'test_engine', 'paint_frame', 'install_engine', 'install_wheel', 'put_fuel', and 'install_seat'). All the others may or may not take place depending on the specific process enactment. Some tasks and transitions are carried out more than once in the same case (e.g., 'install_wheel' is always carried out twice; 'fix_engine' is carried out 0, 1 or 2 times depending on the case). Transitions $p10$, $p13$, $p15$ and $p16$ start alternative routes to the process termination: $p13$ rejects the motorcycle; $p10$, $p15$ and $p16$ terminate by installing the plate. Compared to transition $p15$, that directly installs the plate, optional tasks are involved in transitions $p16 - p17$, where also the radio is installed, and $p10 - p11$, where both the radio and the gps are installed. It is possible to note that task fix_engine occurs in only 2 cases (#3, #5) out of 5, and thus has frequency 0.40. Also, it was carried out at most twice in the same case (specifically, in case #3), which can be used as an upper limit to the number of accepted executions of this task in future process enactments. Task 'reject', and associated transitions $p13$ and $p14$, occur in only 1 case out of 5, yielding a probability of 0.20. So, setting a noise threshold of 0.25, they would be ignored in future process enactments.

```
task(stop,[1,2,3,4,5]).
task(reject,[2]).
task(install_plate,[1,2,3,4,5]).
task(install_gps,[1,5]).
task(install_radio,[1,4,5]).
task(put_fuel,[1,2,3,4,5]).
task(install_seat,[1,2,3,5]).
task(install_wheel,[1,1,2,2,3,3,4,4,5,5]).
task(install_engine,[1,2,3,4,5]).
task(paint_frame,[1,1,2,2,2,3,4,4,5,5]).
task(fix_engine,[2,3,3,5]).
task(test_engine,[1,1,2,2,2,3,3,3,3,4,4,5,5,5]).
task(build_frame,[1,2,3,4,5]).
task(build_engine,[1,2,3,4,5]).
task(start,[1,2,3,4,5]).
transition([start],[build_engine,build_frame],p1,[1,2,3,4,5]).
transition([build_engine],[test_engine],p2,[1,2,3,4,5]).
transition([test_engine],[fix_engine],p3,[3,3,5]).
transition([fix_engine],[test_engine],p4,[3,3,5]).
transition([build_frame],[paint_frame],p5,[1,2,3,4,5]).
transition([test_engine,paint_frame],[install_engine,install_wheel,install_wheel],
                                            p6,[1,2,3,4,5]).
transition([install_wheel,install_wheel],[install_seat],p7,[1,2,3,4,5]).
transition([install_engine],[put_fuel],p8,[1,2,3,4,5]).
transition([install_seat,put_fuel],[test_engine],p9,[1,2,3,4,5]).
transition([test_engine],[install_radio,install_gps],p10,[1,5]).
transition([install_radio,install_gps],[install_plate],p11,[1,5]).
transition([install_plate],[stop],p12,[1,3,4,5]).
transition([test_engine],[reject],p13,[2]).
transition([reject],[stop],p14,[2]).
transition([test_engine],[install_plate],p15,[3]).
transition([test_engine],[install_radio],p16,[4]).
transition([install_radio],[install_plate],p17,[4]).
```

Fig. 3. Activity flow part of a workflow model in WoMan formalism

For the sake of brevity, the part concerning agents will not be reported in its entirety. Only two sample fragments are shown, just to give an idea:

task_agent(test_engine,[mechanic,driver]).

saying that activity 'test_engine' can be carried out by a mechanic or a driver;

transition_agent([mechanic],[electrician,electrician],p16,[1],1).
transition_agent([driver],[electrician,electrician],p16,[5],2).

saying that, in transition p16, the input activity 'test_engine' must be carried out by a mechanic in cases that are compliant to sample case #1, or by a driver in cases that are compliant to sample case #5. In all cases, the other two activities, 'install_radio' and 'install_gps', must be both carried out by an electrician.

Finally, WoMan expresses conditions on task/transition execution as FOL rules. The body is built using suitable domain-dependent predicates that allow to describe the context in which the various activities take place (i.e., the objects involved, their properties and the relationships that come into play among objects, among steps, and between objects and steps). Moreover, two reserved predicates are used to describe the flow of activities in a case:

- activity(s,t): at step s task t is executed;
- after(s',s'',[n',n'']): step s'' follows step s' after a number of steps ranging between n' and n''.

where each step is denoted by a unique identifier. Due to concurrency, predicate after/3 induces a partial ordering on the set of steps. WoMan allows to specify both pre- and post-conditions on three kinds of items: tasks, transitions, and tasks within transitions. The former define what must be true for a given task in general. The latter allow to define further constraints for allowing a task to be run in the context of a specific transition (provided that its general conditions are met). Conditions on transitions define when a transition may take place.

Let us show some examples of conditions. E.g., the task precondition:

test_engine(X) :-
    after(Y,X,[1,8]), activity(Y,build_engine), available(X,T), testing_tool(T), status(X,T,S), working(S).

says that, in order to test the engine at step X, the engine must have been built at step Y, which happened between 1 and 8 steps before X. Indeed it is one step for the leftmost occurrence in Fig. 2, while for the rightmost occurrence it is 4 steps (if activity fix_engine was never carried out), or 6 (if fix_engine was carried out just once), or 8 (if it was carried out twice, which is the upper limit for each single case). Moreover, at the time the engine is tested the testing tool T must be available, and its status S must be 'working'. The task post-conditions:

test_engine(X) :-
    after(Y,X,[1,1]), activity(Y,A), build_engine(A), available(X,T), testing_tool(T), status(X,T,S), working(S), test_outcome(X,O), fail(O).
test_engine(X) :-
    after(Y,X,[1,1]), activity(Y,A), build_engine(A), available(X,T), testing_tool(T), status(X,T,S), working(S), test_outcome(X,O), success(O).

say that, after testing the engine, the outcome must be fail or success.

As regards transition preconditions one might have:

p13(X) :-
    after(Y,X,[1,1]), activity(Y,$A1$), test_engine($A1$), test_outcome(Y,$O1$), fail($O1$), after(Z,Y,[3,7]), activity(Z,$A2$), test_engine($A2$), test_outcome(Z,$O2$), fail($O2$), after(Z,W,[1,1]), activity(W,$A3$), fix_engine($A3$).
p15(X) :-
    after(Y,X,[1,1]), activity(Y,A), test_engine(A), test_outcome(Y,O), success(O), after(Z,Y,[6,10]), activity(Z,S), start(S), required_option(S,R), none(R).

p16(X) :-
    after(Y,X,[1,1]),   activity(Y,A),   test_engine(A),   test_outcome(Y,O),
    success(O), after(Z,Y,[6,10]), activity(Z,S), start(S), required_option(S,R),
    radio(R).

p10(X) :-
    after(Y,X,[1,1]),   activity(Y,A),   test_engine(A),   test_outcome(Y,O),
    success(O), after(Z,Y,[6,10]), activity(Z,S), start(S), required_option(S,R),
    full(R).

The first rule means that the 'reject' transition $p13$ can be applied at step X if the rightmost engine test in Fig. 2 fails when the leftmost engine test already failed and the engine was fixed. The other rules mean that, if the rightmost engine test in Fig. 2 succeeds, transitions $p15$, $p16$ and $p10$ can be applied at step X if at the beginning of the process the user required no optional, just the radio, of full optionals, respectively. So, these rules allow to determine, in each case, which path should be followed in the activity flow model.

Finally, as regards the tasks in the context of specific transitions, consider task 'test_engine' in transition $p9$. The precondition might be:

test_engine_p9(X) :-
    after(Y,X,[4,8]),   activity(Y,A),   build_engine(A),   available(X,T),   test-
    ing_tool(T), status(X,T,S), working(S), fuel_level(X,L), sufficient(L).

i.e., to carry out the engine test in transition $p9$, in addition to the usual pre-conditions, the fuel level must be sufficient. The corresponding post-conditions:

test_engine_p9(X) :-
    after(Y,X,[4,4]),   activity(Y,A),   build_engine(A),   available(X,T),   test-
    ing_tool(T),   status(X,T,S),   working(S),   fuel_level(X,L),   sufficient(L),
    test_outcome(X,O), fail(O), after(X,Z,[1,1]), activity(Y,fix_engine).

test_engine_p9(X) :-
    after(Y,X,[4,8]),   activity(Y,A),   build_engine(A),   available(X,T),   test-
    ing_tool(T),   status(X,T,S),   working(S),   fuel_level(X,L),   sufficient(L),
    test_outcome(X,O), success(O), after(X,Z,[1,2]), activity(Y,install_plate).

say that, after the rightmost test of the engine in Fig. 2, if the test failed and the engine was not fixed in the leftmost test (in which case the distance in steps from activity build_engine must be exactly 4), then the immediately following activity (a range [1,1] means exactly 1 step) is fixing it; otherwise, if the outcome was success, then a subsequent activity, occurring in at most 2 steps, must be installing the plate.

4   Relationships to Petri/WF Nets

It is clear that the expressive power of WoMan conditions is greater than that provided by decision trees. So, let us focus on the relationships between WoMan's formalism for expressing activity flow and Petri Nets. In WoMan, given an atom transition*(I,O,p,C)*, each task in I has an AND-split towards

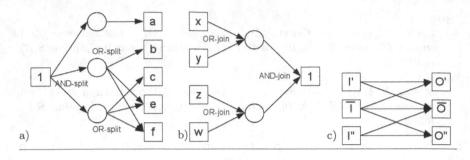

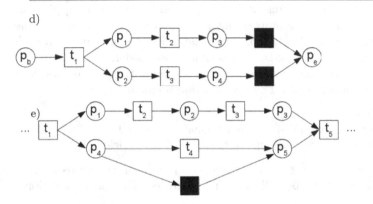

Fig. 4. Sample complex configurations for translation

all tasks in O and each task in O has an AND-join from all tasks in I. Different **transition/4** atoms having the same I represent an (exclusive) alternative choice among the corresponding O's, and, hence, a multiple OR-split. Similarly, different **transition/4** atoms having the same O represent alternative executions that converge to the execution of the same set of tasks, and, hence, a multiple OR-join. So, a **transition/4** atom represents a portion of a Petri net, involving several tasks and places, plus the related incoming and outgoing arcs. With regard to Petri nets, focusing (for compliance with WoMan) on the passage from tasks to tasks, we need to consider fragments involving a set of input Petri Net transitions, a set of intermediate places and a set of output Petri Net transitions. Figure 4(a and b) shows samples of the kinds of patterns that can be found. (a) shows AND/OR-splits; (b) shows OR/AND-joins.

Translating a Petri/WF net $\mathcal{P}$ to a WoMan model $\mathcal{W}$ is straightforward. Each Petri Net transition t yields a **task(t, [0])** atom. Then:

1. each AND-split, say from Petri Net transition t to Petri Net transitions $t_1, \ldots, t_n$, becomes a **transition({t},{$t_1, \ldots, t_n$},p,[0])** atom;

2. each AND-join, say from Petri Net transitions $t_1, \ldots, t_m$ to Petri Net transition t, becomes a `transition`$(\{t_1, \ldots, t_m\}, \{t\}, p, [0])$ atom (the p's being fresh WoMan transition identifiers);
3. each remaining place denotes both an OR-join and an OR-split, say from Petri Net transitions $t_1^I, \ldots, t_m^I$ to Petri Net transitions $t_1^O, \ldots, t_n^O$, and generates $m \cdot n$ atoms `transition`$(\{t_i^I\}, \{t_j^O\}, p_{i,j}, [0])$ (where the $p_{i,j}$'s are fresh WoMan transition identifiers, $i = 1, \ldots, m$ and $j = 1, \ldots, n$).

Note the fictitious case '0' associated to all atoms for compliance with the WoMan representation.

The translation from $\mathcal{W}$ to $\mathcal{P}$ is straightforward if the workflow model involves no duplicate or hidden tasks. Each `task`(t, C) atom in $\mathcal{W}$ generates a Petri Net transition $t \in T$ in $\mathcal{P}$. Each `transition`(I, O, p, C) atom, with $I = \{t_1^I, \ldots, t_m^I\}$ and $O = \{t_1^O, \ldots, t_n^O\}$, generates $m \cdot n$ places $\forall t_i^I \in I, t_j^O \in O : p_{i,j} \in P$ and corresponding arcs $\{(t_i^I, p_{i,j}), (p_{i,j}, t_j^O)\} \subseteq A$. In particular:

- $I = \{i\} \wedge O = \{o\}$ corresponds to a sequential place p in $\mathcal{P}$, having input arc (i, p) and output arc (p, o).
- $I = \{i\} \wedge O = \{t_1, \ldots, t_n\}, n > 1$ is an AND-split in $\mathcal{P}$, involving n places $\{p_j\}_{j=1,\ldots,n}$, each connected in input to Petri Net transition i via an arc (i, p_j) and in output to Petri Net transitions t_j via corresponding arcs (p_j, t_j).
- $O = \{o\} \wedge I = \{t_1, \ldots, t_m\}, m > 1$ is an AND-join in $\mathcal{P}$, involving m places $\{p_i\}_{i=1,\ldots,m}$, each connected in output to Petri Net transition o via an arc (p_i, o) and in input to Petri Net transitions t_i via corresponding arcs (t_i, p_i).

Additionally, WoMan can represent models involving invisible Petri Net transitions (i.e., 'no-action' tasks) or duplicated Petri Net transitions (corresponding to the same task, but in different contexts or under different conditions). Both might be handled only by considering the labeled extension of Petri nets, but this would "complicate matters enormously" [20] and thus is avoided in the process mining literature. Let us prove that WoMan's formalism is more expressive and powerful than basic Petri nets by showing a few examples of models (or fragments of models) that cannot be expressed in Petri net representation without using hidden tasks. Consider Fig. 4.

- In (d), task t_1 causes an AND-split, that concurrently executes tasks t_2 and t_3 and then finishes. Unless a hidden task is added to perform an AND-join before ending, no single sink place is available, so the AND-split has no corresponding AND-join. In WoMan it can be expressed as:
  `transition`$(\{start\}, \{t_1\}, p', [1])$.
  `transition`$(\{t_1\}, \{t_2, t_3\}, p'', [1])$.
  `transition`$(\{t_2, t_3\}, \{stop\}, p''', [1])$.
- In (e), in one branch of the AND-split caused by task t_1, task t_4 is optional. Again, an invisible task must be introduced, because place p_4 cannot be directly connected to place p_5. In WoMan it is easily expressed as:
  `transition`$(\{t_1\}, \{t_2, t_4\}, p', [1])$.
  `transition`$(\{t_1\}, \{t_2\}, p'', [2])$.

```
transition({t_2},{t_3},p''',[1,2]).
transition({t_3,t_4},{t_5},p'''',[1]).
transition({t_3},{t_5},p''''',[2]).
```

- In (c), suppose that I', $\overline{I}$, I'', O', $\overline{O}$ and O'' denote sets of tasks, with $\overline{I} \neq \emptyset$, $\overline{O} \neq \emptyset$, $I' \cup I'' \neq \emptyset$ and $O' \cup O'' \neq \emptyset$. I.e., the tasks in $\overline{I}$ activate tasks $O' \cup \overline{O}$ when combined with I', or tasks $O'' \cup \overline{O}$ when combined with I''. The tasks in $\overline{I}$ should be duplicated in a Petri Net, while in WoMan this can be expressed by two `transition/4` atoms having input and output sets pairwise partially overlapping:

```
transition(I' ∪ Ī,O' ∪ Ō,p',[1]).
transition(I'' ∪ Ī,O'' ∪ Ō,p'',[2]).
```

A logic-based trick proposed in [7] allows to check whether the translation is feasible or not. In this trick, both formalisms are mapped onto logic formulas, where task names are expressed by propositional symbols to be suitably composed by AND ($\wedge$) and XOR ($\oplus$) connectives. This yields formulas in conjunctive normal form (CNF) for Petri net fragments and formulas in disjunctive normal form (DNF) for WoMan ones. Thus, applying standard logistic manipulation to turn a formula from CNF (resp., DNF) to an equivalent one in DNF (resp., CNF) is a trick to translate a WoMan model fragment into an equivalent Petri net one. Specifically, for each `task(t,C)` in $\mathcal{W}$:

- represent as a DNF formula all the O's from atoms `transition(I,O,p,C)` s.t. $t \in I$ and turn the formula into CNF. The resulting formula identifies a place for each of the conjuncts, to be connected by outgoing arcs to the tasks associated to the constants in the disjunction and by an incoming arc to t.
- represent as a DNF formula all the I's from atoms `transition(I,O,p,C)` s.t. $t \in O$ and turn the formula into CNF. The resulting formula identifies a place for each of the disjuncts, to be connected by an outgoing arc to t and by incoming arcs to the tasks associated to the constants in the disjunction.

If the resulting formula is a contradiction, due to the presence of $p \oplus p$ patterns (that are always false) as sub-formulas of some conjunct, then the translation is impossible, unless each conflicting task p is duplicated for its various occurrences in different `transition/4` atoms.

Let us show some examples. Consider Fig. 4(b):

$$(\{x,z\},\{1\}), \qquad (\{x,w\},\{1\}), \qquad (\{y,z\},\{1\}), \qquad (\{t,w\},\{1\}),$$

the input sets $(\{x,z\}, \{x,w\}, \{y,z\}, \{y,w\})$ yield:

$$(x \wedge z) \oplus (x \wedge w) \oplus (y \wedge z) \oplus (y \wedge w) = \cdots = (x \oplus y) \wedge (z \oplus w)$$

which indeed matches the schema in Fig. 4(b): 1 is the target of an AND-join (denoted in the first formula by the outer conjunction) from two places, involved respectively in two OR-joins (the two inner disjuncts in that formula). For

$$(\{1\},\{a,b,c\}), \qquad (\{1\},\{a,e\}), \qquad (\{1\},\{a,f\})$$

one obtains:

$$(a \wedge c \wedge f) \oplus (a \wedge d) \oplus (b \wedge f) = \cdots = a \wedge (b \oplus e \oplus f) \wedge (c \oplus e \oplus f),$$

which indeed matches the schema in Fig. 4(a): 1 is the source of an AND-split (the outer conjunction in the second formula) towards three places, two of which are involved in OR-splits (corresponding to the two inner disjuncts in that formula). Finally,

$$(\{1\}, \{a, c, f\}), \qquad (\{1\}, \{a, d\}), \qquad (\{1\}, \{b, f\})$$

yields:

$$(a \wedge c \wedge f) \oplus (a \wedge d) \oplus (b \wedge f) = \cdots =$$
$$(a \oplus b) \wedge (c \oplus d \oplus b) \wedge (f \oplus d \oplus b) \wedge (a \oplus f) \wedge (c \oplus d \oplus f) \wedge (f \oplus d \oplus f)$$

where $(f \oplus d \oplus f)$ raises a contradiction (and indeed the corresponding Petri net would be invalid). Duplicating f into f' and f'' would yield:

$$(a \wedge c \wedge f') \oplus (a \wedge d) \oplus (b \wedge f'') = \cdots =$$
$$(a \oplus b) \wedge (c \oplus d \oplus b) \wedge (f' \oplus d \oplus b) \wedge (a \oplus f'') \wedge (c \oplus d \oplus f'') \wedge (f' \oplus d \oplus f''),$$

that solves the problem. Also for case (c) in Fig. 4 we have:

$$(O' \wedge \overline{O}) \oplus (O'' \wedge \overline{O}) = ((O' \wedge \overline{O}) \oplus O'') \wedge ((O' \wedge \overline{O}) \oplus \overline{O}) =$$
$$(O' \oplus O'') \wedge (\overline{O}) \oplus O'') \wedge (O' \oplus \overline{O}) \wedge (\overline{O} \oplus \overline{O}),$$

which is a contradiction (due to $(\overline{O} \oplus \overline{O})$), denoting an impossible translation.

5   Comments

The previous section showed that WoMan's formalism is more powerful than Petri nets, the current standard formalism in the Process Mining literature. Indeed, it can smoothly express situations involving invisible or duplicated tasks, that plain Petri nets cannot handle. This is obtained using different transition/4 atoms that combine the same task in different ways with other tasks, or ignore a task when it is optional. However, there are also other reasons why WoMan's formalism may deserve attention.

A process model should be **complete** (i.e. able to generate all event sequences in C), **irredundant** (able to generate as few event sequences not in C as possible) and **minimal** (as simple and compact as possible) [1,2,11,21]. *Accuracy* (i.e. completeness and irredundancy [2]) is typically in contrast with minimality (more compact models are more general and thus tend to cover more cases). Other approaches in the Process Mining literature, imposing a single node for each task, route on this node all occurrences of that task in a process, which introduces combinations that were not intended or foreseen when building the model. Conversely, allowing the use of, in Petri Net terminology, invisible or duplicated tasks, WoMan models significantly improve irredundancy, even at the cost of introducing more nodes into the graph, and, hence, possibly at the expense of minimality. Minimality in WoMan comes into play only for avoiding nodes that would be redundant and whose removal would not introduce additional unseen processes. In a Process Mining setting this is justified, because the aim is constraining future process enactment to pre-defined specifications, not having a predictive model that accounts for new behaviors.

Storing the specific cases in which the various transitions were encountered additionally avoids mixing execution flows from different cases, which ensures even stricter adherence to the correct cases represented by the model. Additionally, this also allow to naturally handle noisy data. Task and transition probabilities are in fact proportional to the number of training cases in which they occurred, which is reflected by the cardinality of the associated multiset of cases.

Finally, WoMan models may specify which agents are allowed to carry out the various tasks, and may impose specific pre- and post-conditions on transitions and tasks (both in general and in the context of a specific transition).

6   Conclusions

Workflow management is fundamental to efficiently, effectively and economically carry out complex processes. The formalism used for expressing process models in process mining systems is in turn fundamental to enforce proper workflow management. Among its innovative peculiarities, the WoMAN framework for workflow management, based on First-Order Logic, includes a formalism for representing the models that ensures strict adherence to the observed practices, also allowing to express conditions and to support noise handling. This paper discussed in details such a formalism, highlighting its most outstanding strengths and comparing it to Petri nets, that are the current standard in state-of-the-art process mining systems. The comparison shows that WoMan is more powerful than standard Petri Nets, and that it can handle naturally and straightforwardly cases that would require complex patterns in Petri Nets. Techniques for translating Petri nets into Woman formalism are also provided, to ensure compatibility with existing systems.

Future work will further investigate the relationships between the WoMan formalism and Petri Nets, to assess more precisely when and how components of the former can be translated into the latter, and which extensions to the latter are required to cover the additional expressive power of the former.

Acknowledgments. This work was partially funded by the Italian PON 2007–2013 project PON02_00563_3489339 'Puglia@Service'.

References

1. Agrawal, R., Gunopulos, D., Leymann, F.: Mining process models from workflow logs. In: Schek, H.-J., Saltor, F., Ramos, I., Alonso, G. (eds.) EDBT 1998. LNCS, vol. 1377, pp. 467–483. Springer, Heidelberg (1998)
2. Cook, J.E., Wolf, A.L.: Discovering models of software processes from event-based data. Technical Report CU-CS-819-96, Department of Computer Science, University of Colorado (1996)
3. Cook, J.E., Wolf, A.L.: Event-based detection of concurrency. Technical report CU-CS-860-98, Department of Computer Science, University of Colorado (1998)
4. de Medeiros, A.K.A., van der Aalst, W.M.P., Weijters, A.J.M.M.T.: Workflow mining: current status and future directions. In: Meersman, R., Schmidt, D.C. (eds.) CoopIS 2003, DOA 2003, and ODBASE 2003. LNCS, vol. 2888, pp. 389–406. Springer, Heidelberg (2003)
5. de Medeiros, A.K.A., van Dongen, B.F., van der Aalst, W.M.P., Weijters, A.: Process mining: extending the α-algorithm to mine short loops. In: WP 113, BETA Working Paper Series. Eindhoven University of Technology (2004)
6. de Medeiros, A.K.A., Weijters, A.J.M.M., van der Aalst, W.M.P.: Genetic process mining: an experimental evaluation. Data Min. Knowl. Discov. **14**, 245–304 (2007)
7. Ferilli, S.: WoMan: logic-based workflow learning and management. IEEE Trans. Syst. Man Cybern. Syst. **44**, 744–756 (2014)
8. Ferilli, S., Esposito, F.: A logic framework for incremental learning of process models. Fundamenta Informaticae **128**(4), 413–443 (2013). IOS Press
9. Herbst, J.: Dealing with concurrency in workflow induction. In: Proceedings of European Concurrent Engineering Conference, pp. 175–182. SCS Europe (2000)
10. Herbst, J., Karagiannis, D.: Integrating machine learning and workflow management to support acquisition and adaptation of workflow models. In: Proceedings of 9th International Workshop on Database and Expert Systems Applications, pp. 745–752. IEEE (1998)
11. Herbst, J., Karagiannis, D.: An inductive approach to the acquisition and adaptation of workflow models. In Proceedings of IJCAI 1999 Workshop on Intelligent Workflow and Process Management: The New Frontier for AI in Business, pp. 52–57 (1999)
12. Herbst, J.: A machine learning approach to workflow management. In: Lopez de Mantaras, R., Plaza, E. (eds.) ECML 2000. LNCS (LNAI), vol. 1810, pp. 183–194. Springer, Heidelberg (2000)
13. van der Aalst, W., et al.: Process mining manifesto. In: Daniel, F., Barkaoui, K., Dustdar, S. (eds.) BPM Workshops 2011, Part I. LNBIP, vol. 99, pp. 169–194. Springer, Heidelberg (2012)
14. Maggi, F.M., Bose, R.P.J.C., van der Aalst, W.M.P.: Efficient discovery of understandable declarative process models from event logs. In: Ralyté, J., Franch, X., Brinkkemper, S., Wrycza, S. (eds.) CAiSE 2012. LNCS, vol. 7328, pp. 270–285. Springer, Heidelberg (2012)
15. Pesic, M., van der Aalst, W.M.P.: A declarative approach for flexible business processes management. In: Eder, J., Dustdar, S. (eds.) BPM Workshops 2006. LNCS, vol. 4103, pp. 169–180. Springer, Heidelberg (2006)
16. Rozinat, A., van der Aalst, W.M.P.: Decision mining in business processes. In: WP 164, BETA Working Paper Series. Eindhoven University of Technology (2006)
17. van der Aalst, W.M.P.: The application of petri nets to workflow management. J. Circ. Syst. Comput. **8**, 21–66 (1998)

18. van der Aalst, W.M.P.: Process mining: overview and opportunities. ACM Trans. Manag. Inf. Syst. **3**, 7.1–7.17 (2012)
19. van der Aalst, W.M.P., Dustdar, S.: Process mining put into context. IEEE Internet Comput. **16**, 82–86 (2012)
20. van der Aalst, W.M.P., Weijters, T., Maruster, L.: Workflow mining: discovering process models from event logs. IEEE Trans. Knowl. Data Eng. **16**, 1128–1142 (2004)
21. Weijters, A., van der Aalst, W.M.P.: Rediscovering workflow models from event-based data. In: Proceedings of 11th Dutch-Belgian Conference of Machine Learning (Benelearn 2001), pp. 93–100 (2001)
22. Wen, L., Wang, J., Sun, J.: Detecting implicit dependencies between tasks from event logs. In: Zhou, X., Li, J., Shen, H.T., Kitsuregawa, M., Zhang, Y. (eds.) APWeb 2006. LNCS, vol. 3841, pp. 591–603. Springer, Heidelberg (2006)

Duration-Aware Alignment of Process Traces

Sen Yang[1]([⊠]), Moliang Zhou[1], Rachel Webman[2], JaeWon Yang[2],
Aleksandra Sarcevic[3], Ivan Marsic[1], and Randall S. Burd[2]

[1] Department of Electrical and Computer Engineering,
Rutgers University, Piscataway, NJ, USA
Sy358@scarletmail.rutgers.edu
[2] Division of Trauma and Burn Surgery, Children's Nat'l Medical Center,
Washington, D.C., USA
[3] College of Information Science and Technology, Drexel University,
Philadelphia, PA, USA

Abstract. Objective: To develop an algorithm for aligning process traces that considers activity duration during alignment and helps derive data-driven insights from workflow data. Methods: We developed a duration-aware trace alignment algorithm as part of a Java application that provides visualization of the alignment. The relative weight of the activity type vs. activity duration during the alignment is an adjustable parameter. We evaluated proportional and logarithmic weights for activity duration. Results: We used duration-aware trace alignment on two real-world medical datasets. Compared with existing context-based alignment algorithm, our results show that duration-aware alignment algorithm achieves higher alignment accuracy and provides more intuitive insights for deviation detection and data visualization. Conclusion: Duration-aware trace alignment improves upon an existing trace alignment approach and offers better alignment accuracy and visualization.

1 Introduction

Many contemporary information systems record activity logs, including online shopping habits, personal calendars and electronic health records (EHR). Process mining techniques aim to extract knowledge and insights from these types of logs [1]. Most research in process mining has focused on workflow discovery using process model analysis (e.g., conformance checking [1]) and deviation discovery based on these models. Bose and Van der Aalst [2] derived a trace alignment from the multiple sequences alignment (MSA) algorithm in bioinformatics [3] and used it to gain insights from activity logs. Trace alignment places the same or similar activities in the same column of the alignment matrix. If a matching activity cannot be found, a gap symbol "-" is inserted. For example, given a trace with activities {a,c} and a trace with activities {a,b,a,c}, two possible alignments are:

$$\begin{bmatrix} a & - & - & c \\ a & b & a & c \end{bmatrix} \text{ and } \begin{bmatrix} - & - & a & c \\ a & b & a & c \end{bmatrix}$$

© Springer International Publishing Switzerland 2016
P. Perner (Ed.): ICDM 2016, LNAI 9728, pp. 379–393, 2016.
DOI: 10.1007/978-3-319-41561-1_28

Trace alignment can be used to find similarities within a group of traces and to determine how a given trace differs from the well-established work practice. Trace alignment can also help identify conserved patterns of activities and deviations from the norm.

Existing trace alignment approaches consider only the sequential order of activities and ignore activity duration. Activity duration may, however, be an important parameter for some processes. For example, consideration of duration helped understand that nurses routinely switch between tasks, spending less time on interrupted tasks that are later resumed [21]. To consider activity duration during alignment when finding a matching activity in another trace, the duration may need to be part of similarity calculation. Activities of the same type and comparable durations should be assigned high similarity. Markedly different durations of the same activity type may indicate deviations and can be expected to have a lower similarity [23]. Activity duration can also help improve alignment accuracy by identifying deviations in the process execution. Activities with a typical duration suggest normal operation while activities with unusual durations indicate difficulties or atypical performance [24]. Additionally, activity duration may be linked with activity importance, i.e. task importance can vary based on situation and may require more time and effort when more integral to the process. Finally, visualizing the trace alignment result incorporating durations helps users detect duration-related deviations in activity performance that otherwise would not be evident.

To consider activity duration as a parameter in trace alignment, we extended the classic Needleman Wunsch algorithm (NW) [4] used for aligning biological sequences, to include a cost for activity. Because a process execution can be considered as a sequential data on the timeline, one may expect that the alignment of traces with activity durations can be solved by Dynamic Time Warping (DTW) [5, 6, 7], an algorithm used in signal processing and pattern recognition to align time-series sequences. The standard DTW algorithm, however, does not explicitly penalize the difference in activity durations between aligned activities of the same type. A variation of DTW called Distortion Penalized DTW or Variable Penalty DTW [8, 25], was developed to address this issue by introducing the time-distortion penalty to penalize the expansion and contraction of the original sequence. Our preliminary experimental results, however, showed that even the distortion-penalized DTW could not produce the alignment results as expected. To perform the distortion-penalized DTW, we discretized the time axis and displayed each activity as function of time over one or more time steps. This uniform slicing produces multiple contiguous segments for each activity, each segment being one unit of time long, that are treated as independent points during the DTW time-warping. Consider a trace T1 with activities {a, c} where a lasted three time units and c lasted one time unit, and another trace T2 with activities {a, b, a, c} where all lasted one time unit (Fig. 1(a)). After time discretization, activity a in T1 is split in three discrete-time segments {a1, a2, a3} (Fig. 1(b)). Alignment of segments from one trace is attempted with segments from the other trace, without considering that adjacent segments may be part of the same activity. As a result, segments of one long-duration activity in one trace may be aligned with several short-duration activities in the other trace. In our example, after the alignment, the first and third segments of activity a in T1 will be separately aligned with the two different activities of type a in T2. This observation occurs because the minimum

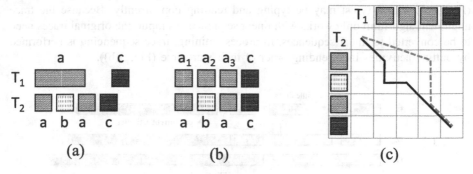

Fig. 1. Example of aligning two traces using distortion-penalized DTW. (a) Original traces with activity durations. (b) Representing activities on discrete timeline where activity a in T_1 is three time units long. (c) Alignment of the discrete time-points of the two traces. An optimal path found by the distortion-penalized DTW is shown by the solid line, which does not keep together the segments of activity a in T_1. The expected alignment is shown by the dashed line.

warping distance results from warping around activity b in T2 rather than keeping a1, a2, and a3 together (Fig. 1(c)) [5, 6, 7]. Our modified trace alignment algorithm addresses this problem by introducing a time-scale distortion penalty into the alignment algorithm. In addition to modifying the trace alignment algorithm, we also customized several existing metrics for evaluating alignment algorithms, such as the sum-of-pairs score for evaluating alignment accuracy [17].

The rest of the paper is organized as follows. Section 2 introduces our duration-aware trace alignment algorithm. Section 3 evaluates the performance of this algorithm and compares the results to previous trace alignment work using two case studies. Section 4 concludes the paper and presents the limitations of our current work.

2   Methodology

Our duration-aware trace alignment algorithm (Fig. 2) works by: (1) sequencing of traces with concurrent activities, (2) computing duration-aware pairwise alignment, (3) building a guide tree, and (4) performing progressive alignment of multiple traces.

2.1   Sequencing of Process Traces

Activities in a process mining dataset (or activity log) are usually coded with timestamps indicating the start time and end time for each activity. Idle time may exist between activities and some activities may be executed concurrently with each other

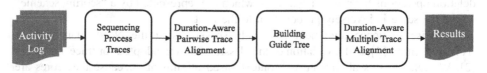

Fig. 2. Duration-aware trace alignment flowchart.

(Fig. 3(a)), e.g. a typist may be typing and reading concurrently. Because the trace alignment algorithm only works with linear sequences as input, the original traces need to be converted to linear sequences. In process mining, trace sequencing is performed by putting activities in ascending order of their start time (Fig. 3(b)).

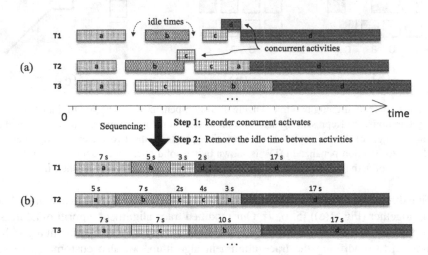

Fig. 3. Two steps of sequencing the traces with concurrent activities (such as d in T_1 and c in T_2) and idle times (white spaces in all three traces). (a) Example process traces before sequencing. (b) The same process traces after sequencing.

2.2 Duration-Aware Pairwise Trace Alignment

Consider example traces $T_1 = \{abcdd\}$ and $T_2 = \{abccad\}$. Example alignment is:

	1	2	3	4	5	6	7
T_1 :	a	b	c	-	-	d	d
T_2:	a	b	c	c	a	d	-

The alignment includes three types of operations: substitution, insertion and deletion. The insertion and deletion operations are usually referred as indel operations because an insertion could be considered a deletion in another trace. For example, columns 1, 2, 3 and 6 show that substituting an activity in one trace with the same type of activity in the other trace; columns 4 and 5 show insertions of activities in T_2 that do not exist T_1; column 7 shows deletion of a T_1 activity in T_2. Substitution, insertion and deletion operations have associated costs, which are represented by a "scoring scheme" where the score is inversely proportional to the cost.

Pairwise sequence alignment is usually solved using the Needleman Wunsch (NW) algorithm [4], and a variation of the classic NW was adopted for trace alignment [2]. We refer to this alignment as "context-based alignment" because the scores are derived from the sequential order of adjacent activities, i.e., the process context.

Both the NW alignment and context-based alignment algorithm align sequences without considering activity duration. To incorporate time information into alignment calculation, we introduce a time-scale distortion penalty into the trace alignment algorithm, which originates from Dynamic Time Warping [8]. The time distortion penalty is applied whenever a sequence is expanded or contracted during alignment.

The accumulated score matrix for our duration-aware trace alignment is defined as:

$$F(i,j) = \max \begin{cases} F(i-1,j-1) + S(T_1(i), T_2(j)) * ddp(T_1(i), T_2(j)) & \text{Substitute} \\ F(i-1,j) + g * hdp(T_1(i)) & \text{Insert} \quad (1) \\ F(i,j-1) + g * vdp(T_2(j)) & \text{Delete} \end{cases}$$

where $S(T_1(i), T_2(j))$ is the score for substituting element $T_1(i)$ with $T_2(j)$, and g is the score for indel operations. $ddp(x,y), hdp(x), and\ vdp(x)$ are time-scale distortion-penalty functions. The initial conditions are: $F(0,0) = 0, F(i,0) = F(i-1,0) + g * hdp(T_1(i), F(0,j) = F(0,j) + g * vdp(T_2(j))$. The diagonal distortion penalty (ddp) penalizes the time distortion generated in the substitution operation. The horizontal distortion penalty (hdp) penalizes the time distortion generated in the horizontal direction, viz. deletion operation. The vertical distortion penalty (vdp) penalizes the time distortion generated in the vertical direction, viz. insertion operation. These three distortion penalty functions are defined as:

$$hdp(T_1(i)) = \varphi(d(T_1(i))) \tag{2}$$

$$vdp(T_2(j)) = \varphi(d(T_2(j))) \tag{3}$$

$$ddp(T_1(i), T_2(j)) = \begin{cases} Min(\varphi(d(T_1(i))), \varphi(d(T_2(j)))) - |\varphi(d(T_1(i))) - \varphi(d(T_2(j)))|, S(T_1(i), T_2(j)) \geq 0 \\ \varphi(d(T_1(i))) + \varphi(d(T_2(j))), S(T_1(i), T_2(j)) < 0 \end{cases}$$

$$(4)$$

where $\varphi(t)$ is defined as the time-weighting function used to control the influence of activity duration, and $d(activity)$ is the duration of an activity. The duration unit is the same as the unit used for recording the activity. For ddp, we need to consider two scenarios. First, $S(T_1(i), T_2(j)) > 0$ means a "match" between activities $T_1(i)$ and $T_2(j)$, i.e. activities are of the same type or substitutable. In this scenario, ddp is decided by rewarding the extent to which the durations overlap and penalizing the extent to which the durations differ. Second, $S(T_1(i), T_2(j)) < 0$ means a "mismatch" between activities $T_1(i)$ and $T_2(j)$, i.e. $T_1(i)$ and $T_2(j)$ are incompatible and should not be aligned. Instead, this substitution operation should be decomposed to a deletion and an insertion. In this scenario, ddp equals to the sum of hdp and vdp.

Our group and others (e.g. [21]) have observed that activity duration has a distribution where extremely long-duration activities occur rarely. These rare long duration activates, however, can have a significant effect on the distortion penalty. We use time weighting to control the influence of activity duration. Choosing a proper time-weighting method is critical for the performance of our algorithm. We analyzed two weighting methods: linear and logarithmic weighting:

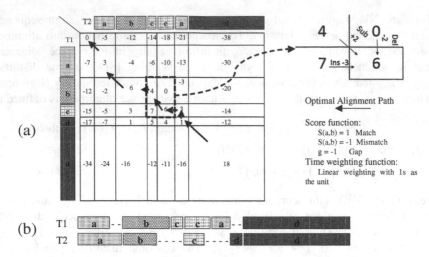

(a)

(b)

Fig. 4. Pairwise alignment with linear duration-weighting function of traces T_1, T_2 from Fig. 3. (a) Scoring matrix for the alignment. (b) The alignment result.

$$\begin{cases} \varphi_{Linear}(d(event)) = c * d(event) \\ \varphi_{Log}(d(event)) = \log_b(d(event)) \end{cases} \tag{5}$$

where $c = 1$ and $b = e$ (base of natural logarithm). Linear weighting preserves the original duration information, but a problem arises when the difference between activity durations is large. The duration becomes the driving factor for alignment and the trace context is mostly ignored. In other words, long-duration activities "overpower" short-duration activities and dominate the alignment. Logarithmic weighting can mitigate this problem.

The scoring scheme, or matrix that specifies the cost of the three operations during alignment in score matrix F can be customized based on domain knowledge. If domain knowledge is unavailable, a simple choice would be the unit score scheme (match = 1, mismatch = −1, indel = −1). Once the score matrix $F(i,j)$ is constructed, the optimal alignment can be deduced by tracing back from lower right corner of the score matrix to the upper left corner choosing the neighboring cell that leads to the maximum score at each step (Fig. 4).

2.3   Building the Guide Tree

A guide tree needs to be constructed to determine the order of trace pairs to be aligned in the progressive iteration of multiple traces. Pairwise alignment is then performed from guide tree's leaves to the root. In our approach, the guide tree is generated based on the hierarchical clustering algorithm (AHC) [9] with Ward's method [10, 11] (Fig. 5). To measure the proximity of traces, we define a new distance named "Duration-Aware Edit Distance", which is derived from Edit Distance [12] (also called Levenshtein Distance) and includes dissimilarity between activity durations. Given two traces α and β, their duration-aware edit distance $t_{\alpha,\beta}$ can be calculated progressively as:

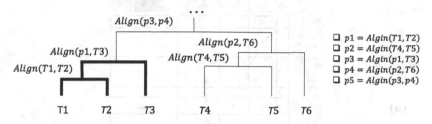

Fig. 5. Guide tree that involves traces T1, T2 and T3 (thick lines) from Fig. 3.

$$
t_{\alpha,\beta}(i,j) = \begin{cases} \max(i,j) & \text{if } \min(i,j) = 0 \\ \min \begin{cases} t_{\alpha,\beta}(i-1,j-1) + \begin{cases} d_\alpha(i) + d_\beta(j) & \text{if } \alpha_i \neq \beta_j \\ |d_\alpha(i) - d_\beta(j)| & \text{if } \alpha_i = \beta_j \end{cases} \\ t_{\alpha,\beta}(i,j-1) + d_\beta(j) \\ t_{\alpha,\beta}(i-1,j) + d_\alpha(i) \end{cases} \end{cases} \quad (6)
$$

where $d_\alpha(i)$ and $d_\beta(j)$ are the duration of ith activity in α and j^{th} activity in β.

2.4 Duration-Aware Multiple Trace Alignment

Multiple trace alignment is essentially a progressive pairwise alignment process that aligns pairs of individual traces, as well as a trace and a *profile* (a temporary alignment result at intermediate stage of the alignment process) (Fig. 6) or two profiles. Because a column in a profile contains more than one activity, the substitution score in multiple sequence alignment is redefined as:

$$
S(C_A^i, C_B^j) = \sum_{a,b \in \mathbb{A}} n_A^i(a) \cdot n_B^j(b) \cdot S(a,b) \cdot \overline{ddp(a,b)}. \quad (7)
$$

where $S(a,b)$ denotes the substitution score of activities a, b; $\mathbb{A}$ denotes the activity set of activity log; $n_A^i(a)$ denotes the frequency of activity a in the column i of profile A; C_A^i denotes the contents of ith column of profile A. $\overline{ddp(a,b)}$ denotes the diagonal distortion penalty calculated based on the average duration of activity types a and b. Similarly, the indel score of a column is redefined as:

$$
\begin{cases} I(C_A^i) = \sum_{a \in \Sigma} f_A^i(a) \cdot g \cdot \overline{hdp(a)} & \text{deletion} \\ I(C_A^i) = \sum_{a \in \Sigma} f_A^i(a) \cdot g \cdot \overline{vdp(a)} & \text{insertion} \end{cases} \quad (8)
$$

where g is the indel score of activity a; $f_A^i(a)$ denotes the frequency of a in column i of profile A; $\overline{hdp(a)}$ and $\overline{vdp(a)}$ are the horizontal or vertical distortion penalties calculated based on the average duration of a deleted or inserted activity.

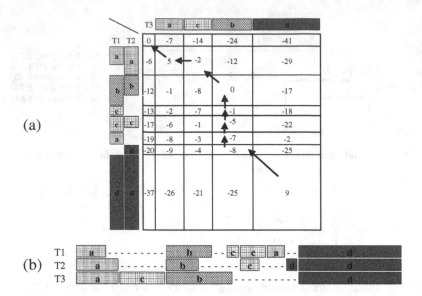

Fig. 6. Duration-aware multiple trace alignment with linear weighting of traces T_1, T_2, T_3 from Fig. 3. (a) Scoring matrix for the alignment. (b) The alignment result.

3   Experimental Results

We implemented our duration-aware trace alignment algorithm in the Java programming language. Due to the space limitations, we do not discuss the computational complexity of our algorithm, but we evaluated its performance on real logs and an artificially generated log. The results show that duration-aware trace alignment algorithm can compute and visualize an activity log of 50,000 activities ($\sim 1,000$ traces and ~ 50 activities for each trace) within 25.5 ± 1.5 s (mean value and standard deviation over 20 different runs).

Although there are many quality-assessment approaches or metrics for biological sequence alignment (TCS [14], Heads-or-Tails [15], GUIDANCE [16], etc.), very few such metrics exist for trace alignment. We evaluated the performance of the duration-aware alignment using two case studies and compared the results with that of previous research on trace alignment.

3.1   Evaluation Criteria

Our evaluation of alignment algorithm performance is based on the following criteria:

1. *Sum-of-pairs Score (SPS):* This metric is widely used to measure the alignment accuracy of multiple sequences in bioinformatics [17]. To our knowledge, it has not been used in the context of process mining. SPS for biological sequences is defined as: $SPS = n/N$ where n is the number of correctly aligned residue pairs found in the test alignment and N is the total number of aligned residue pairs in the reference alignment. In the context of trace alignment, the elements are process activities

instead of biological residues. Our ground-truth alignment (reference alignment) was generated by medical experts who aligned the traces manually.

2. *Average Information Score:* The information score is defined for each column of the alignment matrix as [2]: $1 - E/E_{max}$ where E_{max} is the maximum entropy of a column, equal to $\log_2(|\mathbb{A}| + 1)$. E is the entropy of activities in the column, defined as: $E = \sum_{a \in \mathbb{A} \cup \{-\}} -p_a \log_2(p_a)$ where a is an activity; $\mathbb{A}$ is the set of activity types in the activity log; "-" denotes the gap symbol and p_a is the probability of a's occurrence in this column. Lower information score indicates sparser distribution and higher diversity of activities in a column. Because the purpose of alignment is to find significant information with strong confidence, high diversity of activities in one column is not expected. For this reason, higher information score indicates higher alignment quality. To reflect the quality of the whole alignment, we used the mean value of information scores of all columns.

3. *Consensus Sequence (CS):* The concept of consensus sequence comes from bioinformatics, where it denotes a sequence of most frequent residues found in each column of the alignment. In process mining, the consensus sequence captures the most frequent activity in each column [2]. A gap could also be included in the consensus sequence if the corresponding column is mostly filled with gaps, but we are more interested in non-gap activities in the consensus sequence. The consensus sequence measures the alignment quality because good alignment algorithms should be able to discover the common activity sequences in a process.

4. *Alignment Matrix Length:* Alignment matrix length could also reflect the quality of alignment. Longer alignment matrix indicates that more gaps are introduced into the alignment, which tends to be sparse. Good alignment is expected be dense with only necessary gaps included and unnecessary gaps avoided.

5. *Deviation Detection Ability:* A major objective of trace alignment is to help diagnose the process executions, which includes the ability to identify deviations from common practice. These deviations were previously classified into two main categories, viz. omission and commission [2]. "Omission" denotes an activity that should exist at certain position in a trace but is missing. "Commission" denotes an activity that should not exist but is inserted. In addition, in duration-aware trace alignment we are able to observe one more type of deviation, viz. "abnormal duration". Abnormal duration is present when an activity is either much shorter or much longer when compared to other activities in the same column. We did not quantify this metric but rather illustrate it by examples.

Among these metrics, the sum-of-pairs score reflects alignment accuracy because it performs direct comparison to the reference alignment (ground truth) [17]. The average information score is associated with the alignment matrix length. Including more gaps into the alignment will not only increase the alignment matrix length, but may also increase the average information score. The reason is that the average information score increases for every column mostly filled with gaps. Based on the definition of information score, the information score of such columns will be high.

Legend at the bottom shows the color-coding of activities.

The boxes inside the alignment matrices labeled with numbers are discussed in text.

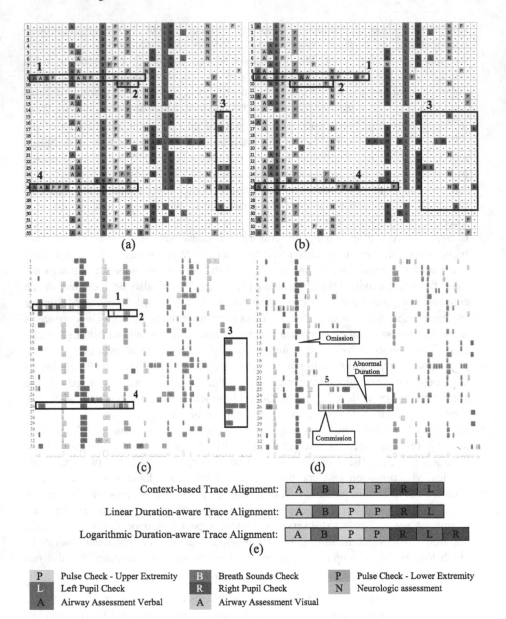

Fig. 7. (a) Reference alignment created by medical experts. (b) Context-based trace alignment. (c) Duration-aware trace alignment with logarithmic weights. (d) Duration-aware trace alignment with linear weights. Rows of the alignment were originally ordered according to the guide tree but we reordered them by trace-ID for easier comparison. (e) Consensus sequences of different alignments. (Color figure online)

3.2 Case Study 1: Trauma Resuscitation Process

Input Dataset Information. This dataset was obtained from trauma resuscitations performed at Children's National Medical Center in Washington, DC. We videotaped and coded the start and end times of 8 different activity types during the initial evaluation of 33 injured children [18]. Resuscitation traces contained on average 7.4 activities and lasted on average of 3.9 min (between 0.5 and 13 min). Our ground-truth reference alignment of activities was created manually by medical experts (Fig. 7(a)).

Generating Alignment Matrix. Context-based trace alignment (Fig. 7(b)) was generated in ProM (http://www.promtools.org/) using the Trace Alignment Plugin with the recommended settings. The guide tree was constructed using maximal repeats as the feature set with Euclidean distance as the proximity measure and minimum variance as the join criteria [2]. The scoring scheme was derived based on the trace context [19]. The duration-aware trace alignment results (Fig. 7(c), (d)) were computed with the following settings. The guide tree was built using duration-aware edit distance to measure the trace proximity (Eq. 6) and minimum variance as the join criteria. We assume that users may not have domain knowledge for generating scoring metrics, so we adopted the unit score function (Sect. 2.2) as our scoring scheme for computing alignment. The alignment results (Table 1) were compared based on the criteria described in Sect. 3.1.

Table 1. Performance comparison of context-based alignments and duration-aware alignments on trauma resuscitation dataset. The best results in each row are shown in bold font.

Algorithm: Metrics:	Duration-aware (linear)	Duration-aware (logarithmic)	Context-based
Sum-of-pairs score	0.617	**0.807**	0.731
Avg. information score	**0.870**	0.863	0.848
No. non-gap activities in CS	6	**7**	6
Alignment matrix length	49	39	**36**

Discussion of Performance Comparison Metrics. Our results (Table 1) show that the duration-aware trace alignment with logarithmic weights has the highest alignment accuracy with the sum-of-pairs score (0.807) based on reference alignment (Fig. 7(a)). This finding shows that activity duration can influence the alignment results and proper weighting of the duration can improve the alignment accuracy. We believe this improvement occurs for the following reasons:

1. *Repeated activities aligned based on typical duration for the column:* When faced with multiple options (because of a repeated activity), duration-aware trace alignment chooses to align the activities with similar durations. We observed that most resuscitation activities had a "typical" duration. When an activity is repeated within a trace, the instance with the typical duration was likely performed similarly as the same activity in other traces. For example, several activities in trace 9 were

repeated, so different alignments are possible. Context-based alignment aligns the first set of repeated activities in trace 9 with the same activities in other traces (Fig. 7(b), box "1"), while duration-aware alignment aligns the second set of activities (Fig. 7(c), box "1"). *Medical explanation:* The second set of repeated activities in trace 9 is more similar to the typical performance as it was done by the physician assigned to this portion of the evaluation, rather than the first set which was done by a substituting team member. Another example is the two instances of "Pulse Check-Lower Extremity" in trace 10 (Fig. 7(b), (c), box 2). Again, context-based alignment aligns the first set, while duration-aware alignment aligns the second set. *Medical explanation:* the task "Pulse Check-Lower Extremity" was repeated when the resident prematurely advanced to the secondary survey. A senior physician intervened and instructed the resident to return to the initial assessment where the "Pulse Check-Lower Extremity" task was repeated.

2. *Better guide tree:* Another factor that influences alignment accuracy is misalignment, the incorrect positioning of activities. The possibility of misalignments increases if dissimilar traces are aligned early in the progressive alignment. These misalignments cannot be corrected later and can propagate into more alignment errors [20, 22]. Because the guide tree algorithm determines the order of traces being aligned, the quality of guide tree is directly associated with the occurrence of misalignment. Compared with the feature (maximal repeats [2]) based distance, duration-aware edit distance, which is based on both activity type and activity duration, can do better in capturing the similarity of traces and producing a guide tree. As a result, misalignment is reduced and alignment accuracy improves. For example, the activity "Breath Sounds Check" in traces 15, 17, 23, 26, 29 is well aligned in duration-aware trace alignment, but it is poorly aligned in the context-based alignment (Fig. 7(a), (b), (c), block "3"). Because of similarity, repeat performance of "Breath Sounds Check" is captured and properly weighted by our duration-aware edit distance and these five traces are aligned early in the algorithm, with a lower risk of being misaligned according to the guide tree.

In some scenarios duration-aware trace alignment may not perform well:

1. *Activity duration does not always accurately predict an anomaly:* From an activity duration perspective, the first performance of "Airway Assessment Visual and Breath Sounds Check" in 26 (Fig. 7(b), (c), box "4") has a more typical duration than the second performance. The reference alignment, however shows that the second performance rather than the first should be aligned. The context-based alignment also made a mistake in this case (Fig. 7(b), box "4"). *Medical explanation:* Video review showed these tasks initially being performed by a substituting clinician and the second performed by the regular team member.

2. *Long-duration activities can dominate:* A single long-duration activity might dominate over several short-duration activities which should be aligned if only activity type were considered (Fig. 7(d), box "5"). This situation explains why linear duration-aware trace alignment performs worse in some cases. This problem is mitigated by using logarithmic time weighting function. The logarithmic duration-aware alignment achieves higher accuracy than the context-based one.

Linear duration-aware alignment had the highest average information score, but it also had the longest matrix (49 columns). This matrix length is much larger than the other two alignment matrices (39 and 36 columns). The large matrix indicates that many unnecessary gaps are included into the alignment and can make visual analysis more difficult.

The consensus sequences for context-based trace alignment and the two duration-aware alignment results were similar (Fig. 7(e)). Context-based alignment had a consensus sequence with six non-gap activities. Linear and logarithmic duration-aware trace alignment had six non-gap activities and seven non-gap activities respectively in the consensus sequence. The additional activity discovered by logarithmic duration-aware alignment is the second check of "Right Pupil". *Medical explanation*: The reason for this interesting finding is unclear, but it may be because the examining clinicians stand on the right side of the patient and leading them to check the right pupil, then the left pupil, then the right pupil again on their way back.

Deviation Detection. Like the context-based trace alignment, our duration-aware algorithms can discover commission and omission deviations (Fig. 7(d)). Duration-aware alignment can also provide additional insights based on activity duration and can identify duration anomalies. For example, the third "Breath Sounds Check" in trace 26 had an abnormally long duration (Fig. 7(d)). This fact could indicate patient disease or clinician's error. *Medical explanation*: In trace 26 the patient had an injury to the lungs and the physician spent extra time examining the chest to be sure the breath sounds were normal.

3.3   Case Study 2: Endotracheal Intubation Process

Input Dataset Information. The endotracheal intubation (breathing tube insertion) process was also reviewed using videos from Children's National Medical. This dataset contained 31 cases with a total of 602 activities of 21 different types.

Table 2. Performance comparison between context-based alignment and duration-aware alignment for endotracheal intubation process. The best results in each row are shown in bold.

Algorithm: Metrics:	Duration-aware (linear)	Duration-aware (logarithmic)	Context-based
Sum-of-pairs score	0.731	**0.843**	0.721
Avg. information score	0.918	**0.919**	0.899
No. non-gap activities in CS	10	12	**13**
Alignment matrix length	134	119	**115**

Discussion of Performance Comparison Metrics. The results show that logarithmic duration-aware trace alignment achieved the highest sum-of-pairs score on this dataset (Table 2). The duration-aware alignment with a logarithmic time weight function had similar alignment matrix length but higher average information score compared to context-based alignment. The context-based alignment had one more non-gap activity "Passive Oxygen Placement" in the consensus sequence than the logarithmic duration-aware alignment. The linear duration-aware alignment performed the worst. The difference in performance between linear duration-aware and logarithmic duration-aware alignments shows that the logarithmic weighting strategy is better than linear weighting strategy in this context.

4 Conclusions

We implemented a novel trace alignment approach using activity duration to improve trace alignment accuracy. We also introduced a set of criteria to quantify trace alignment performance. Based on these criteria and case studies, we compared our algorithms with an existing trace alignment algorithm. The results showed that our duration-aware trace alignment achieved better alignment accuracy and provided more insights into deviations. Our algorithm has important limitations. First, this algorithm cannot handle concurrent activities. When two or more activities occur simultaneously, the algorithm rearranges them in a chronological order based on their start times. Alignment of concurrent activities is still an open research. Second, our duration-aware trace alignment algorithm needs activity durations as the input and cannot be applied on a dataset without activity duration information. Our future work will extend the concepts presented in this paper to improve process trace alignment by considering more information from the time dimension, e.g. activity start/end time, idle time.

Acknowledgments. This paper is based on research supported by National Institutes of Health under grant number 1R01LM011834-01A1.

References

1. Van Der Aalst, W.: Process Mining: Discovery Conformance and Enhancement of Business Processes. Springer, Heidelberg (2011)
2. Bose, R.J.C., van der Aalst, W.M.: Process diagnostics using trace alignment: opportunities, issues, and challenges. Inf. Syst. **37**(2), 117–141 (2012)
3. Waterman, M.S.: Introduction to Computational Biology: Maps. Sequences and Genomes. CRC Press, Boca Raton (1995)
4. Needleman, S.B., Wunsch, C.D.: A general method applicable to the search for similarities in the amino acid sequence of two proteins. J. Mol. Biol. **48**(3), 443–453 (1970)
5. Rakthanmanon, T., et al.: Searching and mining trillions of time series subsequences under dynamic time warping. In: Proceedings of the 18th ACM SIGKDD International Conference on Knowledge Discovery and Data Mining. ACM (2012)
6. Forestier, G., et al.: Classification of surgical processes using dynamic time warping. J. Biomed. Inf. **45**(2), 255–264 (2012)
7. Müller, M.: Dynamic time warping. In: Information Retrieval for Music and Motion, pp. 69–84. Springer, Heidelberg (2007)
8. Holmes, W.: Speech Synthesis and Recognition. CRC Press, Boca Raton (2001)

9. Jain, A.K., Richard, C.D.: Algorithms for Clustering Data. Prentice-Hall Inc., Upper Saddle River (1988)

10. Ward Jr., J.H.: Hierarchical grouping to optimize an objective function. J. Am. Stat. Assoc. **58**(301), 236–244 (1963)

11. Murtagh, F., Legendre, P.: Ward's hierarchical clustering method: clustering criterion and agglomerative algorithm. arXiv preprint arXiv:1111.6285 (2011)

12. Levenshtein, V.I.: Binary codes capable of correcting deletions, insertions, and reversals. Soviet Phys. Dokl. **10**(8), 707–710 (1966)

13. Edgar, Robert C. "MUSCLE: a multiple sequence alignment method with reduced time and space complexity." BMC Bioinformatics 5.1 (2004): 113

14. Chang, J.-M., Di Tommaso, P., Notredame, C.: TCS: a new multiple sequences alignment reliability measure to estimate alignment accuracy and improve phylogenetic tree reconstruction. Mol. Biol. Evol. (2014). msu117

15. Landan, G., Graur, D.: Heads or tails: a simple reliability check for multiple sequence alignments. Mol. Biol. Evol. **24**(6), 1380–1383 (2007)

16. Osnat, P., et al.: GUIDANCE: a web server for assessing alignment confidence scores. Nucleic Acids Res. **38**(2), W23–W28 (2010)

17. Thompson, J.D., Plewniak, F., Poch, O.: A comprehensive comparison of multiple sequence alignment programs. Nucleic Acids Res. **27**(13), 2682–2690 (1999)

18. Kelleher, D.C., et al.: Effect of a checklist on advanced trauma life support task performance during pediatric trauma resuscitation. Acad. Emerg. Med. **21**(10), 1129–1134 (2014)

19. Bose, R.P.J.C., van der Aalst, W.M.P.: Context aware trace clustering: towards improving process mining results. In: SDM (2009)

20. Feng, D.-F., Doolittle, R.F.: Progressive sequence alignment as a prerequisite to correct phylogenetic trees. J. Mol. Evol. **25**(4), 351–360 (1987)

21. Cornell, P., et al.: Transforming nursing workflow, part 1: the chaotic nature of nurse activities. J. Nurs. Adm. **40**(9), 366–373 (2010)

22. Chakrabarti, S., et al.: Refining multiple sequence alignments with conserved core regions. Nucleic Acids Res. **34**(9), 2598–2606 (2006)

23. Bashford, H., et al.: Workflow analysis in production homebuilding. In: 43rd Annual International Conference on Associated Schools of Construction (2007)

24. Rogge-Solti, A., Kasneci, G.: Temporal anomaly detection in business processes. In: Sadiq, S., Soffer, P., Völzer, H. (eds.) BPM 2014. LNCS, vol. 8659, pp. 234–249. Springer, Heidelberg (2014)

25. Clifford, D., et al.: Alignment using variable penalty dynamic time warping. Anal. Chem. **81**(3), 1000–1007 (2009)

A Matching Approach Based on Term Clusters for eRecruitment

Gülşen Bal, Aşkın Karakaş, Tunga Güngör, Fatmagül Süzen,
and Kemal Can Kara$^{(\boxtimes)}$

Computer Engineering Department, Boğaziçi University, Istanbul, Turkey
{gulsenb,askin,fatmaguls,cank}@kariyer.net,
gungort@boun.edu.tr

Abstract. As the Internet occupies our daily lives in all aspects, finding jobs/employees online has an important role for job seekers and companies that hire. However, it is difficult for a job applicant to find the best job that matches his/her qualifications and also it is difficult for a company to find the best qualified candidates based on the company's job advertisement. In this paper, we propose a system that extracts data from free-structured job advertisements in an ontological way in Turkish language. We describe a system that extracts data from resumés and jobs to generate a matching system that provides job applicants with the best jobs to match their qualifications. Moreover, the system also provides companies to find the best fit for their job advertisement.

1 Introduction

The Internet has affected our daily life activities in several ways and it enables us to perform many things easily, including shopping, reading news, online banking, etc. In the past years, seeking for a job used to require interviews and several visits to the offices and companies. However, currently online recruitment websites make it possible for job seekers to look for jobs based on their criteria. Additionally, companies are able to find appropriate candidates based on their job advertisement. However, it takes too much time for a company to examine hundreds of resumes and find the most qualified candidates among them. Moreover, it is also difficult for a candidate to find jobs that meet his/her education, qualifications and work experiences.

There are some similar works that match resumés and job advertisements in English language. Crow and DeSanto proposed a system that identifies the basic parts of the resumés and creates search methods based on these identified parts [1]. They create ontologies for job applications and use rule-based approaches to extract concepts.

Another similar study performs semantic search for job advertisements in online recruitment websites [2]. The system creates ontologies with the help of other ontologies that were built in the related area before. The contents of the texts are clustered thematically and the similarities between concepts are observed. Another study based on matching resumés and job applications aims at generating a system which is based on description logics [3]. This study just focuses on skills in resumés and tries to analyze and match skills. A Concept-based algorithm that gives an output

© Springer International Publishing Switzerland 2016
P. Perner (Ed.): ICDM 2016, LNAI 9728, pp. 394–404, 2016.
DOI: 10.1007/978-3-319-41561-1_29

of potential-match and partial-match is developed. In the literature, there are other studies that aim to match resumes and job advertisements and prepare ontologies with semi-automatic methods [4–6].

Another study examines ontology matching approaches in different ways [7]. These approaches include terminological, structural and extensional matching techniques. In terminological matching, the system calculates the similarity between texts using names, labels etc. and matches them. In structural matching technique, the system compares entity descriptions for each ontology (internal structure) or the corresponding points that each entity may have with others (external structure). In Extensional matching technique, the system compares the instance/extension or the length of the classes of ontologies: in other terms, class instantiations or objects. This study determines whether the matching process composed of text based string sequences followed by terminological matching technique present more accurate results than the other techniques. The basic approach of these studies is to figure out concepts in a semantic way and analyze different matching techniques. Because of that, instead of 'word based' analysis, concept-based approaches or items that consist of word groups and ontologies should be used. In this paper, the proposed system embraces the same thought as well.

2 Methodology

2.1 Creation of the Lexicon of Terms

Before the process of matching resumés and job advertisements, the system extracts information from resumés and job advertisements and to create a lexicon of terms for the positions in job advertisements. Since the information in the resumé is saved in the database, when job seekers sign up the system, there is no need for an extra process to extract information. However, companies use free format texts for job advertisements and there is no standard format for it. Job Advertisements in Kariyer.net system are composed of two parts; General Qualifications and Job Description. General Qualification is used to specify the qualifications of a position; like required skills, educational background, foreign language and work experience information. Job Description is used for indicating responsibilities and duties of the jobs. Since the proposed system is interested only in qualifications required for the position, the job description part will not be analysed for information extraction. For analysing, the system first creates sentences from free format text. We analyzed more than 100 job advertisements and tried to find the most common sentence structure and define the rules. At the final phase of the analysis, we generate 'the Lexicon of endings' to identify the end of the sentences. This lexicon contains the words or word groups that are used at the end of the sentences so we can extract them from the free format texts. In English, verbs are at the middle of the sentences, whereas in Turkish, they are at the end of the sentences. That's why we labeled it as "the Lexicon of the Endings"

Examples of 'endings' in English: *Experienced in, Having knowledge of, Hands-on experience with, Having experience in, Experienced in* etc.

Examples of 'endings' in Turkish: *bilgili (well informed), tecrübeli (exprerienced), bilgi sahibi (has knowledge of), deneyimli (Has hands-on experience), yeteneğine sahip (has the ability)* etc.

After the sentences are found, we analyzed them and focused on how to indentify the terms in the sentences. Based on this analysis, we created some pattern rules to find the terms. Examples of some pattern rules are shown below:

- T + and/or + T + {specialWords} + {EndingWords}
- T, T, T {EndingWords}
- T(3) + {specialWords} + {EndingWords}
- T, T(2) + and + T + {EndingWords}
- T(3) + {specialWords} + ... + {EndingWords}
- T + {specialWords} + (T,T) + {specialWords} + {EndingWords}

T: Term, T(3): Terms composed of three words
T(2): Terms composed of two words
specialWords: {about, in the field of, upon...}
specialWords in Turkish: {konusunda(about), konularında(in the field of), üzerinde (upon)...}
EndingWords Words: {Experienced in, Having knowledge of, Hands-on experience with...}

We use the conjunctions "and & or" as well as commas to find the terms. Terms can be composed of one or more words. Moreover, we use the data containing specialWords and EndingWords, which are determined by analyzing the job advertisements. Based on these rules, we extract words or word groups and label them as terms.

The system finds the terms based on pattern rules, but we do not know whether these terms are related to the position in the job advertisement or not. To eliminate the unnecessary terms, the system implements the 'Morphological Analysis' process to the terms. The morphological analysis process is implemented in order to reveal the words' stems and paragoges. With the help of this process, the words that derive from same the root can be noticed. Generally, using word stems gives better results for agglutinative languages like Turkish. After the morphological analysis, morphological disambiguation process is implemented. Morphological disambiguation is applied in order to find the right usage the word among many other ways to use it. In this system, we used Turkish Morphological Analysis and Morphological Disambiguation programme [7].

Table 1 shows examples of extracted terms from a job advertisement for software engineering and their representation in morphological analysis.

To detect the accurate terms among the extracted terms, first, we try to find which terms are accurate. To achieve this, we have analyzed job advertisements that are looking for a software engineer and found the terms manually. Secondly, we extract the manually found terms' morphological analysis representation and find the most used morphological analysis representation of the software engineering terms. Finally, we define the most used representation of the software engineering terms as the rules to eliminate unnecessary terms.

Table 1. Morphological representation of words

Term	Morphological meaning
ASP	ASP[Noun] + [Acro] + [A3sg] + [Pnon] + [Nom]
C	C[Noun] + [Acro] + [A3sg] + [Pnon] + [Nom]
Microsoft.NET	Microsoft[Noun] + [Prop] + [A3sg] + [Pnon] + [Nom].Net [Unknown]
problem	problem[Noun] + [A3sg] + [Pnon] + [Nom]
yazılım	yazılım[Unknown]
Java J2EE	java[Noun] + [A3sg] + [Pnon] + [Nom] J2EE[Unknown]
Yazılım Mühendisliği	Yazılım[Unknown] Mühendisliği[Unknown]
ASP.NET tecrübeye	ASP.NET[Unknown] tecrübeye[Unknown]
Microsoft.NET C	Microsoft[Noun] + [Prop] + [A3sg] + [Pnon] + [Nom].NET [Unknown] C[Noun] + [Acro] + [A3sg] + [Pnon] + [Nom]

Software engineering position's morphological analysis rules and its explanation are shown in Table 2.

Table 2. Meaning of morphologic analysis rules

Rule	Rule meaning
[Unknown]	Unknown word
[Noun] + [A3sg] + [Pnon] + [Nom]	Noun
[Noun] + [Acro] + [A3sg] + [Pnon] + [Nom]	Acronym Nominative
[Noun] + [Prop] + [A3sg] + [Pnon] + [Nom]	Proper Noun Nominative
[Noun] + [A3sg] + [Pnon] + [Nom], [Noun] + [A3sg] + [Pnon] + [Nom]	Singular Noun Nominative, Singular Noun Nominative
[Unknown], [Unknown]	Unknown word, Unknown word

This term extraction process is applied to 21 selected positions. These positions are: software engineer, accounting specialist, mechanical engineer, architect, electrical engineer, production engineer, graphic designer, lawyer, electrical and electronic engineer, project engineer, business analyst, quality engineer, planning engineer, financier, interior architect, research and development engineer, environmental engineer, technical service engineer, project manager and industrial engineer. We also write morphologic analysis rules for each position by analyzing each position's job advertisements. After the elimination of terms with morphological analysis rules is completed, we implement three steps in order to determine which terms belong to the position's domain [8, 9].

The first step is the Domain Relevance (DR) process. DR is the amount of information captured within the target corpus with respect to the entire collection of corpora. More precisely, given a set of n domains (D1, …, Dn) the domain relevance of a term t (t is now a single word or multiword term) is computed as:

$$DR_D(t) = \frac{\hat{P}(\frac{t}{D_i})}{max_j(\hat{P}(\frac{t}{D_j}))} = \frac{freq(t, D_i)}{max_j(freq(t, D_j))}$$

where

P(t|D$_i$) = The probability of term t is in the domain of Di.
Freq(t,D$_i$) = (how many times term t is in the domain of Di)/(how many times all the terms are in the domain of Di)
For every domain Dj:
max$_j$(freq(t,Dj)) = max (how many times the term t is in the all domains of Dj)/ (how many times all the terms are in the domain of Di).

The second step is the Domain Consensus (DC) process. DC measures the distributed use of a term in a domain Di. The distribution of a term t in documents dj can be taken as a stochastic variable estimated throughout all dj ∈ Di. The entropy H of this distribution expresses the degree of consensus of t in Di . More precisely, the domain consensus is expressed as follows:

$$DR_{Di}(t) = -\sum \hat{P}d_k \in D_i\left(\frac{t}{d_k}\right) \log(\hat{P}(\frac{t}{d_k})) = -\sum norm\, d_k$$
$$\in D_i - freq(t, d_k) \log(norm - freg(t, d_k))$$

where

P(t|D$_k$) = The probability of term t is in the domain of Dk.
Freq(t,D$_k$) = (how many times term t is in the domain of Dk)/(how many times all the terms are in the domain of Dk).

The third step is the Lexicon Cohesion (LC) process. In this step, it is used to determine whether the words in the term T occur in the documents separately or together.

$$LCD_i = \frac{n \cdot freq(t, D_i) \cdot \log(freq(t, D_i))}{\sum freqw_j(w_j, D_i)}$$

where

n: number of terms that t has
freq(t,D$_i$) = (The probability of term t is in the domain of D$_i$.)
wj = jth word of the term, 1 <= j <=n
Freq(w$_j$,D$_i$) = (how many times term w$_j$ is in the domain of D$_i$).

The system implements all of DR, DC and LC processes and every process returns between 0 and 1. To assembly these results, we use the formula that is shown below.

$$DomainResult(T, D) = \alpha1DR + \alpha2DC + \alpha3LC$$
$$\alpha1 + \alpha2 + \alpha3 = 1/3$$

where T denotes a term and D denotes the domain.

Afterwards the system eliminates terms based on the value of their results; we have analyzed and defined cut off values for every domain to eliminate terms. Then, we manually check and delete the unnecessary terms and create a lexicon of terms for every domain. To illustrate, we have a lexicon of terms for software engineering position, which is composed of 887 terms.

2.2   Creation of Term Clusters

The system finds terms related to the positions available in the job advertisements in the previous section. In order to use these terms for the matching process, the system analyzes and sorts the terms which are used together so many times.

To determine the term groups, the system performs the process that is shown below

- How many times the term used in job advertisements
- How many job advertisements have the term
- Terms which are used together and their frequency

Based on these rules, the system finds the terms that are used together in the frequency of %30 or more. And from these terms, we create the term clusters that composed of 2, 3, 4, 5, 6 or 7 terms. Table 3 shows examples of the term clusters.

Table 3. Example of term clusters

Length	Terms
7	ajax, asp.net, CSS, HTML, Javascript, XML, Web
6	asp.net, C#, Javascript, SQL, XML, Web
5	.NET, asp.net, C#, Web, SQL
5	afnetworking, CoreData, CoreGraphics, CoreLocation, QuartzCore
5	ajax.CSS, HTML, Jquery, JavaScript
3	Amazon AWS, Bamboo, Microsoft Azure
3	Hibernate, J2EE, Spring
3	Cassandra, Hbase, Hadoop
3	JSP, struts, servlet
2	java, Oracle
2	MongoDB, noSQL
2	MS Visio, Ms Project
2	android, ios
2	ABAP, SAP

2.3   The Matching Method of Resumés and Job Advertisements

First of all, to match resumés and job advertisements, the system finds related terms. Afterwards, the system uses extracted terms from job advertisements and resumés as well as the term clusters to include term relations to the matching process. The system implements the Cosine Similarity method to evaluate the similarity between job advertisements and resumés. To evaluate the similarity, the system creates vectors for each job advertisement and resumé. For example, in order to match the resumés and the job advertisements in the domain of software engineering position, the system creates vectors that are composed of the items in the lexicon of terms concerning software engineering (887 terms).

Cosine Similarity calculation formula is shown below.

$$similarity = \cos(0) = \frac{A \cdot B}{||A|| ||B||} = \frac{\sum_{i=1}^{N} A_i \times B_i}{\sqrt{\sum_{i=1}^{n} (a_i)^2} \times \sqrt{\sum_{i=1}^{n} (B_i)^2}}$$

Similarity results are between 0 and 1. 0 means two vectors are completely different while 1 means they are completely the same. If the result is approaching to 1, it means the similarities between the resumé and the job advertisement are increasing. The system calculates cosine similarity in two different ways.

2.3.1   Boolean Weighting
The vectors of the resumé and the job advertisement have values that are based on their domain's lexicon of terms. When assigning values to the vectors of resumés and job advertisements, if either one of them has a lexicon of terms, the system initializes the vectors. The initialization of the values of the vectors are shown below.

If the resumé/job advertisement has the lexicon of terms = 1, if not = 0

Examples of the values of the job advertisement and resumé vectors and the lexicon of terms are shown below. If the resumé or the job advertisement has a value in the lexicon of terms, the value of their vector is assigned to be 1. If not, it is 0.

Term lexicon	Job advertisement vector	Resume vector
.NET	1	0
C#	1	0
ASP.NET	0	0
HTML	0	1
JAVA	0	1
Hibernate	0	0
XML	1	1
MS SQL Server	1	0
PL/SQL	0	1
iOS	0	0
Ireport	0	0
J2EE	0	1
J2ME	0	1

Based on these vectors of the resumé and the job advertisement, the resumé has HTML, JAVA, XML, PL/SQL, J2MEE, J2ME terms while the job advertisement has .NET, C#, XML ve MS SQL Server terms. According to these values, the similarity between job advertisement and resume is calculated by cosine similarity.

2.3.2 Assigning Vector Values with Term Clusters

In addition to the concern of the term being in the resumé or the job advertisement, the relationships between the terms are also important, adding that it can affect matching results. To include term clusters in the matching process, the system assumes that the terms which are in the same group are related. In this way, if the resume/job description has a term which is in the same group with a term that the resume/job advertisement has, instead of assigning the value of 0 to the vector, the frequency of terms that used together (0.3, 0.4, 0.5.. etc.) is assigned. Thereby, the vectors of the job advertisement and the resumé come closer and the distance between vectors decreases.

3 Experiments

The experiments are applied to the positions of Software Engineering and Accounting Specialist. We randomly select 10 job advertisements and 100 resumés for each domain. Then the system calculates the matching with Cosine Similarity. The system selects 5 highest scoring resumés for each one of the 10 job advertisements. Table 4 shows some examples of job advertisement – resumé matching results.

Table 4. Matching results

ID	Method	JA_ID	ResumeID	Result	Index
1	1-01	112	106890930	0.3651483	1
2	1-01	112	102318815	0.2637521	2
3	1-01	112	110299507	0.2581988	3
4	1-01	112	9463976	0.2581988	4
5	1-01	112	11777808	0.2306328	5
6	1-01	156	215689	0.3481553	1
7	1-01	156	102318815	0.3405026	2
8	1-01	156	2563345	0.3333333	3
9	1-01	156	108766850	0.3333333	4
10	1-01	156	105445360	0.3187883	5
11	1-01	163	102318815	0.4865336	1
12	1-01	163	108766850	0.4082482	2
13	1-01	163	8624239	0.3644054	3
14	1-01	163	215689	0.3553345	4
15	1-01	163	11777808	0.3403516	5
16	1-W	112	1307469	0.5207212	1

(Continued)

Table 4. (*Continued*)

ID	Method	JA_ID	ResumeID	Result	Index
17	1-W	112	308945	0.3693846	2
18	1-W	112	106890930	0.3651483	3
19	1-W	112	7134122	0.3132371	4
20	1-W	112	102318815	0.2637521	5
21	1-W	156	7134122	0.4235080	1
22	1-W	156	215689	0.3481553	2
23	1-W	156	102318815	0.3405026	3
24	1-W	156	2563345	0.3333333	4
25	1-W	156	108766850	0.3333333	5
26	1-W	163	102318815	0.4865336	1
27	1-W	163	108766850	0.4082482	2
28	1-W	163	8624239	0.3644054	3
29	1-W	163	215689	0.3553345	4
30	1-W	163	11777808	0.3403516	5

In Table 4, the method represents the Cosine Similarity method (1-01: calculation without term clusters, 1: W calculation with term clusters). JA_ID means Job Advertisement ID and Index means the sequence of the resumés at top 5. To measure the accuracy of our system, we define a gold standard which was determined by a Human Resources Specialist. We asked for the specialist to select 5 resumés that are the most suitable for the selected job advertisements. Table 5 shows the results of the resumé selection of the specialist for the job advertisements 112,156 and 163.

Table 5. Human resource specialist matching results

JA_ID	CV1	CV2	CV3	CV4	CV5
112	1321224	163083	41816	106890930	308945
156	41816	2563345	215689	102318815	106890930
163	102318815	41816	168821	215689	11777808

To find the accuracy of our system, we compare our results with the results of the Human Resources Specialist's performance, as shown in Table 6.

Table 6. Performance results

Method	Similarity
1-01	0.36
1-W	0.42

As seen, the matching method that uses the relationship between terms gives a more successful result than the Boolean weighting method. Thus, we can say that analysing the relationship between terms gets the system closer to finding the appropriate match.

4 Conclusions

Finding the best candidates that fit the requirements of the job is time consuming and necessitates human effort. Also, finding the best jobs that meet the qualifications and work experiences of the candidate has been an important research issue in recent years. The proposed system is the first system that works in Turkish. The system extracts the terms from job advertisements and creates a lexicon of terms. Afterwards, proposed system implements resume and job advertisement matching with Cosine Similarity. Based on performance results, the matching method that uses the term clusters gives better results. For future work, we are planning to calculate the performance results with the indexes as seen in Table 4. By providing this, the system can evaluate the performance in a more detailed way.

Acknowledgements. This study is supported by TÜBİTAK TEYDEB programme with the project number 3130841.

References

1. Crow, D., DeSanto, J.: A hybrid approach to concept extraction and recognition-based matching in the domain of human resources. In: Proceedings of the 16th IEEE International Conference on Tools with Artificial Intelligence (ICTAI). IEEE (2004)
2. Mochol, M., Wache, H., Nixon, L.J.: Improving the accuracy of job search with semantic techniques. In: Abramowicz, W. (ed.) BIS 2007. LNCS, vol. 4439, pp. 301–313. Springer, Heidelberg (2007)
3. Colucci, S., Di Noia, T., Di Sciascio, E., Donini, F.M., Mongiello, M., ve Mottola, M.: A formal approach to ontology-based semantic match of skills descriptions. J. Univ. Comput. Sci. 9(12), 1437–1454 (2003)
4. Mochol, M., Paslaru, E., ve Simperl, B.: Practical guidelines for building semantic eRecruitment applications. In: Proceedings of International Conference on Knowledge Management, Special Track: Advanced Semantic Technologies (AST) (2006). Gonzàlez, E., Fuentes, M.: A new lexical chain algorithm used for automatic summarization. In: Proceedings of the 12th International Congress of the Catalan Association of Artificial Intelligence (CCIA) (2009)
5. Le, B.T., Dieng-Kuntz, R., Ve Gandon, F.: On ontology matching problems for building a corporate semantic web in a multi-communities organization. In: Proceedings of the Sixth International Conference on Enterprise Information Systems, Kluwer, Porto (2005)
6. Ehrig, M., Sure, Y.: Ontology mapping - an integrated approach. In: Bussler, C.J., Davies, J., Fensel, D., Studer, R. (eds.) ESWS 2004. LNCS, vol. 3053, pp. 76–91. Springer, Heidelberg (2004)

7. Hassan, F., Ghani, I., Faheem, M., Hajji, A.: Ontology matching approaches for eRecruitment. In: Proceedings of ESWS, LNCS, vol. 3053, pp. 76–91. Springer (2004). International Journal of Computer Applications (2012)
8. Navigli, R., Velardi, P.: Learning domain ontologies from document warehouses and dedicated web sites. Association of the Computational Linguistics (2004)
9. Sclano, F., Velardi, P.: Term extractor: a web application to learn the common terminology of interest groups and research communities. In: TIA 2007

Splitting Algorithm for Detecting Structural Changes in Predictive Relationships

Olga Gorskikh[✉]

Department of Information and Service Economy, Aalto School of Business,
Runeberginkatu 14-16, 00100 Helsinki, Finland
olga.gorskikh@aalto.fi

Abstract. Change point analysis is crucial in many different fields of science because real world data are full of instability. In this paper, we introduce a new parametric technique that allows to perform multiple structural change point analysis in a single-dependent variable relationship. The main idea in the splitting method is a heuristic smart search for structural breaks with identification of corresponding significant variables at each stable period.

Keywords: Change point analysis · Predictive relationship · Parametric technique

1 Introduction

The goal of performing change point analysis is to detect distributional changes within observed data. A change point is defined as a point separating the data into parts where predictive relationships have different structure. Detecting such changes is important in many different areas such as climatology (Gallagher et al. [7]), bioinformatic applications (Mihin [15]) and finance (Fryzlewicz et al. [17]). For example, the parameters of financial models are generally highly dependent on time (prices of materials, information in the press). Thus, when the time of change is unknown, the problem of its detection becomes critically important.

Due to the relevance of the topic plenty of studies were conducted in this area. The history of change point analysis started sixty years ago (Page [16]). The first developed method allowed to detect changes in the mean of a sequence of independent random variables which was later extended also to dependent sequences. There are two main categories in which change point analysis is considered: classical change point detection (identifying changes in distribution) and structural change point detection (identifying changes in predictive structure). In general, an analysis of changes may be performed in either parametric or nonparametric settings. The main assumption of a parametric case is that the underlying distribution belongs to some known family. In this field, plenty of methods were developed for single variable settings, such as Davis et al. [5] analysing nonlinear processes, Lavielle and Moulines [12] and Lebarbier [13] both estimating changes in the means of time series, Harchaoui and Levy-Leduc [8] dealing

© Springer International Publishing Switzerland 2016
P. Perner (Ed.): ICDM 2016, LNAI 9728, pp. 405–419, 2016.
DOI: 10.1007/978-3-319-41561-1_30

with piecewise constant signals. At the same time some methods focus on multivariate data. Cho and Fryzlewicz [3] designed a solution for high-dimensional time series having second-order structure. Kolar et al. [11] introduced a technique allowing to estimate a changing structure of a varying-coefficient varying-structure model. A few years earlier, Bai and Perron [1] proposed an efficient algorithm for detecting multiple structural changes in the multivariate linear model. Another way to group change point analysis techniques in a regression study is to consider informational and Bayesian approaches. For instance, Ferreira [6] developed a method to identify a known number of change locations from a Bayesian point of view. The informational approach assumes using some criterion (for ex. Schwartz information criterion [18]) to reduce the number of computations in the likelihood-ratio procedure approaches. Brown et al. [2] studied series where regression coefficients were allowed to be polynomials in time. Overall, although the field of change point analysis in predictive relationships has a considerably long history, the most important part of methods focuses on finding a single change location or on detecting a known number of turning points. Multiple structural search in a high-dimensional space was briefly mentioned in a comparatively smaller number of works, and it almost always contains a comparison of different possible splitting combinations which might increase calculation time significantly. The main goal of this research is to introduce a new parametric method that allows to estimate:

1. the number of change-points
2. the locations of the change-points
3. significant variables at each of the stable periods
4. variables' weights at each of the stable periods

The splitting method is designed to analyse relationships having a single response variable and many explanatory variables. The main benefits of the proposed approach are: (i) reduction of computational costs: we consider consequent time splits from large to small in order to narrow down the search which significantly increases the method's efficiency; (ii) variable selection: we exclude insignificant variables from the analysis which also leads to a reduction in computational costs and a better understanding of models in general. The novelty of the proposed method is in considering the possible regions of change points instead of suspecting every point to be a turning one.

Having such information, it becomes relatively easy to analyse the data, to view its structure, and to understand its underlying trends. Such abilities are highly important, for example, in the field of economic behaviour, since it has a wide variety of data which needs to be analysed from different angles.

2   Preliminaries

The method proposed in this paper uses widely known techniques and approaches such as Lasso regularization and the Chow Test.

2.1   Lasso Regularization

In the case when there are a lot of insignificant features in the set, it is critically important to consider only valuable ones. Otherwise, taking into account all independent variables may lead to an enormous loss in efficiency. The Lasso regularization algorithm (Tibshirani [19]) addresses this issue. It is a shrinkage estimator: it "minimizes the residual sum of squares subject to the sum of the absolute value of the coefficients being less than a constant. Because of the nature of this constraint it tends to produce some coefficients that are exactly zero". Suppose we have the following data $\{x_{i1}...x_{ip}, y_i\}_{i=1}^{N}$, where N is the number of observations, p is the number of features, x_{ij} corresponds to the independent features, and y_i is the response variable. Let $\{\beta_j\}_{j=1}^{p}$ be a vector of feature weights. For a non-negative turning parameter λ, Lasso solves the problem:

$$\{\hat{\beta}_j\}_{j=1}^{p} = \text{argmin} \left\{ \sum_{i=1}^{N}(y_i - \sum_{j=1}^{p} x_{ij}\beta_j)^2 + \lambda \sum_{j=1}^{p} |\beta_j| \right\}, \tag{1}$$

As λ increases, the number of non-zero components of β decreases. Hence, all insignificant features might be detected and excluded from further analysis. There are many different techniques in optimization theory which allow to solve the stated minimization problem, and in the current research we chose stochastic gradient descent (Zinkevich et al. [20]), since it is one of the simplest and most popular stochastic optimization methods.

2.2   Chow Test

G.C. Chow [4] proposed a method defining whether two sets of observations belong to the same linear regression or not. Although initially the author introduced this estimator in the field of economic relationships measurement, it still might be used in any field where issues related to structure comparison between different datasets exist. The main hypothesis in the Chow test is that the coefficients are equal for both sub-samples. The regression test statistic is given as

$$F(p, N - 2p) = \frac{(RSS_c - (RSS_1 + RSS_2))(N - 2p)}{(RSS_1 + RSS_2)p}, \tag{2}$$

where p is the number of features, N is the total number of observations, RSS_c is the residual sum of squares for the whole sample, RSS_1 is the residual sum of squares for the first data set, and RSS_2 is the residual sum of squares for the second data set. In the Chow test, if there is no significant statistical difference between two sets of observations, then the regression test statistic follows an $F(p, N - 2p)$ distribution.

3   Problem Setup

In the following section, we introduce the setup of a problem to be solved. A general model of the data looks the following way:

$$y_t = x_t'\beta(t) + \epsilon_t, \tag{3}$$

where y_t is a response variable, $t = 1..N$ is a time stamp, $x'_t \in \mathbb{R}^p$ is a vector of predictors, and regression coefficients $\beta(t) \in \mathbb{R}^p$ are modelled as functions of t. We consider a model with a fixed nature of structural changes (Fig. 1):

$$y_t = \sum_{m=1}^{M} x'_t \beta_m \chi_{T_m}(t) + \epsilon_t, \tag{4}$$

$$\chi_{T_m}(t) = \begin{cases} 1, & t \in T_m \\ 0, & \text{otherwise} \end{cases}$$

where $\{T_m\}_{m=1}^{M}$ is the partitioning of timeline T. Here $T = \bigcup_{m=1}^{M} T_m$ and $T_m \cap T_{m'} = \emptyset$ for every $m \neq m'$.

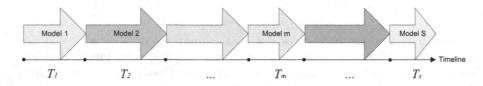

Fig. 1. Structural changes illustration.

Each model is described with its own set of coefficients for each variable. Our key interest is to solve the following estimation problem:

$$\{\hat{T}_m, \hat{\beta}_m\}_{m=1}^{M} = \underset{\{T_m, \beta_m\}_{m=1}^{M}}{\operatorname{argmin}} \left\| \sum_{t=1}^{N} (y_t - \sum_{m=1}^{M} x'_t \beta_m \chi_{T_m}(t)) \right\|_{L_2}, \tag{5}$$

Thus we are trying to find such regression coefficients $\{\beta_m\}_{m=1}^{M}$ and time partitioning $\{T_m\}_{m=1}^{M}$ which minimize the difference between the predicted value of a response variable and its actual value.

The overall task is to find:

1. the number of structural change points $M - 1$
2. the positions of structural change points $\{K_j\}_{j=1}^{M-1}$
3. S linear models described by the matrix $\mathbb{B} \sim M \times n$
4. variables contributing at each period, i.e. those with $\beta_j^k \neq 0$

4   Splitting Algorithm

In this section we introduce the splitting algorithm performing change point analysis. First, we discuss its assumptions and input parameters. After that describe the algorithm itself.

4.1 Main Characteristics

The **parameters of the algorithm** are presented in Table 1. All the parameters are needed to perform the first- and second-round splits as well as to properly estimate the models at each step.

Table 1. Splitting method parameters

Parameter notation	Parameter meaning
Θ	Overall number of variables.
θ	The maximum number of variables. Assumed to be smaller than Θ.
Δ	Number of points needed to evaluate a model with Θ variables well.
δ	Number of points needed to evaluate a model with θ variables well. Assumed to be smaller than Δ.

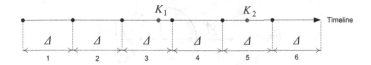

Fig. 2. Assumption 1 illustration.

Assumptions of the algorithm:

1. **Normality assumptions:**
   (a) Features within the dataset are assumed to be independent and normally distributed.
   (b) Error term ϵ_t is also normally distributed.
2. **Change point location assumptions:**
   (a) The distance between two structural change points should be big enough, particularly if we split the data into blocks of size Δ, then the block containing a jump point should be surrounded by blocks not containing such points. For instance (see Fig. 2), for the first change K_1, this means that periods 2 and 4 shouldn't contain change points. This assumption is required by the construction of the algorithm: we should be able to detect accurately sets of features contributing to the left and to the right of the change point.
   (b) First and last Δ points of the data shouldn't contain structural jump points. In Fig. 2, this means that periods 1 and 6 should be clear. This is a natural assumption since it's not possible to estimate a model correctly without having enough data.
3. **Other assumptions:**
   (a) Δ should be a multiple of δ. This assumption is needed due to the construction of the algorithm.

4.2   Detailed Description

The designed method is purely empirical. It is based on three steps of data splitting, which represent a so-called "smart search" of the jump points. Using such a technique leads to a significant increase in its efficiency. Instead of considering each point as a possible change point, at first we are trying to define large areas of suspected changes and then narrow them down until the exact points are found. More specifically, the splitting steps might be described as follows:

1. **First estimation: identifying Δ-areas of structural changes.** Dividing all the data into Δ-sized periods and analyzing those to find suspected ones.
2. **Second estimation: identifying δ-areas of suspected change points.** Extending selected periods by adding $\delta(\lfloor \Delta/(2\delta) \rfloor + 1)$-sized areas on each side, dividing the obtained points into δ periods and finding suspected ones.
3. **Identifying exact jump points.** Extending found δ periods by adding δ areas on each side and analyzing to find the exact jump points.

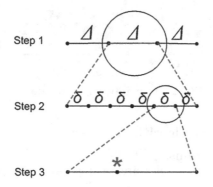

Fig. 3. Splitting algorithm general approach.

One of the main approaches used in the splitting method is the application of the Chow Test. We use it to compare neighbor periods in order to find out whether they have a similar structure. This knowledge allows us to detect suspicious behavior for further analysis. A detailed description of the splitting method is presented below (Fig. 3).

1. **First estimation: identifying Δ-areas of structural changes.** The aim of the first estimation is to define a set of current partitioning indexes S_1, corresponding to the Δ-periods possibly containing jump points and to determine variables' significance for a possible data split. At this step several actions are performed (Fig. 4):
   (a) First, the initial dataset is divided into Δ-sized parts. We assume that any of the parts may contain a potential structural change point. Thus, we get partitioning $\{T_i'\}_1^{N^*}$, $N^* = \frac{N}{\Delta}$.

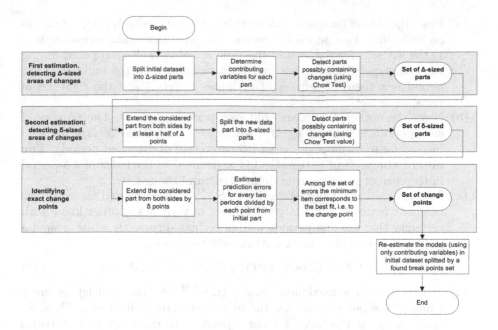

Fig. 4. Splitting algorithm general schema.

(b) We evaluate contributing variables at each period using the Lasso algorithm for feature selection. As a result, we have a set of different variable sets $\{Set_i\}_1^{N^*}$.

(c) After that, we estimate which of the periods might have a change based on the Chow Test statistic. In this test for a linear regression we use all the variables because we know that Δ samples is enough to assess a model with Θ variables. The regression test statistic follows an $F(\Theta, 2\Delta - 2\Theta)$ distribution. Based on the set of test indicators $\{CT_i\}_1^{N^*-1}$, we define the periods containing suspected jump points (if $CT_i = 1$, then block number $i + 1$ is the one we are looking for). The final set of suspected block numbers is denoted as S_1.

$$CT_i = \text{ChowTest}(T_i', T_{i+1}'), \quad i = 1..N^* - 1 \tag{6}$$

$$S_1 = \{i \in \{2, N^*\} | CT_{i-1} = true\} \tag{7}$$

2. **Second estimation: identifying δ-areas of suspected change points.** This step is performed for each of the periods in S_1 separately. The goal in this step is to narrow down the search to smaller areas around the change points. While at the previous step we found the Δ-areas, an output of the current method phase will be the δ-areas of possible points. Second step estimation includes the following sequence of operations:

(a) First, the period (suspected Δ-area) is extended by $\delta(\lfloor\Delta/(2\delta)\rfloor+1)$ areas on both sides. This period has correctly defined (according to assumption 1) the left set of variables and the right set of variables computed during the last stage (from the left and right neighbors). If the number of the period is i, then the left set of variables is Set_{i-1} and the right set of variables is Set_{i+1}.

(b) Next, the obtained data are split into periods of size δ. The number of such periods is denoted by M^*, where $M^* = \Delta/(2\delta) + 2(\lfloor\Delta/(2\delta)\rfloor + 1)$ (according to assumption 3a and construction of extended period). **The aim of the second estimation** is to define $\imath^*$ as the index of a δ-period in the current partition containing structural change.

(c) Then, the value of Chow statistics $\{CS_i\}_1^{M^*-1}$ is calculated for every two neighbor periods (using only the left set of variables in order know with certainty that the first periods will be evaluated correctly while around a potential jump point the evaluation will be wrong).

$$CS_i = \text{ChowValue}(T_i'', T_{i+1}''), \quad i = 1..M^* - 1 \qquad (8)$$

(d) Lastly, we find a maximum among $\{CS_i\}_1^{M^*-1}$. This will let us know which of the period pairs has the biggest structural difference. Thus, we are looking for the index i^{**} corresponding to the maximum statistics value:

$$i^{**} = \underset{i\in[1,M^*-1]}{\text{argmax}} \{CS_i\} \qquad (9)$$

In order to decide which of the periods we should choose as the final area to be found, we simply calculate the squared error (SSE) to the left and to the right of i^{**}. The period with the maximum error contains a jump point since a bigger error indicates a bigger loss in estimation which means that this period is more likely to contain a change.

$$i^* = \text{argmax}\{SSE(T_{i^{**}}''), SSE(T_{i^{**}+1}'')\} \qquad (10)$$

3. **Identifying exact jump points.** The last phase of the search is a straightforward estimation of each point within the selected δ-period T_{i^*}''. At the current stage, we extend the period by a δ-area on both sides. We denote the total interval considered in this step as T''', $|T'''| = 3\delta$. This period has the same left set of variables and right set of variables computed during the first estimation step (from the left and right neighbors), already used in the second estimation.

(a) First, we calculate δ pairs of models, each time using left variables for left models $\text{Model}_{\text{left}}$ and right variables for right models $\text{Model}_{\text{right}}$. Each time we use data of size δ for linear regression (LR). The aim of using different variable sets is natural since the estimation error and estimation time will be the smallest if we consider only contributing variables from each side.

$$\begin{pmatrix} \text{Model}_{\text{left}} \\ \text{Model}_{\text{right}} \end{pmatrix}_i^T = \begin{pmatrix} \text{LR}(T'''[i, i+\delta] \\ \text{LR}(T'''[i+\delta+1, i+2\delta+1]) \end{pmatrix}^T, i = 1..\delta \qquad (11)$$

(b) Then we calculate the array $\{error_i\}_1^\delta$ containing δ items, each of which corresponds to a pair and equals $|SSE_{left}| + |SSE_{right}|$ for each pair. The smallest item corresponds to a pair that produced the smallest squared error, which means that the splitting point between the data for this pair is the exact change point.

$$error_i = |SSE(\text{Model}_{\text{left}})_i| + |SSE(\text{Model}_{\text{right}})_i|, \quad i = 1..\delta \qquad (12)$$

$$i^{***} = \underset{i}{\arg\min}\{error_i\} \qquad (13)$$

(c) Thus, we find an index of a change point i^{***} within the observed period; we might easily convert this index to the one with respect to initial data.

Algorithm 1. Splitting algorithm

1: **Begin First stage**:
2: Split the data into Δ-sized parts
3: Using Lasso, calculate contributing variables for each part
4: For each two neighbor parts calculate Chow Test

5: **if** Chow Test showed difference **then**
6:   Extend second period by $(\lfloor \Delta/(2*\delta) \rfloor + 1)*\delta$-sized on both sides
7:   Split the extended period into δ-sized parts

8:   **Begin Second stage**:
9:   For each two neighbor parts calculate Chow Test statistic value (using left set of variables)
10:   Find a part with maximum Chow Test value
11:   **End Second stage**.

12:   **Begin Third stage**:
13:   Extend found period by δ-sized parts on both sides
14:   Calculate neighbor pairs of models using left and right sets of variables
15:   Calculate sum of absolute squared errors of the models
16:   **The point of a minimum sum is the change point**
17:   **End Third stage**.

18: **End First stage**.

Having the whole set of jump points we may re-estimate the models within each of the found periods to find the final weights and contributing variables. The main benefit of the described method is efficiency. The splitting procedure of finding change regions from larger to smaller allows to skip the excess estimations and point considerations. The described algorithm is simple and intuitively natural because in general it is based on the comparison of regression errors.

5   Experiments and Results

5.1   Evaluation Preliminaries

To evaluate the efficiency and correctness of the proposed algorithm we need to determine a naive baseline as well as exact formulation of the changes magnitude.

1. **Naive baseline.** To compare the computational cost of the splitting method we introduce a simple baseline for detecting a single change point. Assume we have the same natural assumption as for the proposed algorithm: first and last Δ points cannot contain a change. The idea of the naive method is to consider every point as a possible jump point by minimizing the sum of absolute squared errors obtained from the left and right part estimations. Assume we have N observations. The naive method might be described as follows:
   As it is clear from the outlined method, a linear regression estimation is run $2 * (N - \Delta)$ times, which is not critical if the size of the data is small but enormously expensive otherwise.

Algorithm 2. Naive algorithm

1: point = Δ
2: **while point!=(datasize-Δ+1): do**
3:        LeftModel = linear model of a period [0; point-1]
4:        RightModel = linear model of a period [point, N-1]
5:        Error = |SSE(LeftModel)|+|SSE(RightModel)|
6:        point = point + 1
7: Change point = point with minimum Error

---

2. **Magnitude of a change.** One of the representative features in change point analysis is the magnitude sensitivity. In order to compare how well the splitting method works with different changes, we need to define the magnitude. Assume we have two models with weight vectors $\boldsymbol{\beta_1}, \boldsymbol{\beta_2}$, where p is the total number of explaining variables. Then the magnitude of a change is given by

$$M(\boldsymbol{\beta_1}, \boldsymbol{\beta_2}) = \sqrt{\sum_{j=1}^{p}(\beta_{1j} - \beta_{2j})^2}. \tag{14}$$

5.2   Results

While evaluating the effectiveness of the splitting method, we assessed the data set with one change in order to be able to apply also a naive method for comparison. The evaluation setup was as follows:

1. We ran 12 simulations with different data size N from 360 to 1680
2. $\Theta = 10$, $\theta = 5$, $\Delta = 120$, $\delta = 40$

3. Magnitude of a change: $M = 2.5$, more precisely:

$$y_i = \begin{cases} 2x_{i1} + x_{i2} + 4x_{i3} + x_{i4} + 3x_{i5}, & \text{if } i \in [1, \frac{N}{3} - 1] \\ 2x_{i1} + x_{i2} + 4x_{i3} + (1 + M)x_{i4} + 3x_{i5}, & \text{if } i \in [\frac{N}{3}, N] \end{cases}$$

4. $(x_6, x_7, x_8, x_9, x_{10})_i$ do not contribute to y_i for $i = 1..N$
5. $X \sim i.i.d, N(0, 1)$ and $\epsilon_t \sim i.i.d, N(0, 0.1)$
6. The change occurs at $\frac{N}{3}$

Basically, for each N from the chosen range [360,1680] we generated a dataset containing 10 independent dimensions and one dependent, having one change point occurred at $\frac{N}{3}$ causing a difference of a magnitude $M = 2.5$.

All estimations were run on the same machine having an Intel Core i5-4310CPU 2.00 GHz processor with 8.00 GB installed memory (RAM). The resulting curves are presented in Fig. 5. While the naive method's time costs increase dramatically with growth in data size, the splitting method's costs remain nearly the same and stay under the five-minute rate. These results are natural due to the construction of both algorithms.

To assess the accuracy of the proposed splitting method we varied the magnitude of changes. This approach detects how sensitive the algorithm is to different changes. The experiment's setup was the following:

1. We varied magnitude M from 0.10 to 4.00
2. $\Theta = 10$, $\theta = 5$, $\Delta = 120$, $\delta = 40$
3. $N = 1560$
4. $X \sim i.i.d, N(0, 1)$ and $\epsilon_t \sim i.i.d, N(0, 0.1)$
5. The change points are $\{138, 405, 701, 911, 1395\}$

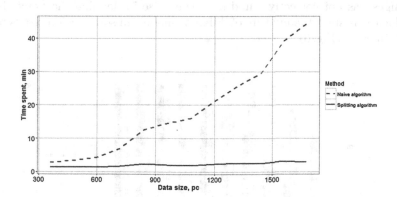

Fig. 5. Computational costs. 12 simulations with different data size from 360 to 1680, $\Theta = 10$, $\theta = 5$, $\Delta = 120$, $\delta = 40$. Magnitude of a change is $M = 2.5$, $X \sim i.i.d, N(0, 1)$, $\epsilon_t \sim i.i.d, N(0, 0.1)$, the change occurs at $\frac{N}{3}$.

Under these settings for each $M_j \in \{0.1, 0.2, 0.3, 0.5, 1, 1.5, 2, 2.5, ..., 4\}$ we generated a separate dataset based on the following linear model:

$$y_i = \begin{cases} 2x_{i1} + x_{i2} + 4x_{i3} + x_{i4} + 3x_{i5}, & \text{if } i \in [1, 137] \\ 2x_{i1} + x_{i2} + 4x_{i3} + (1 + M_j)x_{i4} + 3x_{i5}, & \text{if } i \in [138, 404] \\ (2 + M_j)x_{i1} + x_{i2} + 4x_{i3} + (1 + M_j)x_{i4} + 3x_{i5}, & \text{if } i \in [405, 700] \\ (2 + M_j)x_{i1} + x_{i2} + 4x_{i3} + (1 + 2M_j)x_{i4} + 3x_{i5}, & \text{if } i \in [701, 910] \\ (2 + 2M_j)x_{i1} + x_{i2} + 4x_{i3} + (1 + 2M_j)x_{i4} + 3x_{i5}, & \text{if } i \in [911, 1394] \\ (2 + 2M_j)x_{i1} + (1 + M_j)x_{i2} + 4x_{i3} + \\ \quad + (1 + 2M_j)x_{i4} + 3x_{i5}, & \text{if } i \in [1395, 1560] \end{cases}$$

At each of the stable periods only first five variables contribute to the dependent y_i while the rest $(x_6, x_7, x_8, x_9, x_{10})_i$ do not effect on the response. It is crucial to mention that the magnitude is the same for every change in a single simulation set. For instance, if we consider $M_3 = 0.3$ then this means that for the third dataset each of five structural changes has the magnitude M_3. After all the datasets are generated, we analyze them with the splitting algorithm. The aim is to assess the method's accuracy as well as to estimate the effect of the changes' magnitude on the final result.

The histograms below show the results of the second round simulations. The splitting method is unable to detect the changes of magnitude $M = 0.1$. This is predictable since such a small change is not detectable by the Chow Test. However, the method performs already better with a slightly bigger magnitude of 0.3. Even though the changes are not detected precisely, 60 % of the them were found nonetheless. Naturally, since accuracy grows with the magnitude, it is under 100 % for the first half of the simulations. The cases when the number of changes was not correctly found have the noise in defining the exact change point locations since the mixture of different models in some regions leads to the growth in error (Fig. 6).

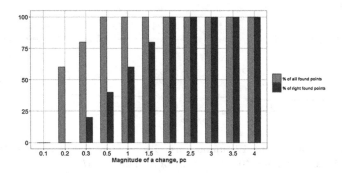

Fig. 6. Percentage of points detected. Magnitude varies from 0.10 to 4.00, $\Theta = 10$, $\theta = 5$, $\Delta = 120$, $\delta = 40$, $N = 1560, X \sim i.i.d, N(0, 1)$, $\epsilon_t \sim i.i.d, N(0, 0.1)$. The change points are $\{138, 405, 701, 911, 1395\}$.

The exact errors in detected change locations are presented in Fig. 7. As the magnitude grows, the difference between found and predicted points decreases. While for $M = 0.20$ none of the points were defined correctly, already for $M = 0.30$ this value became 25 %, and for the biggest magnitude of $M = 2.00$, it reached 100 %. The observed highest errors 6-20 and 20-35 correspond to the smallest M, which was also expected as the low difference between models' effects' prediction accuracy.

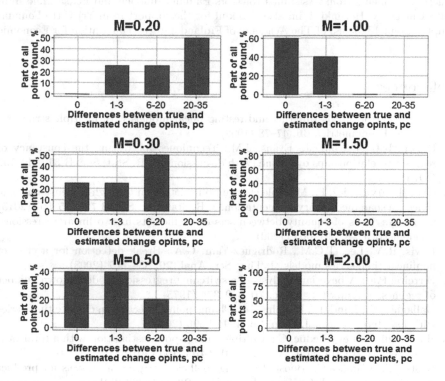

Fig. 7. Errors of points detected. Magnitude varies from 0.20 to 2.00, $\Theta = 10$, $\theta = 5$, $\Delta = 120$, $\delta = 40$, $N = 1560$, $X \sim i.i.d, N(0, 1)$, $\epsilon_t \sim i.i.d, N(0, 0.1)$. The change points are $\{138, 405, 701, 911, 1395\}$.

6 Summary

The field of structural change point analysis has a comparatively long history, and there are many different ways in which it may be performed. The known methods vary from purely parametric to nonparametric, from univariate settings to multivariate. In the current paper, we proposed a new efficient technique allowing to estimate structural changes in high-dimensional predictive relationships. Its approach is intuitively simple due to its natural construction. By using the Chow Test and least square linear regression we bind the application of the method by distributional assumptions. A few studies were conducted in the field of nonparametric and/or purely heuristic change point analysis such as Wang and Hastie [9], Davis et al. [5], Khaleghi and Ryabko [10]. Making as few assumptions as possible is crucial in such type of problems, hence the future developments will focus on transforming the method from parametric to a nonparametric.

Acknowledgements. I would gratefully acknowledge the help, support and very constructive comments from Assistant Professors Pauliina Ilmonen and Pekka Malo from Aalto University, Helsinki. I am also thankful for the support from Yrj Uitto Foundation, the grant 13253583 of The Academy of Finland and the Foundation for Economic Education.

References

1. Bai, J., Perron, P.: Estimating and testing linear models with multiple structural changes. Econometrica **66**, 47–78 (1998)
2. Brown, R.L., Durbin, J., Evans, J.M.: Techniques for testing the constancy of regression relationships over time (with discussion). J. R. Stat. Soc. B **37**, 149–192 (1975)
3. Cho, H., Fryzlewicz, P.: Multiple change-point detection for high-dimensional time series via sparsified binary segmentation. J. R. Stat. Soc. Ser. B **77**, 475–507 (2015)
4. Chow, G.: Tests of equality between sets of coefficients in two linear regressions. Econometrica **28**, 591–605 (1960)
5. Davis, R.A., Lee, T.C.M., Rodriguez-Yam, G.A.: Break detection for a class of nonlinear time series models. J. Time Ser. Anal. **29**, 834–867 (2008)
6. Ferreira, P.E.: A bayesian analysis of a switching regression model: known number of regimes. J. Am. Stat. Assoc. **70**, 370–374 (1975)
7. Gallagher, C., Lund, R., Robbins, M.: Changepoint detection in climate time series with long-term trends. J. Clim. **26**, 4994–5006 (2013)
8. Harchaoui, Z., Levy-Leduc, C.: Multiple change-point estimation with a total variation penalty. J. Am. Stat. Assoc. **1480–1493**, 105 (2010)
9. Hastie, T., Wang, J.C.: Boosted varying-coefficient regression models for product demand prediction. J. Comput. Graph. Stat. **23**, 361–382 (2014)
10. Khaleghi, A., Ryabko, D.: Nonparametric multiple change point estimation in highly dependent time series. In: Jain, S., Munos, R., Stephan, F., Zeugmann, T. (eds.) ALT 2013. LNCS, vol. 8139, pp. 382–396. Springer, Heidelberg (2013)
11. Kolar, M., Song, L., Xing, E.P.: Sparsistent learning of varying-coefficient models with structural changes. In: Advances in Neural Information Processing Systems, pp. 1006–1014 (2009)
12. Lavielle, M., Moulines, E.: Least-squares estimation of an unknown number of shifts in a time series. J. Time Ser. Anal. **21**, 33–59 (2000)
13. Lebarbier, E.: Detecting multiple change-points in the mean of gaussian process by model selection. Sig. Process. **85**, 717–736 (2005)
14. Matteson, D.S., James, N.A.: A nonparametric approach for multiple change point analysis of multivariate data. J. Am. Stat. Assoc. **109**, 334–345 (2014)
15. Minin, V.N., Dorman, K.S., Fang, F., Suchard, M.A.: Dual multiple change-point model leads to more accurate recombination detection. Bioinformatics **21**, 3034–3042 (2005)

16. Page, E.S.: Continuous inspection schemes. Biometrika **41**, 100–115 (1954)
17. Schrder, A.L., Fryzlewicz, P.: Adaptive trend estimation in financial time series via multiscale change-point-induced basis recovery. Stat. Interface **6**, 449–461 (2013)
18. Schwartz, U.: Estimating the dimension of a model. Ann. Stat. **6**, 461–464 (1978)
19. Tibshirani, R.: Regression shrinkage and selection via the lasso. J. R. Stat. Soc. Ser. B (Methodological) **58**, 267–288 (1996)
20. Zinkevich, M., Weimer, M., Li, L., Smola, A.J.: Parallelized stochastic gradient descent. In: Advances in Neural Information Processing Systems, pp. 2595–2603 (2010)

Early Predictive System for Diabetes Mellitus Disease

Karim M. Orabi[1(✉)], Yasser M. Kamal[2], and Thanaa M. Rabah[3]

[1] Information System Department,
Arab Academy for Science Technology and Marine, Cairo, Egypt
Karim_orabi@hotmail.com
[2] College of Computing and Information Technology,
Arab Academy for Science Technology and Marine, Cairo, Egypt
dr_yaser_omar@yahoo.com
[3] Public Health and Preventive Medicine, National Research Center,
Cairo, Egypt
thanarabah@yahoo.com

Abstract. Diabetes is a menacing disease, which can cause death without any cautions. In this paper we introduce a way to assist people by raising an alert for precautions. It is a prediction system for the diabetes disease, which will predict whether to be a candidate and at what age. The datasets are for Egyptian diabetes patients, 2/3 will be used for training and 1/3 will be used for testing. This system is based on the machine learning concept, by using decision tree technique. This paper introduces a new idea in prediction and differs from previous papers, which focused on classification prediction to answer a yes or no question only. This contribution was new in the prediction system, by adding a regression technique with a randomization code to predict the age. The results were promising, the system predicts diabetes incidents at what age, with accuracy 84 %.

Keywords: Component · Diabetes · Diabetic · Decision tree · Machine learning · CDC · Prediction

1 Introduction

Diabetes Mellitus is a long standing disease that affects people either when hormone of the insulin is produced by the pancreas and the body cells cannot use it, or the pancreas fails to produce enough insulin [1]. There were 1.5 million deaths caused in 2012 by the diabetic disease. Also World Health Organization released statistics proved about 9 % of adults and older people got diabetes in 2014 [2].

Most of the previous research used the data mining techniques and the machine learning in order to build systems to predict whether any person could be a diabetic or not. These systems were built in order to raise an alert for anyone who is about to be a diabetic. But in this paper we introduce a new contribution, as in the previous works all the predictions are either based on yes or no question. The new contribution is to give a regression prediction, as to predict whether a person could be a candidate of diabetes

© Springer International Publishing Switzerland 2016
P. Perner (Ed.): ICDM 2016, LNAI 9728, pp. 420–427, 2016.
DOI: 10.1007/978-3-319-41561-1_31

and at what age. Rotation mechanism is used in order to randomize the process of learning and testing, and then the average is calculated for each iteration to get the accuracy for comparing the final results.

2 Related Works

The prediction of Diabetes was adopted in much research in the past. The first research was held in 2008 and it was the diabetes Data Analysis and Prediction Model Discovery Using RapidMiner. The techniques used in this research were rapid miner and ID3 Algorithm. The results reached in this research were plasma-glucose which is main attribute that leads to diabetes and the system predicts about 200 patients develop diabetes and 523 that do not, of those 200 patients, 19 are wrong this means that the system has about 72 % of accuracy. Furthermore another algorithm ID3, which predicts 231 patients develop diabetes and 492 that do not. This means ID3 algorithm has 80 % of accuracy [3].

The second research adopted the idea of prediction in 2011 and was the prediction on diabetes using the data mining approach. In this research the data mining approach used and resulted in people with high blood pressure diagnosis with value 2 or −1, then the probability of not being affected by diabetes is 98 % and in the other case the probability to be affected is 2 % [4].

In 2012, an international conference posted the following research Study of Type 2 Diabetes Risk Factors Using Neural Network for Thai People and Tuning Neural Network Parameters, the author in this research studied the risk factors on Thai people datasets. He also mentioned research that used Logistic regression and the risk factors they reached [5–12]. The used method in this research is neural network and different experiments were used to compare the result. Finally the best fitting number of the hidden nodes is 50 since, it means of RMSE.

In 2014, A Survey on Data-Mining technologies for prediction and diagnosis of diabetes [13] was made. In this research the author did a survey on the techniques used on diabetes prediction. These are samples from the Techniques used in this survey:

Patil, generated association rules on numeric data. They used pre-processing to improve the quality of data by handling the missing values and applied an equal interval with approximate values based on medical expert's advice to Pima Indian diabetes data [14].

Nuwangi used advanced and reliable data mining techniques to identify different risk factors behind the diabetes and the relationship between diabetes and other diseases [15].

The conclusion reached that, mechanical life style and obesity have contributed to the sudden rise of type 2 diabetes worldwide. Furthermore, the analysis of different research papers proved, evident that the occurrence of diabetes has a strong relation with diseases like, Wheeze, Edema, and Oral disease. So the use of data mining techniques can be helpful in the early detection for the diabetes disease.

3 Datasets Preparation

The Datasets were collected by the Egyptian National Research Center based on a standard format. This standard format is a questionnaire form, this form is questions regarding the risk factors of the diabetes disease for diabetes patients. These questions were asked for diabetic patients. After that these forms were extracted into a statistical tool called SPSS, for doing statistical analysis on these data and finally were exported into an excel sheet, in order to be used in our experiments. These datasets were collected from 6-years. The Dataset comprises 23 features; these features are summarized in Table 1.

Table 1. Description of dataset feature

No.	Feature name	Type	Range
1	Diabetes age	Numeric	Real values
2	Gender	Numeric	Categorical
3	Education	Numeric	Categorical
4	Diabetic family member	Numeric	Categorical
5	Smoker	Numeric	Categorical
6	Cigarette number	Numeric	Real values
7	Smoking start date	Date	Real values
8	Exercising status	Numeric	Categorical
9	Frequent exercise per week	Numeric	Real values
10	Exercise type	Numeric	Categorical
11	Food type	Numeric	Categorical
12	Healthy food status	Numeric	Categorical
13	No of basic meals	Numeric	Real values
14	Snacks status	Numeric	Categorical
15	Snacks number	Numeric	Real values
16	Snack type	Numeric	Categorical
17	Regime status	Numeric	Categorical
18	Blood pressure status	Numeric	Categorical
19	Blood fat status	Numeric	Categorical
20	Foot complications	Numeric	Categorical
21	Neuro complications	Numeric	Categorical
22	Low vision status	Numeric	Categorical
23	Wound recovery status	Numeric	Categorical

4 Preprocessing

Most of the collected data is highly liable to be influenced unreliable, missing and insufficient due to the size or the transformed way from manually to computerized data. In that case the data will be unreliable leading to low quality results. So the data should be preprocessed in order to enhance the quality of data and to reach the best accuracy in prediction. On this section we will use different methods: First data cleaning to remove the null values and inconsistencies data. Second, data reduction, by removing the redundant features or features will not affect on the results of prediction. Third, data transformation or normalization by scaling and categorizing high range numeric features, to assist in improving efficiency and accuracy especially for the age prediction [16].

4.1 Data Cleaning

Data Cleaning is one of the major and important tasks in the preprocessing, routines filling or discard the null values. This task will increase the trust of the result accuracy, furthermore unreliable or uncleaned data can lead to confusion and unreliable results.

Ignoring diabetes 1 records. In our Case we are focusing on the patients of diabetes 2, the first rows were deducted for the peoples whose age was less than 19 years, are affected with diabetes 1 per medical reference which will be out of our scope.

Discard null values. Some of the values were not completed for some features and we don't have the contacts for these patients to correct the missing values. So to retain the consistency and accuracy we had to remove the whole record for the missing values.

Discard Discrepancy values. There are some values for the same records that are not inconsistence. For instance, the smoking feature with value smoking = 1 means he is smoking and the number of cigarettes = 0 which is inconvenient as he is smoking but the number of the cigarettes is zero. In that case we discard the whole row as we can't rely on this row.

4.2 Data Integration

We have only one source of data, so we didn't face the issue of data integration. The data was collected then inserted into the SPSS tool and exported to one excel sheet. The SPSS tool has feature to export the data to different programs. One of these programs is Microsoft office Excel.

4.3 Data Reduction

Some features were useless and will not add anything in the prediction. These features were eliminated from the data sets, based on a medical prescription where these features will be resulted in the complication phase, which is a late phase of the disease. In our case we are looking for the feature that assists us in the early prediction.

4.4 Data Transformation and Data Discretization

Normalization. When representing the data into a smaller value, it helps to avoid the dependence on the unit's selection [17]. The used transformation was min max normalization, Eq. 1:

$$(Vi - minA) / (maxA - minA) \qquad (1)$$

The Vi is the current value of the feature, minA is the minimum value in the column and the maxA represent the maximum value for that feature.

Data discretization. Transform the categorical data into a numerical value. By giving a weight value based on a medical scoring for example, the number of family members for the patients. This weight was based on the scoring sheet for International Diabetes federation; here is a sample of the score sheet:

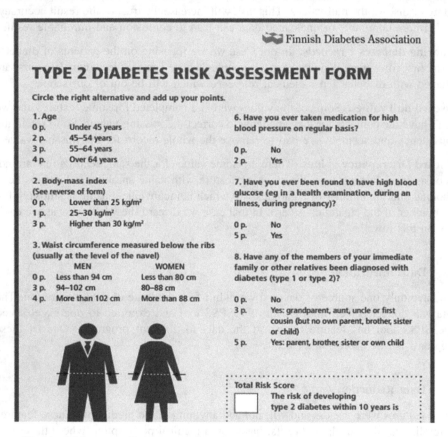

Fig. 1. Diabetes prediction score sheet (source: International Diabetes Federation)

On Fig. 1, is the diabetes Type two risk assessments, based on patient questionnaire by the international Diabetes Federation. It should be used, as the basis for developing national questionnaires, which take into account local factors.

The score of the test will result for a score; this score should represent the probability of developing type 2 diabetes over the following 10 years.

5 Experiment and Results

The patients' dataset is used twice, the first time where the age of the patients numerical and the second time by classifying the ages into 9 classes based on the statistical model, frequency distribution, where the class interval for the ages was calculated as follow: $60 - 31 = 29$, then the magnitude was calculated based on Sturge's rule, the number of the class can be calculated using the following Eq. 2:

$$NC = 1 + 3.222\text{Log } 300 = 9.251 = 9 \tag{2}$$

Where NC is the number of classes, log N which in our case 300 is the logarithm of the total number of observation [18]. Finally, the class intervals were determined. In order to calculate the precision, sensitivity and specificity the following equations were used to calculate the classification experiment:

True Positive [TP] = Condition Present + Positive result (1)
False Positive [FP] = Condition absent + Positive result [Type I error] (2)
False (invalid) Negative [FN] = Condition present + Negative result [Type II error] (3)
True (accurate) Negative [TN] = Condition absent + Negative result (4)
Precision (class) = TP (class)/(TP (class) + FP (class)) (5)
Sensitivity (class) = Recall (class) = TruePositiveRate (class) = TP (class)/(TP (class) + FN (class)) (6)
Specificity (mostly used in 2 class problems) = TrueNegativeRate(class) = TN (class)/(TN(class) + FP(class)) (7)

Moreover, the decision tree mechanism was applied on both data sets, in addition to adding a customized rotation mechanism, this rotation mechanism trains each portion of the data with random rotation to ensure that all the data used for both testing and training process. Another reason for using the rotation mechanism is to provide us with the smallest percentage of error and gets the best accuracy. The result of the experiment on the classified Age was as follow Table 2:

Table 2. Results of prediction

	Actual						
	Class	1	2	3	4	5	Total
Predict	1	6	8	0	1	1	16
	2	11	25	12	6	6	60
	3	1	1	6	2	0	10
	4	0	4	6	2	1	13
	5	0	0	1	0	0	1
Total		18	38	25	11	8	100

Then from this data we were able to calculate the precision, sensitivity and specificity as shown on the Table 3:

Table 3. Prediction accuracy

	Class				
	1	2	3	4	5
TP	6	25	6	2	0
FP	10	35	4	11	1
FN	12	13	19	9	8
TN	72	27	71	78	91
Precision (%)	37.5	42	60	15	0
Sensitivity (%)	33	65	24	18	0
Specificity (%)	88	43.5	95	88	99

The result of the second experiment on numerical age data was promising as the best rotation predicted with average accuracy 84 %, for the 100 case in the test phase. This indicates that if we keep working in applying the preprocessing and the association rules, we can reach to a better accuracy. The association rules are planned to be used in the next phase, in order to increase the percentage of the accuracy. So if we worked on increasing the percentage, we can give hope to reduce the number of patients and trying to get rid of the diabetes disease.

The following equation was used to calculate the average error percentage for the 100 predicted cases: Eq. 3

$$X = 100 \sum_{l=r}^{l=1} \frac{\max\left(\frac{Absolute(Y-Z)}{Z}\right)}{T})$$ (3)

X present the percentage of the error resulted from the accuracy. Y is the value of the predicted age resulted from the system. Z represents the actual value of the patient's age and finally T is the total number of the testing cases which is in our case would be 100, the 1/3 of the datasets. Summation the average value of each case.

6   Conclusion

The goal of this system is to add a new parameter in the prediction for the diabetes disease, this parameter is the age. The old studies and papers build the prediction system on whether this person would be a candidate of the disease or not, which is not enough to save him in a good and precise time; furthermore, to give people a warning alert before reaching the dangerous phase and lead to deadly complications. That is why we used in building this system randomization and classification, in order to reach the best accuracy from these data. The results of the system show that the accuracy of the experiment is promising, where the percentage reached 84 % by using the age as numeric. In the future we will add the association rules to get used of the critical risk factors, as we suspect the percentage of the accuracy may increase and lead us to better results.

References

1. World Health Organization: Global Health Estimates: Deaths by Cause, Age, Sex and Country, 2000–2012. WHO, Geneva (2014)
2. Global status report on noncommunicable diseases 2014. World Health Organization, Geneva (2012)
3. Han, J., Rodriguze, J.C., Beheshti, M.: Diabetes Data Analysis and Prediction Model Discovery Using RapidMiner (2008)
4. Repalli, P.: Prediction on Diabetes Using Data Mining Approach (2011)
5. Aekphakorn, W.: Diabetes risk score: Office of Health Information System, pp. 28–31 (2005)
6. Baa, C.A., et al.: Performance of a predictive model to identify undiagnosed diabetes in a health care setting. Diabetes Care **22**(2), 213–219 (1999)
7. Glumer, C., et al.: A Danish diabetes risk score for targeted screening: the Inter99 study. Diabetes Care **27**(3), 727–733 (2003)
8. Stern, M.P., et al.: Does the metabolic syndrome improve identification of individuals at risk of type 2 diabetes and/or cardiovascular disease? Diabetes Care **27**(11), 2676–2680 (2004)
9. Schmidt, M.I., et al.: Identifying individuals at high risk for diabetes: the Atherosclerosis Risk in Communities study. Diabetes Care **28**(8), 2013–2018 (2005)
10. Mohan, V., et al.: A simplified Indian Diabetes Risk Score for screening for undiagnosed diabetic subjects. J. Assoc. Physicians India **53**, 759–763 (2005)
11. Wilson, P.W., et al.: Prediction of incident diabetes mellitus in middle-aged adults: the Framingham Offspring Study. Arch. Intern. Med. **167**, 1068–1074 (2007)
12. Visang, K., Chialchanwattana, S., Sunat, K.: Risk factor analysis of diabetes mellitus diagnosis. Master's thesis in Computer Science, Konkan University, pp. 798–805 (2009)
13. Alby, S.: A survey on data-mining technologies for prediction and diagnosis of diabetes (2014)
14. Patil, B.M., Joshi, R.C., Toshniwal, D.: Association rule for classification of type-2 diabetic patients. In: Proceedings of the Second International Conference on Machine Learning and Computing, pp. 330–334 (2010)
15. Nuwangi, S.M., Oruthotaarachchi, C.R., Tilakaratna, J.M.P.P., Caldera, H.A.: Usage of association rules and classification techniques in knowledge
16. Han, J., Kamber, M., Pei, J.: Concepts and Techniques in Data Mining, 3rd edn. Morgan Kaufmann Publishers, Burlington (2012)
17. De Muth, J.E.: Basic Statistics and Pharmaceutical Statistical Application, 3rd edn. CRC Press, Boca Raton (2014)

Data Quality Visualization for Preprocessing

Markus Vattulainen[✉]

University of Tampere, Tampere, Finland
markus.vattulainen@gmail.com

Abstract. Preprocessing is often the most time-consuming phase in data analysis and interdependent data quality issues a cause of suboptimal modelling results. The design problem addressed in this paper is: what kind of framework can support visualization of data quality issue interdependencies for faster and more effective preprocessing? An object framework was designed that uses constructed features as a basis of visualizations. Six real datasets from business performance measurement system domain were acquired to demonstrate the implementation. The framework was found to be a viable preprocessing analysis supplement to both industry practice of exploratory data analysis and research benchmark of preprocessing combinations.

Keywords: Preprocessing · Data quality visualization · Feature construction

1 Introduction

Preprocessing can take as much as 85 % of the duration of a knowledge discovery project [35]. Although the importance of good data quality practices have long since been recognized in the businesses [46], preprocessing a given dataset and its possible data quality issues such as missing values, outliers, noise, insufficient variance, duplicates, class imbalance etc. often requires laborious manual work. Preprocessing can also be a cause of suboptimal results in knowledge discovery tasks. Several papers [12, 24, 45, 48, 49] envision that data preprocessing should have tools specifically designed for it, which would report data quality observations, guide preprocessing steps needed and evaluate preprocessing outcomes.

From the knowledge discovery research point of view, unrecognized or unresolved data quality problems add to the uncertainty of research findings. Complexity of data production, acquisition and integration processes demands recognition of multiple data quality dimensions simultaneously. Only a few of the dimensions such as incompleteness have computational operationalizations (e.g. labeling missing values, see [3]) and there is no solid theoretical guidance for operationalization of data quality dimensions for preprocessing. Most importantly, data quality issues can be interdependent and a method for empirical identification of these interdependencies is a gap in current knowledge.

The design problem addressed in this paper is: what kind of framework can support visualization of data quality issue interdependencies for faster and more effective preprocessing? The design problem is operationalized through its elements. First, focus is on visualization. Secondly, data quality is understood either as compliance to basic

© Springer International Publishing Switzerland 2016
P. Perner (Ed.): ICDM 2016, LNAI 9728, pp. 428–437, 2016.
DOI: 10.1007/978-3-319-41561-1_32

data quality requirements (e.g. regarding acceptable amount of missing values) or fitness for purpose in knowledge discovery tasks (e.g. classification accuracy achieved with the data). Data quality measurement is operationalized exclusively in the context of preprocessing as constructed data quality features and conceptual, organizational or information system related issues of data quality (see [38]) are not further elaborated.

The framework presented is demonstrated with six real datasets from the business performance measurement system domain and was found to be a viable preprocessing analysis supplement to both industry practice of exploratory data analysis and research benchmark of preprocessing combinations. The current study contributes to the existing body of knowledge on preprocessing by establishing data quality issues as objects, presenting a data issue visualization system as a system model, presenting an implementation of the system model and describing example relations between data quality constructs.

2 Related Research

There are proprietary and open source data profiling and preprocessing solutions embedded in database management systems (e.g. MS SQL Server data quality services), data integration systems (Talend, BizTalk, Informatica), big data platforms (Hadoop, Azure, Amazon) and data analysis systems (e.g. SAS, SPSS, RapidMiner, Knime). An earlier literature review [43] did not show, however, solutions that conceptualize data quality problems as objects and focus explicitly on data quality problem interdependencies.

Preprocessing consists of data cleaning, data integration, data reduction and data transformation [17] and includes several techniques [1, 27, 29]. The majority of the preprocessing papers have studied single data problems and competing corrections to them [43]. The defining feature of these studies is that a single data problem is isolated and transformed with techniques ranging from simple such as omitting rows with missing values to complex such as imputation of missing values by modelling. Combination studies [10, 11, 42] recognize the interdependence of preprocessing techniques. Contrary to single techniques and combination approaches that both transform data, incorporation studies do not aim to transform data problems but to incorporate them. Data issue interdependence visualizations and incorporation both build on feature construction.

Feature construction [16] is related to feature selection and extraction. Single preprocessing techniques can be modified to compute constructed features. Single data problem correction techniques include outlier detection [8, 13, 19, 25], missing values [20], class imbalance [31], feature selection [16, 28], low variance [26] and duplicates [50]. A survey on feature construction [40] and the conducted literature review pointed that there is a gap in knowledge in construction of data quality features. The current paper expands and applies constructed features developed by the author in [43] from knowledge discovery task of prediction to that of visualization and provides an R application to use constructed features.

3 Method

The method consist of the steps taken to solve a question [22]. Data quality issues as objects is established as a fundamental concept driving method selection and thus object-oriented design methods [39] are used.

For improved understandability and extensibility, the structure of the solution should reflect the structure of the problem domain. Industry standards such as CRISP-DM [9] build on the idea that there are 'things' in data such as missing values and outliers. They are abstractions of the data in a sense that attention is focused on a thing (e.g. missing values) and other qualities of the data (e.g. outliers) are at that moment ignored. Taking missing value imputation as an example: missing values can be represented in the raw data in several ways, then identified as such and coded as NAs. Number of missing values by row can be counted and constructed as a new feature. Transforming missing values can be done by imputing by mean of the variable and this imputation may be erroneous due to the outliers in the same variable. Thus data problems can intuitively be understood as objects having three distinct member functions: identification, transformation and feature construction.

4 Design

The system architecture is represented as a simplified component diagram highlighting the essential design choices (Fig. 1). Setup component is responsible for storing the feature definitions, which are created by inheriting them from PreprocessorBaseClass in the Feature construction component. Analysis component evaluates the definitions against the data (i.e. computes the numerical values of the constructed features) and Reporting component reports the results. As statistical software should not only allow exploration of the data but also be trustworthy [7], defensive programming [47] was applied to stop execution as early as possible in case of errors. Class, function and interface details are presented in [44].

5 Implementation

The design was implemented by using statistical language R [37], its S4 object system [7] and made freely available as package 'preproviz' [44] in CRAN repository.

5.1 Cases

Current business performance measurement systems aim to predict financial results from non-financial results [23] and rely on metrics design, data gathering and analysis [14]. Realizing the aims have not been as successful as expected [34]. To demonstrate constructed data quality features in business performance measurement system context six real corporate datasets (Toyota Material Handling Finland, Innolink Group, 3StepIt, M-Files, Papua Merchandising and Lempesti) were acquired. The main selection

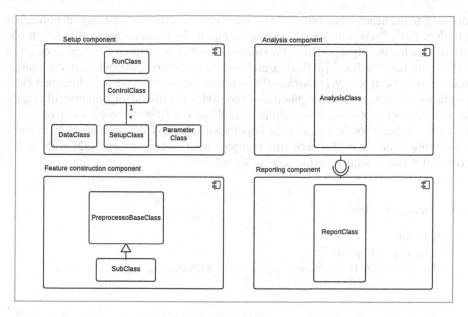

Fig. 1. Component diagram

criteria were the availability of the financial target feature in the data and at least two of the three other scorecard dimensions (customer, process and HR) included.

5.2 Features

Data quality features were constructed for the cases by selecting a set of data quality dimensions and operationalizing them with either supervised or unsupervised computations. Data quality is understood here either as deviation from a basic data quality requirement or fitness for purpose in knowledge discovery task. A data point is considered to be of good quality in basic data quality requirement sense if it has: 1. No or minimal amount of missing values, 2. Low probability of being an outlier, 3. Variance as opposite to flatness (e.g. respondent giving 4 on a 1 to 5 scale to all questions in a survey), 4. Low probability of being a duplicate and 5. Its nearest neighbor or neighborhood is homogenous and does not include changes in class labels. High quality in a sense of fitness for purpose if: 1. Its class probability is high in prediction task, 2. It adds to the clustering tendency of the data, 3. It is close to the cluster center (clustering by continuity ignored).

The constructed data quality features included were (expanding [43]): *MissingValueShare* as 'count the number of missing values in a row and divide it by the total number of features'. *MissingValueToClass* as 'MissingValueShare of a point divided by average MissingValueShare in the class'. *LenghtOfIQR* as 'min-max normalize the data and compute the length of IQR (inter-quartile range) for each row'. *DistanceToNearest* as 'min-max normalize the data and compute the Euclidean distance to the nearest data point'. This is an implementation of an entity identification problem [50]. *ClassificationCentainty* as 'compute the Random forest [4] class probability that has the highest value'. *ClassificationScatter* as 'compute the scatter count in 1/10 point neighborhood [21]'. *NearestPointPurity* as 'compare, whether the nearest point is of same class'. *NeighborhoodDiversity* as 'count the number of dominant class points in 1/10 data set size neighborhood and divide it with the total number of classes in the data set'. *LOFScore* as 'compute LOF scores (density-based local outliers)' [5, 41]. *MahalanobisDistance* as 'compute Mahalanobis distance to class center'. *ClusteringTendency* as 'for each row compute the Hopkins statistic [32] without the row and divide it with Hopkins statistic for all data'.

5.3   Demonstration

A simple run:

```
library(preproviz)
plotDENSITY(preproviz(tdata)) #tdata is R data frame
```

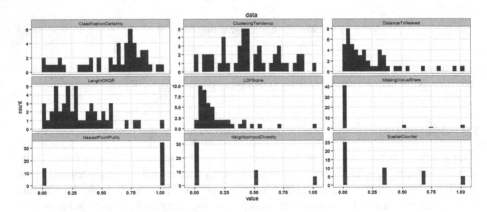

Fig. 2. Bar chart of constructed data quality features in the Toyota dataset

Figure 2 presents bar chart for each of the constructed features. It gives an overall exploration of how frequent the issues are. There are seven points with high LOFScore and roughly three groups of ClassificationCertainty. In LengthOfIQR some of the points in the left have low within-point variance, that is after normalization the values in each variables of the point are close to each other. ClusteringTendency distribution is multimodal. Figure 3. presents multidimensional scaling of constructed data quality features before and after preprocessing. It shows that 1. Data quality can form clear clusters (that is, data quality is not scattered randomly), 2. The cluster shape changes as a result of preprocessing but is still far from random scatter and 3. There are points that are outliers in their data quality profile. Figure 4. presents hierarchical clustering of data quality features by using correlation as a distance measure. It shows how data quality features correlate with each other. Innolink and 3StepIt data quality features are not correlated (exception: two missing value features in the Innolink case). Toyota and M-Files cases show more clear correlation (shorter vertical distances in forming clusters). Lempesti and Papua cases have almost identical data issue clusters. In M-Files case missing value issues are clearly separated from the other issues.

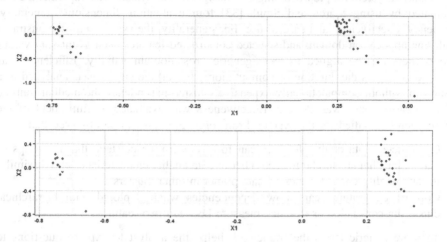

Fig. 3. Multidimensional scaling of constructed data quality features in the Toyota dataset before (top) and after (bottom) preprocessing shows two data quality clusters and only a small change in cluster shapes after preprocessing.

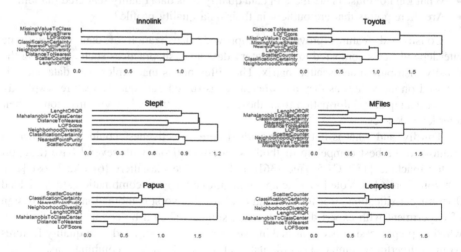

Fig. 4. Hierarchical feature clustering of constructed features for all cases (expanding [43]). Closely related features are shown by short vertical distance when forming a cluster.

6 Conclusions

Data quality issues were conceptually established as objects having three member functions: identification, transformation and feature construction. A data quality issue visualization system for preprocessing was designed. The system was implemented and evaluated by using March and Smith [33] four evaluation dimensions: generality, efficiency, effectiveness and ease of use. For generality, the system was demonstrated with one application domain and selected constructed features such as missing values and outliers can be argued to have generic cross-domain validity. Efficiency was operationalized as the duration of computations. For all the six datasets analyzed in a same run without computationally expensive ClusteringTendency the median runtime of 10 measurements was 10, 8 s. Effectiveness was evaluated as fulfillment of the requirements specified for the system. From Figs. 3 and 4 can be observed that:

- Constructed data quality features can form discernible clusters, that is groups of data points that are similar to each other in data quality characteristics and dissimilar to data quality characteristics of data points in other clusters
- Constructed features can show dependencies when explored with hierarchical clustering correlation as distance measure between two features.

In more generic terms the framework helps the analyst to explore questions to support preprocessing decision making in choices of preprocessing techniques and level of preprocessing achieved:

- Which data quality features are close related (linearly) to each other?
- What kind of clusters are there in data quality or is data quality scattered randomly?
- Are there points that are outliers in their data quality profile?

Constructed features provide an improved visualization of data quality issue interdependencies compared to exploratory data analysis practices such as plotting the density distribution and scatter matrix. The difference is that exploratory data analysis is based on the values as they are whereas constructed data quality features explicitly extract data quality information from those values and thus allow visualization of their interdependencies.

Thirdly, and what is not further studied here, effectiveness could be evaluated in relation to the best competing method. As a comparison, with several base (Support vector machine [18]; C4.5 Tree [36]) and ensemble classifiers [6] (Adaboost [15], Random Forest [4], Vote [30]) the above mentioned Toyota combination study yielded 0,16 misclassification error for best and 0.48 for the worst combination [42]. A design of experiment with expert data analysts as subjects should be conducted to find out, whether preprocessing decision making supported by constructed data quality features would outperform completely computational approach such as combinations.

The practical implications are that the framework can be used as basis to build production level systems for faster and more effective preprocessing. The research implications are that constructed data quality feature visualizations as shown in the demonstrations is a relevant research topic and can shed light to data quality interdependencies.

There are three limitations and further needs identified. First, currently the system deals with only one of five fundamental knowledge discovery task, namely characterization as visualization. The system should be extended to embed constructed features for prediction, clustering, association rules and outliers. Of these there is already evidence that prediction can benefit from constructed data quality features at least in the business performance measurement system domain (i.e. counterintuitively, the class to be predicted relates to data quality issues) [43]. Obviously, the effect of adding more dimensions [2] should be taken into consideration. Secondly, the current system does not take into account the competitive perspective of preprocessing combinations. Visualization of data quality interdependencies may help to provide insight into combinations, which are essentially black boxes as far understanding of interdependencies is concerned. In contrary to combinations visualizations do not, however, in themselves resolve the data issues. Thirdly, currently the system does not provide numerical statistics and their reliability estimates. As one of the goals of preprocessing is transform systematic data quality patterns to random, there is a need to develop numerical tools such as mutual information of constructed data quality features.

References

1. Abdul-Rahmana, S., Abu Bakara, A., Hussein, B., Zeti, A.: An intelligent data pre-processing of complex datasets. Intell. Data Anal. **16**, 305–325 (2012)
2. Bellman, R.: Dynamic Programming. Rand Corporation, Princeton University Press (1957)
3. Berthold, M.R., Borgelt, C., Höppner, F., Klawonn, F.: Guide to Intelligent Data Analysis – How to Intelligently Make Sense of Real Data. Springer, London (2010)
4. Breiman, L.: Random forests. Mach. Learn. **45**, 5–32 (2001)
5. Breunig, M.M., Kriegel, H-P., Ng, R.T., Sander, J.: LOF: identifying density-based local outliers. In: Proceedings of the ACM SIGMOD 2000 International Conference on Management of Data, pp. 93–104 (2000)
6. Caruana, R., Niculescu-Mizil, A., Crew, G., Ksikes, A.: Ensemble selection for libraries of models. In: Proceedings of ICML, p. 18 (2004)
7. Chambers, J.: Software for Data Analysis. Springer, New York (2008)
8. Chandola, V., Banerjee, A., Kumar, V.: Anomaly detection: a survey. ACM Comput. Surv. **41**(3), 15 (2009)
9. Chapman, P., Clinton, J., Kerber, R., Khabaza T., Reinartz T., Shearer, C., Wirth R.: Crisp-Dm 1.0 step by step data mining guide, Crisp-DM Consortium (2000)
10. Crone, S.F., Lessmann, S., Stahlbock, R.: The impact of preprocessing on data mining: an evaluation of classifier sensitivity in direct marketing. Eur. J. Oper. Res. **173**(3), 781–800 (2005)
11. Engel, J., Gerretzen, J., Szymanka, E., Jansen, J.J., Downey, G., Blanchet, L., Buydens, L.: Breaking with trends in preprocessing. TrAC Trends Anal. Chem. **50**, 96–106 (2013)
12. Fayyad, U., Piatetsky-Shapiro, G., Smyth, P.: The KDD process for extracting useful knowledge from volumes of data. Commun. ACM **39**(11), 27–34 (1996)
13. Filzmoser, P., Maronna, R., Werner, M.: Outlier identification in high dimensions. Comput. Stat. Data Anal. **52**(3), 1694–1711 (2008)

14. Franco-Santos, M., Kennerley, M., Micheli, P., Martinez, V., Mason, S., Marr, B., Gray, D., Neely, A.: Towards a definition of a business performance measurement system. Int. J. Oper. Prod. Manage. **27**(8), 784–801 (2007)

15. Freund, Y., Schapire, R.E.: A decision-theoretic generalization of on-line learning and an application to boosting. J. Comput. Syst. Sci. **55**(1), 119–139 (1995)

16. Guyon, I., Elisseeff, A.: An introduction to variable and feature selection. J. Mach. Learn. Res. **3**, 1157–1182 (2003)

17. Han, J., Kamber, M., Pei, J.: Data Mining: Concepts and Techniques. Morgan Kaufmann, San Francisco (2012)

18. Hsu, C.-W., Chang, C.-C., Lin, C.-J.: A Practical Guide to Support Vector Classification. Taiwan National University, Taipei (2010)

19. Hodge, V.J., Austin, J.: A survey of outlier detection methodologies. Artif. Intell. Rev. **22**(2), 85–126 (2004)

20. Hu, M-X., Salvucci, S.: A Study of Imputation Algorithms, Institute of Education Science, NCES, U.S. Department of Education (1991)

21. Juhola, M., Siermala, M.: A scatter method for data and variable importance evaluation. Integr. Comput. Aided Eng. **19**(2), 137–149 (2012)

22. Järvinen, P.: On Research Methods. Opinpajan kirja, Tampere (2012)

23. Kaplan, R.S., Norton, D.P.: The balanced scorecard – measures that drive performance. Harvard Bus. Rev. **71**(1), 71–79 (1992)

24. Kriegel, H.-P., Borgwardt, K.M., Kröger, P., Pryakhin, A., Schubert, M., Zimek, A.: Future trends in data mining. Data Min. Knowl. Disc. **15**(1), 87–97 (2007)

25. Kriegel, H-P., Kröger P., Zimek, A.: Outlier detection techniques. In: 16th ACM SIGKDD Conference on Knowledge Discovery and Data Mining, Washington, DC (2010)

26. Kuhn, M., Johnson, K.: Applied Predictive Modeling. Springer, New York (2013)

27. Kochanski, A., Perzyk, M., Klebczyk, M.: Knowledge in imperfect data in advances in knowledge representation. In: Ramirez, C. (ed.) (2012). DOI:10.5772/37714. http://www.intechopen.com/books/advances-inknowledge-representation/knowledge-in-imperfect-data

28. Kohavi, R., John, G.H.: Wrappers for feature subset selection. Artif. Intell. **97**, 273–324 (1997)

29. Kotsiantis, S.B., Kanellopoulos, D., Pintelas, P.E.: Data preprocessing for supervised learning. Int. J. Comput. Sci. **2**, 111–117 (2006)

30. Kuncheva, L.: Combining Pattern Classifiers: Methods and Algorithms. Wiley-Interscience, Hoboken (2004)

31. Longadge, R., Dongre, S.S., Malik, L.: Class imbalance problem in data mining: review. Int. J. Comput. Sci. Netw. **2**(1) (2013)

32. Lawson, R.G., Jurs, P.C.: New index for clustering tendency and its application to chemical problems. J. Chem. Inf. Comput. Sci. **30**(1), 36–41 (1990)

33. March, S., Smith, G.: Design and natural science research on information technology. J. Decis. Support Syst. **15**(4), 251–266 (1995)

34. Nørreklit, H.: The balance on the balanced scorecard—a critical analysis of some of its assumptions. Manage. Account. Res. **11**(1), 65–88 (2000)

35. Pyle, D.: Data Preparation for Data Mining. Morgan Kauffman, San Francisco (2003)

36. Quinlan, J.R.: C4.5: Programs for Machine Learning. Morgan Kaufmann, San Francisco (1993)

37. R Development Core Team: R: A Language and Environment for Statistical Computing. R Foundation for Statistical Computing, Vienna (2008)

38. Sadig, S., Yeganeh, N.K., Induska, M.: 20 years of data quality research: themes, trends and synergies. In: ADC 2011 Proceedings of the Twenty-Second Australasian Database Conference, vol. 115, pp. 153–162 (2011)

39. Somerville, I.: Software Engineering. Pearson, Boston (2015)
40. Sondhi, P.: Feature Construction Methods: A Survey, Jan 2016. sifaka.cs.uiuc.edu
41. Torgo, L.: Data Mining with R: Learning with Case Studies. CRC Press, Boca Raton (2010)
42. Vattulainen, M.: A method to improve the predictive power of a business performance measurement system by data preprocessing combinations: two cases in predictive classification of service sales volume from balanced data. In: Ghazawneh, A., Nørbjerg, J., Pries-Heje, J. (eds.) Proceedings of the 37th Information Systems Research Seminar in Scandinavia (IRIS 37), Ringsted, Denmark, 10–13 August 2014
43. Vattulainen, M.: Improving the predictive power of business performance measurement systems by constructed data quality features - five cases. In: Perner, P. (ed.) ICDM 2015. LNCS, vol. 9165, pp. 3–16. Springer, Heidelberg (2015)
44. Vattulainen, M.: Preproviz: Tools for Visualization of Interdependent Data Quality Issues (2016). https://cran.r-project.org/web/packages/preproviz
45. Wu, X., Kumar, V., Quinlan, J.R., Ghosh, J., Yang, Q., Motoda, H., McLachlan, G., Ng, A., Liu, B., Yu, P.S., Zhou, Z.H., Steinbach, M., Hand, D.J., Steinberg, D.: Top 10 algorithms in data mining. Knowl. Inf. Syst. **14**(1), 1–37 (2008)
46. Wand, Y., Wang, R.: Anchoring data quality dimensions in ontological foundations. Commun. ACM **39**(11), 86–95 (1996)
47. Wickham, H.: Advanced R. Chapman (2014)
48. Wu, X., Zhu, X., Wu, G.-Q., Ding, W.: Data mining with big data. IEEE Trans. Knowl. Discov. Data Eng. **26**(1), 97–107 (2013)
49. Yang, Q., Wu, X.: 10 Challenging problems in data mining research. Int. J. Inf. Technol. Decis. Making **5**(4), 597–604 (2006)
50. Zhao, H., Sudra, R.: Entity identification for heterogenous database integration —a multiple classifier system approach and empirical evaluation. Inf. Syst. **30**(2), 119–132 (2005)

List Price Optimization Using Customized Decision Trees

R. Kiran$^{(\boxtimes)}$, Shashank Shekhar, John Kiran, Raghava Rau,
Sam Pritchett, Anit Bhandari, and Parag Chitalia

VMware Inc., Palo Alto, USA
{rki,sshekhar,kjohn,drau,spritchett,
anitb,pchitalia}@vmware.com

Abstract. There are many data mining solutions in the market which cater to solving pricing problems to various sectors in the business industry. The goal of such solutions is not only to give an optimum pricing but also maximize earnings of the customer. This paper illustrates the application of custom data mining algorithms to the problem of list price optimization in B2B. Decision trees used are mostly binary and pick the right order based on impurity measures like Gini/entropy and mean squared error (for CART). In our study we take a novel approach of non-binary decision trees with order of splits being the choice of business and stopping criteria being the impurity measures. We exploit proxies for list price changes as discount %age and SPF discounting. We calculate transaction thresholds, anchor discounts and elasticity determinants for each SKU segment to arrive at recommended list price which gets used by pricing unit.

Keywords: Non-binary decision tree · List price · SPF · B2B · Anchor discount · Classification · Regression · Entropy · Log-loss

1 Introduction

VMware (VMW) is a virtualization, end user computing and cloud company with annual revenues of USD 6 BB (as of 2014) and a market cap of USD 25 BB [1]. VMware sells products in the Software Defined Data Center (vSphere, VSAN, NSX for computing, storage & network virtualization respectively), end user computing (Airwatch, Horizon, and Fusion/Workstation) and cloud. These are all sold to B2B customers.

The prices of products at VMW have been rarely changed over time. However Sales reps could offer discounts via SPF flag which a special discretionary discount was typically given to large orders. The Pricing Business Unit was keen to figure out a way to get to optimal list prices at VMW.

1.1 Objectives

The Advanced Analytics & Data Sciences team came up with the following objectives with the Pricing Business Unit.

© Springer International Publishing Switzerland 2016
P. Perner (Ed.): ICDM 2016, LNAI 9728, pp. 438–444, 2016.
DOI: 10.1007/978-3-319-41561-1_33

- Analyze historical prices and related volume movements and understand levers of discount % and SPF (sales person specific discount) Flag
- To come up with recommended list price along with level of confidence for all SKUs

2   Solution Framework

2.1   Strategy

A traditional list price optimization would use price changes versus quantity changes, but we never change prices. Discounting practices are an alternative inference method but requires additional steps.

Discount% and SPF requests are useful indicators to infer the customer's assignment of value to a product. Segments will be arrived at for discount % and SPF flag. SPF flag is a discount that can be given by a sales representative on request – mostly given on large order sizes. List price is used to measure VMware's assignment of value to a product. Imbalance between assignments of value then indicate tension in List Price. Following detailed steps were planned

- Understand segments of the order-SKUs based on discount %age and also SPF Usage
- Understand the importance of factors that drive variation between segments and within segment
- Incorporate explicit price-volume elasticity algorithms and come up with pseudo elasticity determinant

2.2   Segmentation Approach

Objective was to arrive at segments of order-SKUs based on discount% and SPF usage. Business required that the decision tree be non-binary in nature and in a particular order. Non-Binary decision trees built to create segments to explain discount% and SPF flag. Product platform, Order Size, Local Currency, Industry Vertical, Partner Tier were used in the same order. Uniqueness of the tree is that non-binary tree implementation that has us determine the order of the input variables based on business perceived improvements.

Working of a normal decision tree – Decision trees [2] in standard software packages decide the list of features and the order of features based on mathematics related to mean squared error, entropy, log-loss, Gini [3] etc.

The customized decision tree made during this study was needed to be built from the leaf node and up using packages in R [4]. It takes the features in the same order required by the business and groups the order-SKUs accordingly. The standard splitting criteria is now replaced by the custom criteria required by the business. The stopping criterion is minimum number of observations. These kind of custom trees are neither available as part of any standard package nor have a custom splitting and stopping criteria. It also uses hand-built mean squared error and log-loss for regression and classification trees respectively. Price elasticity based on above didn't yield significant results and will be trying to arrive at recommendations based on segments of a single product. We will correlate discount % and SPF Usage for segments against their characteristics and use insights from the above segmentation to arrive at list price recommendations at product platform level.

3   Modeling

3.1   Dataset Creation

The dataset for the model was created at an order-SKU level using Greenplum and Hadoop. In Greenplum, which is an enterprise data warehouse we created dataset for a time period of FY13Q1 to FY14Q2 with license only transactions for only partner channel. We also excluded vCloud Air/ELA/VSPP transactions while creating this dataset. In Hadoop, we took last 5 years of dataset related to site made available by IT in a flat tabular structure at order-SKU level.

3.2   Custom Regression Tree for Discount Percentage

Features for the regression tree were chosen in the same order of importance as following: product platform, order size group, industry vertical, currency code, and partner tier.

We found *248* segments with their measures which were used to identify orders where discount % is very different from the segment mean. It was also used to group segments. The tree was a non-binary which used *mean-squared error minimization* [5]. It also can split the dataset using an order of variables/features that business desire. The attributes used to identify segments are mean discount %, proportions of order with SPF flag, # transactions, average booking amount and average unit list price in USD (Table 1).

Table 1. Mock-up representation of regression/classification tree data

Segment	Order number	Product SKU	Actual discount %	Segment mean discount %
VCLOUD SUITE UPGRADED	13499196	CLX-VXXX-STD-UG-AXXX	−20 %	−34 %
VCLOUD SUITE UPGRADED	13471982	CLX-VXXX-STD-UG-BXXX	−20 %	−34 %

3.3   Custom Classification Tree for SPF Flag

The feature order of importance was same as in the custom regression tree. We found *188* segments with their measures which were used to identify orders where proportion of SPF is high/low. It was also used to group segments. The tree was a non-binary which used *entropy for minimizing classification error*. It also can split the dataset using an order of variables/features that business desire. The attributes used to identify segments are mean discount %, proportions of order with SPF flag, # transactions, average booking amount and average unit list price in USD.

The outputs based on the decision tree approach using classification and regression [6] were joined back to our original dataset. Hence, we have two additional fields in our dataset – Discount Segment and SPF Segment. These were utilized in later part of the project.

3.4 Determining Relative Importance

The relative importance of the factors in the trees were determined by running custom regressions [7] on the predictions from the two decision trees. The relative importance of factors in the decision tree based on discount % was determined using linear regression while logistic regression was used for the decision tree based on SPF usage.

4 Utilizing Modeling Outputs

4.1 Identifying Anchor Product

A target SKU was selected based on overall bookings and profitability. This SKU was primarily the top selling SKU. *Selecting anchor SKU was a mutual exercise and business judgements were used to identify the SKU.*

4.2 Generating Model Parameters

Transaction Cutoff: The rationale to have a transaction cutoff was to set a minimum number of transactions required for the model to make a recommendation

$$T(Q_i) = \text{mean(transactions)} - \sigma \tag{1}$$

Capped List Price percentage change: Sets a maximum recommendation as % change from original list price. This was set at 15 % after due consultation with the pricing team on what is an acceptable limit.

List Price Integer rounding: Rounds recommendations to make value appear more similar to common rounding practice. This was done in alignment to the corporate standards. In this particular case, we rounded to the nearest $5/LC5.

4.3 Calculating Anchor Discount

One of the top selling SKUs was selected as the anchor product. The mean of all the discount % was calculated for the selected SKU. The model was evaluated iteratively using the mean of discount %age. An adjustment was made/error was added to the mean to come up with the anchor discount. Adjustment was made to ensure that the anchor discount satisfies the ranges from the pseudo elasticity determinant.

$$\text{Anchor Discount} = \text{Mean of Discount} \% + \text{Adjustment Amount} \qquad (2)$$

4.4 Pseudo Elasticity Determinant

Regression of sum of quantity to gross per unit USD value for different product platform and different pricing and SPF segments. The goal was to arrive at a pseudo elasticity determinant and conform that the change in list price doesn't result in decrease in volume for certain important segments.

4.5 SPF % Intercept and Slope

Expected SPF intercept and slope are used to determine what SPF usage should be expected based on a SKU's list price. The intercept and slope came out of a regression of SPF usage on the list price for selected representative SKUs.

$$\text{SPF Usage} = \alpha + \beta * \text{List Price} \qquad (3)$$

$$\text{SPF Intercept} = \alpha \qquad (4)$$

$$\text{Slope} = \beta \qquad (5)$$

4.6 Expected SPF

This was calculated for each SKU based on the regression equation generated from the regression of SPF usage on the list price for selected representative SKUs (Table 2).

$$\text{Exp. SPF of SKU} = \alpha + \beta * \text{List Price of SKU} \qquad (6)$$

4.7 SPF Flag

This was generated as an indication to whether SPF usage for a SKU is recommended or not. There are three different cases possible for SPF Flag based on following criteria:

$$\text{Flag 'YY'} - \text{if expected SPF} > \text{Actual SPF} + 10\% \qquad (7)$$

$$\text{Flag 'Y'} - \text{if expected SPF} > \text{Actual SPF} + \text{Delta } 10\% \qquad (8)$$

$$\text{Flag 'N'—if expected SPF} < \text{Actual SPF} + \text{Delta 10\%} \tag{9}$$

Delta % has to be entered by the business user and is capped at 10 %. For current exercise, we took it as 0 % (Table 3).

4.8 Suggested List Price

This is the suggested list price without any adjustment and rounding off.

$$\text{Suggested LP} = \frac{\text{LP} \times (1 - \text{Actual Discount})}{(1 - \text{Anchor Discount})} \tag{10}$$

4.9 Recommended List Price

The recommended list price was arrived at after applying the transaction cut-off constraint, capped list price percentage change constraint and the list price integer rounding (Table 4).

Table 2. Model parameter/assumptions for mock-up calculation of recommended list price

Model parameters/assumptions	Values (product 1)
D_A = Anchor discount	35 %
T_C = Transaction cut-off	50
I_{LP} = Capped list price increase	15 %
LP_R = Rounding to LP to the nearest	\$5
SPF_{IN} = SPF % intercept	0.90 %
SPF_S = SPF % slope (per \$1,000)	\$0.0000295
SPF_δ = Delta SPF% (capped at 10 %)	0 %

Table 3. Model parameter/assumptions for mock-up calculation of recommended list price

Model parameters/assumptions	Values (product 1)
Product platform	ABC+
Product SKU	VS-ABC-PL-C
LP_{US} = US list price	\$3,495
Rev_{FLAT} = % platform revenue	83.40 %
Rev_{GROSS} = gross revenue	\$113,636,152
#tx = transactions	9,965
Rev_L = list revenue	\$172,292,318
$Disc_E$ = extended discount	\$58,656,166
Disc % = % discount = $Disc_E/Rev_L$	34.04 %
SPF_w = weighted SPF	6.0 %

Table 4. Mock-up calculation for estimated value of list price

Model parameters/assumptions	Values (product 1)
SPF_{EXP} = Expected SPF % = $LP_{US} \times SPF_S + SPF_{IN}$	11.21 %
SPF Flag	Y
$LP_{Suggested}$ = $LP_{US} \times (1 - Disc\ \%)/(1 - D_A)$	$3,546.37
LP_δ = List Price % Delta = $(LP_{Suggested}/LP_{US}) - 1$	1.47 %
LP_{Capped} = Round(If($LP_\delta > LP_{US}$, $LP_{US} \times (1 + I_{LP})$, $LP_{Suggested}$)	$3,546
$LP_{Recommended}$	$3,545

5 Conclusion

The solution provided for optimizing the list price at VMware Inc. is unique and distinctive because there is hardly any instance of price change at VMware Inc. The custom non-binary decision trees that were built in the first phase of the project are inimitable piece of IP creation as there is no existing software/package which builds non-binary decision trees. Also, the implemented approach was very different from standard price optimization methodologies where historical price changes are utilized to arrive at recommended price changes.

References

1. VMware Inc. Annual Report. http://ir.vmware.com/annuals.cfm
2. Quinlan, J.R.: Induction of decision trees. Mach. Learn. **1**, 81–106 (1986)
3. Shannon, C.E.: A mathematical theory of communication. Bell Syst. Tech. J. **27**(3), 379–423 (1948). doi:10.1002/j.1538-7305.1948.tb01338.x.(PDF)
4. Dowle, M., Srinivasan, A., Short, T., Lianoglou, S., with contributions from Saporta, R., Antonyan, E.: https://cran.r-project.org/web/packages/data.table/data.table.pdf (PDF)
5. Hastie, T., Tibshirani, R., Friedman, J.H.: The Elements of Statistical Learning: Data Mining, Inference, and Prediction. Springer, New York (2001)
6. Breiman, L., Friedman, J.H., Olshen, R.A., Stone, C.J.: Classification and regression trees. Wadsworth & Brooks/Cole Advanced Books & Software, Monterey (1984). ISBN 978-0-412-04841-8
7. Cohen, J., Cohen, P., West, S.G., Aiken, L.S.: Applied Multiple Regression/Correlation Analysis for the Behavioral Sciences, 2nd edn. Lawrence Erlbaum Associates, Hillsdale (2003)

Author Index

Printed in the United States
By Bookmasters